$$\nabla \cdot \rightarrow \begin{bmatrix} \partial/\partial x & 0 & 0 & 0 & \partial/\partial z & \partial/\partial y \\ 0 & \partial/\partial y & 0 & \partial/\partial z & 0 & \partial/\partial x \\ 0 & 0 & \partial/\partial z & \partial/\partial y & \partial/\partial x & 0 \end{bmatrix} \qquad \nabla_s \rightarrow \begin{bmatrix} \partial/\partial x & 0 & 0 \\ 0 & \partial/\partial y & 0 \\ 0 & 0 & \partial/\partial z \\ 0 & \partial/\partial z & \partial/\partial y \\ \partial/\partial z & 0 & \partial/\partial x \\ \partial/\partial y & \partial/\partial x & 0 \end{bmatrix}$$

3. COORDINATE TRANSFORMATIONS

$$[r'] = [a][r] \qquad [a] = \begin{bmatrix} a_{xx} & a_{xy} & a_{xz} \\ a_{yx} & a_{yy} & a_{yz} \\ a_{zx} & a_{zy} & a_{zz} \end{bmatrix}$$

A. Stiffness and Compliance Matrices

$$[c'] = [M][c][\widetilde{M}] \qquad [s'] = [N][s][\widetilde{N}]$$

$$[M] = \begin{bmatrix} a_{xx}^2 & a_{xy}^2 & a_{xz}^2 & 2a_{xy}a_{xz} & 2a_{xz}a_{xx} & 2a_{xx}a_{xy} \\ a_{yx}^2 & a_{yy}^2 & a_{yz}^2 & 2a_{yy}a_{yz} & 2a_{yz}a_{yx} & 2a_{yx}a_{yy} \\ a_{zx}^2 & a_{zy}^2 & a_{zz}^2 & 2a_{zy}a_{zz} & 2a_{zz}a_{zx} & 2a_{zx}a_{zy} \\ a_{yx}a_{zx} & a_{yy}a_{zy} & a_{yz}a_{zz} & a_{yy}a_{zz} + a_{yz}a_{zy} & a_{yx}a_{zz} + a_{yz}a_{zx} & a_{yy}a_{zx} + a_{yx}a_{zy} \\ a_{zx}a_{xx} & a_{zy}a_{xy} & a_{zz}a_{xz} & a_{xy}a_{zz} + a_{xz}a_{zy} & a_{xz}a_{zx} + a_{xx}a_{zz} & a_{xx}a_{zy} + a_{xy}a_{zx} \\ a_{xx}a_{yx} & a_{xy}a_{yy} & a_{xz}a_{yz} & a_{xy}a_{yz} + a_{xz}a_{yy} & a_{xz}a_{yx} + a_{xx}a_{yz} & a_{xx}a_{yy} + a_{xy}a_{yx} \end{bmatrix}$$

$[N]$ is obtained from $[M]$ by shifting the factors 2 into the lower left-hand submatrix.

B. Piezoelectric Matrices

$$[d'] = [a][d][\widetilde{N}] \qquad [e'] = [a][e][\widetilde{M}]$$

C. Permittivity Matrix

$$[\epsilon'] = [a][\epsilon][\widetilde{a}]$$

SECOND EDITION

ACOUSTIC FIELDS
AND WAVES
IN SOLIDS

VOLUME I

SECOND EDITION

ACOUSTIC FIELDS AND WAVES IN SOLIDS

VOLUME I

B.A. AULD
Professor (Research)
of Applied Physics
Stanford University

KRIEGER PUBLISHING COMPANY
MALABAR, FLORIDA

Original Edition 1973
Second Edition 1990

Printed and Published by
KRIEGER PUBLISHING COMPANY, INC.
KRIEGER DRIVE
MALABAR, FLORIDA 32950

Library of Congress Cataloging-in-Publication Data

Auld, B.A. (Bertram Alexander), 1922-
    Acoustic fields and waves in solids / B.A. Auld.—2nd ed.
        p.   cm.
    Bibliography: p.
    Includes index.
    ISBN 0-89874-782-1 (v. 1 : alk. paper).—ISBN 0-89874-783-X (v.
2 : alk. paper)
    1. Solids—Acoustic properties. 2. Sound - waves. 3. Elastic
waves. 4. Wave-motion, Theory of. I. Title.
QC176.8.A3A84   1989
530.4'12—dc20   89-15477
                                                          CIP

10   9   8   7   6

FOREWORD

The fifteen years since publication of the first edition have witnessed an impressive expansion of acoustic device applications—scanned ultrasonic medical imagers, NDE (nondestructive evaluation) systems, acoustic microscopy, convolvers and correlators, SAW (surface acoustic wave) resonators and oscillators, acoustic sensors (for robotics, process control, motion detection, chemical detection), analogue and digital filters, etc. New single crystal, ceramic and, especially, composite materials have also been introduced. At the same time some of the concepts and devices that appeared promising in 1973—for example, equivalent voltage and current concepts for acoustic waveguides, and certain types of SAW transducers—are no longer of vital importance. The intent in revising the first edition has been to eliminate most of the redundant and less useful material; and sections have been added treating the fundamentals underlying new developments judged to be of lasting interest.

The second edition retains the same general character of the first, as an advanced textbook and reference on the fundamentals of acoustic waves in solids. General principles and concepts are emphasized, to provide a base for future device modeling. No attempt has been made to document fully the enormous number of relevant papers published since 1983. A Supplementary Reference List contains citations related to the revised text. It also gives a number of recent books on acoustics, special acoustics issues of journals, and published conference proceedings. These publications have extensive bibliographies of the current literature on acoustic waves in solids. The Proceedings of the IEEE Ultrasonics Symposium, published annually, is an excellent resource for the most recent developments.

I am especially grateful to my secretary, Judith C. Clark, for her skill and effort in word processing the revised sections to match the original edition. Also, to Janet K. Okagaki of the Ginzton Drafting Office for her figure artistry. The assistance of J. Fraser, S. Meeks, G. Laguna, and S. Ayter in reading the manuscript is also acknowledged.

Stanford, California   B. A. Auld

PREFACE

This book has developed from a lecture course on mechanical waves and vibrations in solids for first and second year graduate students in Applied Physics. The descriptive term "acoustic" (rather than "elastic") in the title follows common usage in physics, where ordinary elastic motions in crystals are called acoustic modes. This distinguishes them from optical modes, which involve internal degrees of freedom within a crystal unit cell. The title also reflects common terminology among researchers and engineers engaged in developing elastic wave devices for radar and communication systems. This area of technology has been strongly influenced by the philosophy, concepts, and techniques of microwave electromagnetics and has come to be known as microwave acoustics. In this way, use of the term "acoustic" accurately describes the aim and scope of the book. It is intended to present, in a manner congenial to the disciplines of Applied Physics and Electrical Engineering, a coherent treatment of mechanical wave and vibration theory, starting from fundamentals.

In Volume I the development proceeds step-by-step from the basic principles of mechanics and electricity. There are no other requirements beyond an understanding of elementary calculus, differential equations, vector analysis, and matrix theory; but a preliminary acquaintance with some of the basic concepts of acoustics (as presented, for example, in *Fundamentals of Acoustics*, L. E. Kinsler and A. R. Frey, Wiley, 1962) would be helpful to the student. Background in tensor theory is not needed. Tensor concepts are developed in a simplified manner whenever required, and advanced topics are completely avoided.

Throughout the book, symbolic notation is used for acoustic field quantities; that is, vector and tensor variables are represented by boldface letters rather than by their components. Components are used explicitly only when a specific problem is to be solved. Also, the presentation of the acoustic field equations parallels Maxwell's equations for the electromagnetic field. This approach serves two purposes. First, it emphasizes the similarity of concepts between acoustic engineering and electrical engineering; and, second, it simplifies manipulations of the field equations and clarifies the presentation of basic acoustic field theorems. To further strengthen the connection between acoustics and electromagnetics, MKS units are used. This mode of presentation should be helpful to electronics engineers approaching the subject with a background in microwave electromagnetics. The acoustics theory is, however, completely self-contained and can, if desired, be studied without reference to the electromagnetic analogue.

Since this book is intended in part for classroom use, a large number of examples are included in the text and problems are given at the end of each chapter. Examples in the text, which have been chosen both to illustrate concepts and to demonstrate problem solving methods, are progressive in nature and examples in later chapters often build on results obtained earlier. The problem sets are intended to provide exercise in handling concepts, analyzing new situations and, in some cases, deriving results quoted in the text. Tables of material properties and an extensive catalogue of plane wave solutions are given in the Appendices.

Preparation of this book would have been impossible without the co-operation and assistance of both staff and students at the W. W. Hansen Laboratories of Physics. I am indebted to Walter L. Bond for information on transformation methods, to Gordon S. Kino for his important contributions to the theory of piezoelectric waveguides, to John D. Larson and Donald K. Winslow for their collection of material constants, and to many students for locating errors in the manuscript. I am especially grateful to Iona Williams, Joan Cantor, Tony Vacek and others for their expert and generous assistance in preparing the manuscript. Acknowledgements are also due to friends and colleagues elsewhere for their helpful criticisms and comments.

Stanford, California   **B. A. Auld**

CONTENTS

ix

To Anne

Chapter 1

PARTICLE DISPLACEMENT AND STRAIN

A. PARTICLE DISPLACEMENT AND DISPLACEMENT GRADIENT

Acoustics is the study of time-varying deformations, or vibrations, in material media. All material substances are composed of atoms, which may be forced into vibrational motion about their equilibrium positions. Many different patterns of vibrational motion may exist at this atomic level. However, most of these motional patterns are not relevant to the study of acoustics, which is concerned only with material particles that are small but yet contain many atoms. Within each particle the atoms move in unison. Acoustics theory deals, therefore, only with macroscopic phenomena and is formulated as if matter were a continuum. Structure at the microscopic level is of interest only insofar as it affects the medium's macroscopic properties.

When the particles of a medium are displaced from their equilibrium positions, internal restoring forces arise. It is these elastic restoring forces between particles, combined with inertia of the particles, which lead to

1

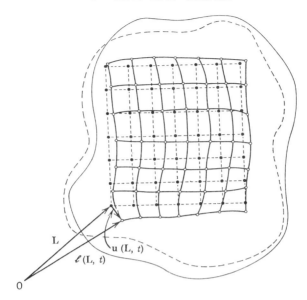

FIGURE 1.1.   **Particle position in the equilibrium and deformed states of a solid body.**

oscillatory motions of the medium. To formulate a mathematical description of these vibrations, which may be either traveling waves or localized oscillations, it is first necessary to introduce quantitative definitions of *particle displacement*, *material deformation*, and *internal restoring forces*.

Displacements of the material particles in a deformed medium may be illustrated by the construction shown in Fig. 1.1. Equilibrium positions of a regular array of selected particles are indicated by dots and the displaced positions of the same particles by circles. Each particle is then assigned an equilibrium position vector $\mathbf{L}$ and a displaced position vector $\ell(\mathbf{L}, t)$, measured from some point of origin O. The displaced position vector ℓ is, in general, a time-varying quantity and is also shown as a function of $\mathbf{L}$. The equilibrium (or reference) position vector $\mathbf{L}$ is used simply as an identifying label for the particles. Both $\mathbf{L}$ and ℓ are continuous variables, and are not restricted to the discrete values shown in the figure.

The displacement of the particle located at $\mathbf{L}$ in the equilibrium state is defined, according to Fig. 1.1, as

$$\mathbf{u}(\mathbf{L}, t) = \ell(\mathbf{L}, t) - \mathbf{L}. \tag{1.1}$$

The *particle displacement field* $\mathbf{u}$ is thus a continuous variable describing the vibrational motions of all particles within a medium. If a vibration is a sinusoidal function of time, with a single frequency ω, there are three possible

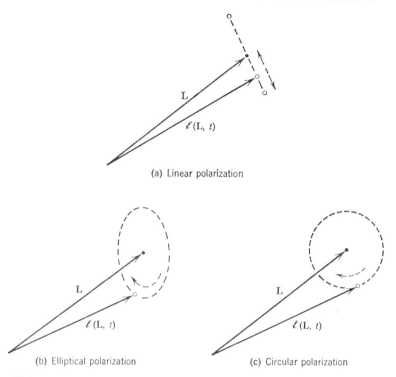

(a) Linear polarization

(b) Elliptical polarization   (c) Circular polarization

FIGURE 1.2.   **Different types of particle displacement polarization.**

patterns of motion for each particle. The particle may move along a linear path, passing through the equilibrium position twice each cycle,

$$\mathbf{u}(\mathbf{L}, t) = \mathbf{A}(\mathbf{L}) \sin \omega t. \tag{1.2}$$

This is called a *linearly polarized* particle displacement (Fig. 1.2a). If the particle simultaneously executes two linear motions with perpendicular polarizations and a 90° difference in time phase (Fig. 1.2b) the displacement field is

$$\mathbf{u}(\mathbf{L}, t) = \mathbf{A}(\mathbf{L}) \sin \omega t + \mathbf{B}(\mathbf{L}) \cos \omega t. \tag{1.3}$$

In this case the particle follows an elliptical path about the equilibrium position,

$$u(\mathbf{L}, t) = \{A^2(\mathbf{L}) \sin^2 \omega t + B^2(\mathbf{L}) \cos^2 \omega t\}^{1/2}, \tag{1.4}$$

and the displacement is said to be *elliptically polarized*. For the special case $A = B$ in (1.4) the particle trajectory is circular (Fig. 1.2c). This is called a *circularly polarized* displacement.

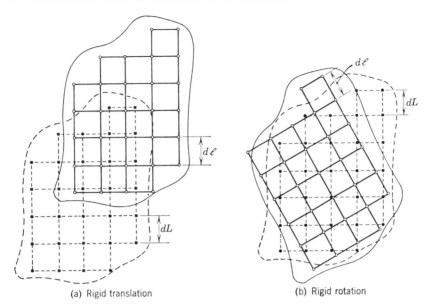

(a) Rigid translation   (b) Rigid rotation

FIGURE 1.3.   **Rigid motions of a solid body.**

The term *material deformation* is understood to apply only when particles of a medium are displaced relative to each other. In rigid translations and rotations, where all particles of a body maintain their relative positions (Fig. 1.3), there is no deformation. Since the particle displacement field **u** in (1.1) is nonzero for all such rigid motions, **u** does not itself provide a satisfactory measure of material deformation. Rigid *translational* displacements may be eliminated by considering the differential form of (1.1) at *constant t*; that is,

$$d\mathbf{u}(\mathbf{L}, t) = d\boldsymbol{\ell}(\mathbf{L}, t) - d\mathbf{L}. \tag{1.5}$$

The physical significance of this relation is illustrated by Fig. 1.4, in which displaced particle positions at a particular instant of time are shown for two neighboring particles *a* and *b*. If the medium experiences a rigid translational motion, the two displacement vectors, $\mathbf{u}(\mathbf{L}, t)$ and $\mathbf{u}(\mathbf{L} + d\mathbf{L}, t)$, in the figure are equal and the differential displacement $d\mathbf{u}$ in (1.5) is therefore zero.

To calculate the differential particle displacement $d\mathbf{u}$ in (1.5) from the particle displacement field $\mathbf{u}(\mathbf{L},t)$, use is made of the partial derivative relation

$$d\mathbf{u}(\mathbf{L}, t) = \frac{\partial}{\partial L_1} \mathbf{u}(\mathbf{L}, t)\, dL_1 + \frac{\partial}{\partial L_2} \mathbf{u}(\mathbf{L}, t)\, dL_2 + \frac{\partial}{\partial L_3} \mathbf{u}(\mathbf{L}, t)\, dL_3, \tag{1.6}$$

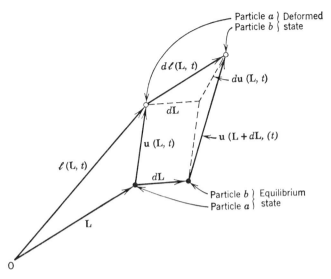

FIGURE 1.4. **Definition of *differential particle displacement* in a deformed medium.**

where dL_1, dL_2, dL_3 are components of $d\mathbf{L}$ with respect to some orthogonal coordinate system and the term

$$\frac{\partial}{\partial t} \mathbf{u}(\mathbf{L}, t) \, dt$$

is omitted because the differential displacement is defined at *constant* time. In rectangular Cartesian coordinates

$$\mathbf{u}(\mathbf{L}, t) = \hat{\mathbf{x}} u_x(\mathbf{L}, t) + \hat{\mathbf{y}} u_y(\mathbf{L}, t) + \hat{\mathbf{z}} u_z(\mathbf{L}, t) \tag{1.7}$$

and since the unit vectors $\hat{\mathbf{x}}$, $\hat{\mathbf{y}}$, $\hat{\mathbf{z}}$ are *constant*†, (1.6) becomes

$$
\begin{aligned}
d\mathbf{u} = \hat{\mathbf{x}} &\left(\frac{\partial u_x}{\partial L_x} dL_x + \frac{\partial u_x}{\partial L_y} dL_y + \frac{\partial u_x}{\partial L_z} dL_z \right) \\
+ \hat{\mathbf{y}} &\left(\frac{\partial u_y}{\partial L_x} dL_x + \frac{\partial u_y}{\partial L_y} dL_y + \frac{\partial u_y}{\partial L_z} dL_z \right) \\
+ \hat{\mathbf{z}} &\left(\frac{\partial u_z}{\partial L_x} dL_x + \frac{\partial u_z}{\partial L_y} dL_y + \frac{\partial u_z}{\partial L_z} dL_z \right).
\end{aligned}
$$

† In cylindrical and spherical coordinate systems, derivatives of the unit coordinate vectors must also be taken into account (Appendix 1).

This may be written in matrix form as

$$
\begin{bmatrix}
du_x(d\mathbf{L}, t) \\[4pt]
du_y(d\mathbf{L}, t) \\[4pt]
du_z(d\mathbf{L}, t)
\end{bmatrix}
=
\begin{bmatrix}
\dfrac{\partial u_x}{\partial L_x} & \dfrac{\partial u_x}{\partial L_y} & \dfrac{\partial u_x}{\partial L_z} \\[8pt]
\dfrac{\partial u_y}{\partial L_x} & \dfrac{\partial u_y}{\partial L_y} & \dfrac{\partial u_y}{\partial L_z} \\[8pt]
\dfrac{\partial u_z}{\partial L_x} & \dfrac{\partial u_z}{\partial L_y} & \dfrac{\partial u_z}{\partial L_z}
\end{bmatrix}
\begin{bmatrix}
dL_x \\[4pt]
dL_y \\[4pt]
dL_z
\end{bmatrix}
\qquad (1.8)
$$

The matrix† in (1.8),

$$
[\mathscr{E}(\mathbf{L}, t)] =
\begin{bmatrix}
\dfrac{\partial u_x(\mathbf{L}, t)}{\partial L_x} & \dfrac{\partial u_x(\mathbf{L}, t)}{\partial L_y} & \dfrac{\partial u_x(\mathbf{L}, t)}{\partial L_z} \\[14pt]
\dfrac{\partial u_y(\mathbf{L}, t)}{\partial L_x} & \dfrac{\partial u_y(\mathbf{L}, t)}{\partial L_y} & \dfrac{\partial u_y(\mathbf{L}, t)}{\partial L_z} \\[14pt]
\dfrac{\partial u_z(\mathbf{L}, t)}{\partial L_x} & \dfrac{\partial u_z(\mathbf{L}, t)}{\partial L_y} & \dfrac{\partial u_z(\mathbf{L}, t)}{\partial L_y}
\end{bmatrix}
\qquad (1.9)
$$

is called the displacement gradient matrix. Using this matrix and relation (1.8), the differential displacements $d\mathbf{u}$ for *any* two neighboring particles may be calculated from their equilibrium spacing $d\mathbf{L}$ in Fig. 1.4. The displacement gradient matrix $[\mathscr{E}(\mathbf{L}, t)]$ is therefore a measure of *differential particle displacement* in a deformed medium.

EXAMPLE 1.   In illustrating the basic concepts of acoustic deformation theory it is often convenient to examine time-independent, or static, field problems. Consider a solid bar that is compressed uniformly along the x direction and constrained so that there are no components of particle displacement along y and z (Fig. 1.5). If the end of the bar at $x = 0$ is fixed, all the particle displacements are zero at that end. At the other end of the bar all particles are displaced by an amount

$$
\mathbf{u}(D) = -\hat{\mathbf{x}}(D - D'),
$$

where D' is the deformed length and D the undeformed length of the bar. For intermediate points, the displacement is linearly proportional to the undeformed distance

---

† Symbols representing matrices are enclosed by square brackets.

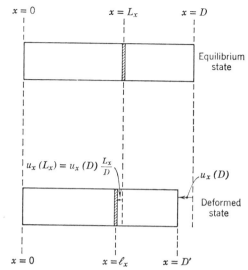

FIGURE 1.5. **Particle displacement in a uniformly compressed bar.**

L_x from the fixed end of the bar, and

$$\mathbf{u}(L_x) = -\hat{\mathbf{x}}(D - D')\frac{L_x}{D} = -\hat{\mathbf{x}}(1 - D'/D)L_x. \qquad (1.10)$$

In this case the displacement gradient matrix (1.9) is independent of x,

$$[\mathscr{E}] = \begin{bmatrix} -(1 - D'/D) & 0 & 0 \\ \\ 0 & 0 & 0 \\ \\ 0 & 0 & 0 \end{bmatrix} \qquad (1.11)$$

and (1.8) reduces to the scalar equation

$$du_x = -(1 - D'/D)\,dL_x. \qquad (1.12)$$

The *differential* particle displacement is thus uniform throughout the bar.

Suppose that the same bar is again compressed from length D to D', but with the midpoint now held rigidly fixed. In this case the particle displacement field is

$$\mathbf{u}(L_x) = -\hat{\mathbf{x}}(1 - D'/D)(L_x - D/2). \qquad (1.13)$$

This differs from (1.10) only by a rigid translation $D/2$ in the positive x direction. Since the *differential* displacement is unchanged, the particle displacement gradient matrix is again given by (1.11).

B. STRAIN

As a measure of material *deformation*, the displacement gradient matrix (1.9) is deficient in one respect. It does not reduce to zero for rigid rotations. This is easily demonstrated. Consider a body that is rigidly rotated through a small angle ϕ about a perpendicular axis passing through O in Fig. 1.6. Positions of two particles, *a* and *b*, are shown before and after the rotation. In this case particle *a* experiences a displacement

$$u = 2L \sin \phi/2$$

and particle *b* experiences a displacement

$$u + du = 2(L + dL) \sin \phi/2.$$

The displacement gradient is therefore nonzero, although the body has not been deformed.

A quantity that does remain zero for the rigid rotation of Fig. 1.6, and for all combinations of rigid rotations and translations, is the *scalar* quantity

$$\Delta = d\ell(\mathbf{L}, t) - dL.$$

For rigid motions Δ is always zero (see Figs. 1.3 and 1.6) and for deformations it is always nonzero. It is therefore a true measure of deformation. However,

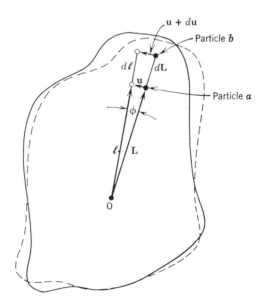

FIGURE 1.6. **Rigid rotation through a small angle ϕ.**

a more convenient quantity that meets the same requirements is

$$\Delta' = d\ell^2(\mathbf{L}, t) - (dL)^2,$$

and this is customarily defined as deformation.

In rectangular Cartesian coordinates the deformation measure Δ' is calculated from $\mathbf{u}(\mathbf{L}, t)$ by using relations

$$d\ell_x = dL_x + du_x = dL_x + \frac{\partial u_x}{\partial L_x} dL_x + \frac{\partial u_x}{\partial L_y} dL_y + \frac{\partial u_x}{\partial L_z} dL_z, \quad (1.14)$$

etc., from (1.5) and (1.8). For simplicity, two-dimensional deformations normal to the z axis† will be considered first. In this case

$$\Delta' = d\ell^2 - dL^2 = (d\ell_x)^2 + (d\ell_y)^2 - (dL_x)^2 - (dL_y)^2$$

$$= \left(2\frac{\partial u_x}{\partial L_x} + \left(\frac{\partial u_x}{\partial L_x}\right)^2 + \left(\frac{\partial u_y}{\partial L_x}\right)^2\right) dL_x^2$$

$$+ \left(2\frac{\partial u_y}{\partial L_y} + \left(\frac{\partial u_x}{\partial L_y}\right)^2 + \left(\frac{\partial u_y}{\partial L_y}\right)^2\right) dL_y^2$$

$$+ \left(2\frac{\partial u_x}{\partial L_y} + 2\frac{\partial u_y}{\partial L_x} + 2\frac{\partial u_x}{\partial L_x}\frac{\partial u_x}{\partial L_y} + 2\frac{\partial u_y}{\partial L_x}\frac{\partial u_y}{\partial L_y}\right) dL_x\, dL_y. \quad (1.15)$$

This is conveniently expressed in matrix notation as

$$(d\ell)^2 - (dL)^2 = 2\left(\begin{bmatrix} dL_x & dL_y \end{bmatrix}\begin{bmatrix} S_{xx} & S_{xy} \\ S_{yx} & S_{yy} \end{bmatrix}\begin{bmatrix} dL_x \\ dL_y \end{bmatrix}\right)$$

$$= 2S_{xx}\, dL_x^2 + 2S_{yy}\, dL_y^2 + 2(S_{xy} + S_{yx})\, dL_x\, dL_y, \quad (1.16)$$

where the matrix elements S_{ij} are evaluated by equating terms in (1.15) and (1.16). Since only the sum of the off-diagonal matrix elements appears in (1.16), the matrix may be chosen to be symmetric ($S_{yx} = S_{xy}$) without loss of generality. That is

$$S_{xx} = \frac{\partial u_x}{\partial L_x} + \frac{1}{2}\left(\frac{\partial u_x}{\partial L_x}\right)^2 + \frac{1}{2}\left(\frac{\partial u_y}{\partial L_x}\right)^2$$

$$S_{yy} = \frac{\partial u_y}{\partial L_y} + \frac{1}{2}\left(\frac{\partial u_x}{\partial L_y}\right)^2 + \frac{1}{2}\left(\frac{\partial u_y}{\partial L_y}\right)^2$$

$$S_{xy} = S_{yx} = \frac{1}{2}\left(\frac{\partial u_x}{\partial L_y} + \frac{\partial u_y}{\partial L_x} + \frac{\partial u_x}{\partial L_x}\frac{\partial u_x}{\partial L_y} + \frac{\partial u_y}{\partial L_x}\frac{\partial u_y}{\partial L_y}\right).$$

† Displacement components u_x and u_y are functions only of x and y; and $u_z = 0$.

For three-dimensional deformations the same argument may be extended to show that[†]

$$\Delta'(\mathbf{L}, t) = 2S_{ij}(\mathbf{L}, t)\, dL_i\, dL_j, \tag{1.17}$$

with

$$S_{ij}(\mathbf{L}, t) = \frac{1}{2}\left(\frac{\partial u_i}{\partial L_j} + \frac{\partial u_j}{\partial L_i} + \frac{\partial u_k}{\partial L_i}\frac{\partial u_k}{\partial L_j}\right)$$

$$i, j, k = x, y, z. \tag{1.18}$$

The matrix elements $S_{ij}(\mathbf{L}, t)$ are termed components of the strain field. The strain field thus determines the deformation $\Delta' = d\ell^2 - dL^2$ in terms of the particle displacement field $\mathbf{u}(\mathbf{L}, t)$, and reduces to zero for all rigid motions.

Solids differ widely in their deformabilities. In some materials, like rubber, displacement gradients greater than unity[‡] are easily reached. With more rigid materials, however, the displacement gradient must be kept below the range 10^{-4} to 10^{-3} if permanent deformation or fracture is to be avoided. *For displacement derivatives much smaller than this range, the quadratic terms in (1.18) are negligible and the linearized strain-displacement relation*

$$S_{ij}(\mathbf{L}, t) = \frac{1}{2}\left(\frac{\partial u_i(\mathbf{L}, t)}{\partial L_j} + \frac{\partial u_j(\mathbf{L}, t)}{\partial L_i}\right)$$

$$i, j = x, y, z \tag{1.19}$$

can usually be used.§

Since $dL_j = d\ell_j - du_j$, the partial derivative of u_i with respect to L_j in (1.19) differs from $\partial u_i/\partial \ell_j$ only by quadratic and higher order terms

$$\frac{\partial u_i}{\partial L_j} = \frac{\partial u_i}{\partial \ell_j}\left(1 - \frac{\partial u_j}{\partial \ell_j}\right)^{-1}.$$

For a linearized theory there is therefore no need to distinguish between components of the deformed position vector ℓ and those of equilibrium position vector $\mathbf{L}$. That is,

$$\mathbf{L} \simeq \ell = \hat{x}x + \hat{y}y + \hat{z}z = \mathbf{r} \tag{1.20}$$

[†] Here and hereafter summation over repeated subscripts is assumed.
[‡] Both displacement gradient and strain are dimensionless quantities.
§ In nonlinear acoustics it is often necessary to use all of the displacement derivatives, rather than the particular combinations of derivatives that appear in (1.19). See, for example, H. F. Tiersten, *J. Math. Phys.* **6**, pp 779–787 (1965); W. F. Brown Jr., *J. Appl. Phys.* **36**, pp. 994–1000 (1965); and D. E. Eastman, *Phys. Rev.* **148**, pp. 530–542 (1966).

in rectangular Cartesian coordinates. The linearized strain-displacement relation therefore takes the form

$$S_{ij}(\mathbf{r}, t) = \frac{1}{2}\left(\frac{\partial u_i}{\partial r_j} + \frac{\partial u_j}{\partial r_i}\right)$$

$$i, j = x, y, z \tag{1.21}$$

in rectangular coordinates.† In the same approximation, the displacement gradient matrix (1.9) becomes

$$\mathscr{E}_{ij}(\mathbf{r}, t) = \frac{\partial u_i}{\partial r_j}. \tag{1.22}$$

EXAMPLE 2.   For the particle displacement field of Example 1,

$$\mathbf{u} = -\hat{\mathbf{x}}(1 - D'/D)x$$

in the linearized approximation. The only nonzero strain component is therefore

$$S_{xx} = \frac{\partial u_x}{\partial x} = -(1 - D'/D)$$

and the strain matrix is

$$[S] = \begin{bmatrix} -(1 - D'/D) & 0 & 0 \\ 0 & 0 & 0 \\ 0 & 0 & 0 \end{bmatrix}$$

Physically, S_{xx} is a change in length per unit length and is negative for a compression, positive for an extension.

In laboratory-scale bodies the maximum particle displacements are minute. If the maximum allowed strain in the bar is

$$S_{xx} = -10^{-4}$$

and $D = 10^{-1}$ m the displacement of the free end at the bar is

$$u_x(D) = -(D - D') = S_{xx}D = 10^{-5} \, \text{m},$$

or 10 microns.

---

† Linearized strain-displacement relations for cylindrical and spherical coordinates are derived in Appendix 1.

EXAMPLE 3. Another common type of deformation is shear strain. Consider a solid cube that is sheared uniformly along the x direction (Fig. 1.7). In this case the particle displacement field is

$$\mathbf{u}(\mathbf{r}) = \hat{\mathbf{x}} \, 2Cy.$$

The nonzero strain components, from (1.21), are therefore

$$S_{xy} = S_{yx} = \frac{1}{2} \frac{\partial u_x}{\partial y} = C,$$

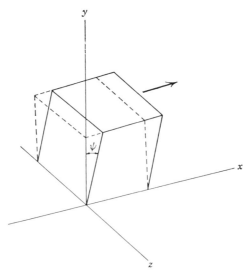

FIGURE 1.7. **Shearing deformation along x. Dashed lines outline the body in its undeformed state and solid lines show the deformed state.**

and the strain matrix is

$$[S] = \begin{bmatrix} 0 & C & 0 \\ C & 0 & 0 \\ 0 & 0 & 0 \end{bmatrix}$$

In this case the strain (called *simple shear*) is a measure of the shearing angle ψ shown in the figure, with

$$\tan \psi = 2C.$$

If a maximum allowed strain of 10^{-4} is again assumed

$$\psi \cong 2 \times 10^{-4} \text{ radians.}$$

C. "LOCAL" ROTATION

In the linearized approximation, a simple and physically significant relationship exists between the strain matrix (1.21) and the displacement gradient matrix (1.22). This may be demonstrated by decomposing the latter into its symmetric and antisymmetric parts,

$$[\mathscr{E}] = \tfrac{1}{2}([\mathscr{E}] + \widetilde{[\mathscr{E}]}) + \tfrac{1}{2}([\mathscr{E}] - \widetilde{[\mathscr{E}]}), \qquad (1.23)$$
symmetric   antisymmetric

where the tilde $(\sim)$ designates a *transposed matrix*.† Comparison with (1.21) and (1.22) shows that the symmetric part of $[\mathscr{E}]$ is identical with the strain matrix. That is

$$[S] = \tfrac{1}{2}([\mathscr{E}] + \widetilde{[\mathscr{E}]}). \qquad (1.24)$$

The antisymmetric part of $[\mathscr{E}]$ also has a simple physical interpretation. It corresponds to a rotation. To see this, consider the purely antisymmetric displacement gradient matrix

$$[\mathscr{E}] = \begin{bmatrix} 0 & \mathscr{E}_{xy} & 0 \\ -\mathscr{E}_{xy} & 0 & 0 \\ 0 & 0 & 0 \end{bmatrix}$$

From (1.8),

$$du_x = \mathscr{E}_{xy}\, dL_y$$
$$du_y = -\mathscr{E}_{xy}\, dL_x$$
$$du_z = 0$$

and therefore

$$d\mathbf{u} \cdot d\mathbf{L} = du_x\, dL_x + du_y\, dL_y = 0.$$

That is, $d\mathbf{u}$ is at right angles to $d\mathbf{L}$. This is certainly true for the rigid rotation of Fig. 1.6, where the linearized approximation implies that ϕ must now be small. It also applies to a "local" rotation. Figure 1.8 gives an example. A rectangular block of material, held rigidly at its boundaries, is deformed by a localized torque applied about the z axis. As in the previous figures, positions are again designated by solid dots for the equilibrium state and by circles for the deformed state. At the application point of the torque the medium experiences a pure "local" rotation, since $d\ell = dL$; and the strain is therefore zero. Elsewhere both strain and rotation are present. Local rotations are frequently encountered in acoustic vibration problems. It will be seen in the next chapter,

† Interchange of rows and columns.

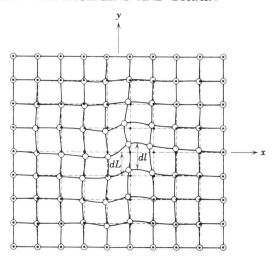

FIGURE 1.8.   An illustration of local rotation about the z axis.

however, that these "local" rotations do not enter into the dynamics of the vibration problem.[†] The antisymmetric part of the displacement gradient matrix does not, therefore, appear in the acoustic field equations.

EXAMPLE 4.   Since the strain matrix is not sensitive to either rigid or local rotations of a medium, strain components do not uniquely characterize the particle displacement field. Consider for example, the two-dimensional static displacement field

$$\mathbf{u}(\mathbf{r}) = \hat{x}Cy + \hat{y}Cx$$

From (1.21), the strain matrix for this case has the same nonzero strain components as Example 3. That is,

$$S_{xy} = S_{yx} = C.$$

In this case, however, the displacement gradient matrix is purely symmetric,

$$[\mathscr{E}] = [S] = \begin{bmatrix} 0 & C & 0 \\ C & 0 & 0 \\ 0 & 0 & 0 \end{bmatrix}$$

There is therefore no rotation, and the deformation is called *pure shear*. By contrast, the *simple shear* displacement field of Example 3,

$$\mathbf{u}(\mathbf{r}) = \hat{x}2Cy,$$

[†] These local rotations *are*, however, significant in other physical phenomena such as acousto-optic scattering. See D. F. Nelson and M. Lax, *Phys. Rev. Lett.* **25**, pp. 379–380 (1970); D. F. Nelson and P. D. Lazay, *Phys. Rev. Lett.* **25**, pp. 1187–1190 (1970).

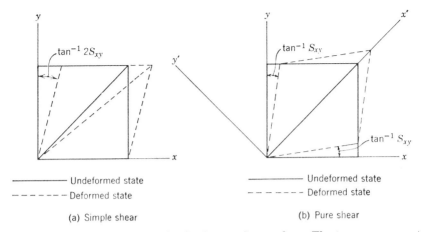

FIGURE 1.9. Comparison of simple shear and pure shear. The two cases cannot be distinguished by the strain field alone because one is transformed into the other by a rigid rotation.

does have an antisymmetric component of displacement gradient,

$$\tfrac{1}{2}([\mathscr{E}] - [\widetilde{\mathscr{E}}]) = \begin{bmatrix} 0 & C & 0 \\ -C & 0 & 0 \\ 0 & 0 & 0 \end{bmatrix}$$

The simple shear particle motion of Example 3 thus differs from that of the pure shear case by a clockwise rotation through an angle $\tan^{-1} S_{xy}$ about the z axis (Fig. 1.9).

D. PICTORIAL REPRESENTATIONS

In electromagnetic theory it has for many years been common practice to represent electric and magnetic field distributions by means of field line diagrams. This kind of pictorial representation, which often provides valuable physical insights in problem solving, is constructed by drawing a family of smooth lines following the direction of the field vector at every point. Variations in field intensity are then represented by the density of field lines, with closely spaced lines corresponding to regions of high intensity and widely spaced lines corresponding to regions of low intensity.

For the same reasons, pictorial representations are helpful in acoustic field theory and they have already been used to this purpose in Fig. 1.1 and Fig. 1.8. In these illustrations the deformation pattern is visualized by first considering a regular array of particles in the equilibrium state of a medium

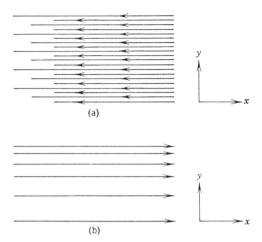

FIGURE 1.10.   **Field line diagrams for particle displacement fields (a) Uniformly compressed bar (Fig. 1.5). (b) Uniformly sheared block (Fig. 1.7).**

and then drawing a grid of lines through the displaced positions of the same particles in the strained medium. An alternative technique is to use the field line method of electromagnetism to represent the vector particle displacement $\mathbf{u}(\mathbf{r}, t)$. Figure 1.10 shows field diagrams of this kind for the static displacement fields of Examples 2 and 3.

EXAMPLE 5.   Up to this point, only static examples with spatially uniform strains have been considered. For time-varying vibration problems the strains are always spatially nonuniform. The simplest examples of this kind are uniform (or straight-crested) plane waves. In a uniform plane wave, there is propagation along a particular direction but uniformity of the field in planes perpendicular to the propagation direction. For example,

$$\mathbf{u}(\mathbf{r}, t) = \hat{\mathbf{x}} \cos (\omega t - ky) \qquad (1.25)$$

is a uniform plane wave of x-polarized particle displacement propagating along the y axis. Here, the quantity

$$k = \frac{2\pi}{\lambda}, \qquad (1.26)$$

where λ is the wavelength, is called the *wave vector*. The propagation velocity of a point of constant phase ($\omega t - ky =$ constant) is called the phase velocity,

$$V_p = \frac{\omega}{k}. \qquad (1.27)$$

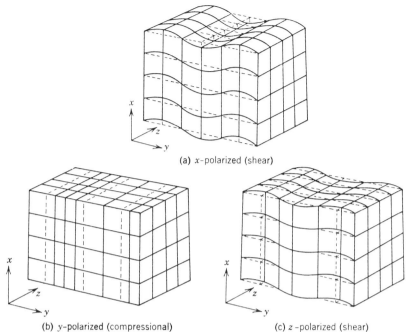

(a) x-polarized (shear)

(b) y-polarized (compressional)

(c) z-polarized (shear)

FIGURE 1.11. **Grid diagrams for plane uniform particle displacement waves propagating along** y.

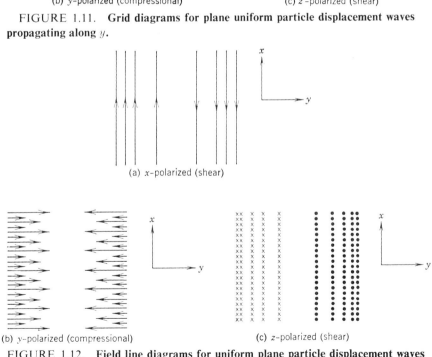

(a) x-polarized (shear)

(b) y-polarized (compressional)

(c) z-polarized (shear)

FIGURE 1.12. **Field line diagrams for uniform plane particle displacement waves propagating along** y.

From (1.21), the strain corresponding to (1.25) is

$$S_{yx} = S_{xy} = \frac{1}{2}\frac{\partial u_x}{\partial y} = \frac{k}{2}\sin{(\omega t - ky)}. \tag{1.28}$$

This is a case of simple shear, as in Example 3, and the wave is called an *x-polarized y-propagating shear wave*. Grid and line representations of the deformation pattern at $t = 0$ are shown in Figs. 1.11 and 1.12. It has not yet been proved that (1.25) and (1.28) represent a permissible acoustic vibration pattern, but this will be verified in Chapter 3. Other uniform plane waves propagating along y are the z-polarized wave

$$\mathbf{u} = \hat{\mathbf{z}}\cos{(\omega t - ky)} \tag{1.29}$$

and the y-polarized wave

$$\mathbf{u} = \hat{\mathbf{y}}\cos{(\omega t - ky)}. \tag{1.30}$$

The first of these has a simple shear strain field

$$S_{zy} = S_{yz} = \frac{1}{2}\frac{\partial u_z}{\partial y} = \frac{k}{2}\sin{(\omega t - ky)} \tag{1.31}$$

and is a *z-polarized y-propagating shear wave*, while the second has a compressional-extensional strain field

$$S_{yy} = \frac{\partial u_y}{\partial y} = k\sin{(\omega t - ky)} \tag{1.32}$$

and is called a *y-propagating compressional wave*. Field patterns for both of these waves are also illustrated in Figs. 1.11 and 1.12.

E.   TRANSFORMATION PROPERTIES

In Sections 1.A and 1.B the particle displacement field $\mathbf{u}(\mathbf{r}, t)$ and the strain field $S_{ij}(\mathbf{r}, t)$ were defined with respect to a set of rectangular Cartesian coordinate axes. In acoustic vibration and wave propagation analysis, it is often necessary to transform the acoustic field into a coordinate system which is suited to the geometry of a particular problem. This may involve either a rotation of the rectangular coordinate axes or a transformation from rectangular coordinates to either cylindrical or spherical coordinates. Only rectangular coordinates will be discussed here,† and only right-handed coordinate systems will be considered. The old and new coordinates are taken to be x, y, z and x', y', z', respectively, where the relative orientation of the two sets of axes is described by the direction cosines a_{ij} defined in Fig. 1.13a. That is a_{ij} is the direction cosine of the angle between the new axis $\hat{\mathbf{i}}'$ and the old axis $\hat{\mathbf{j}}$.

---

† Transformations to cylindrical and spherical coordinates are given in Part D of Appendix 1.

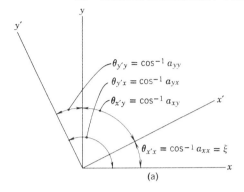

(a)

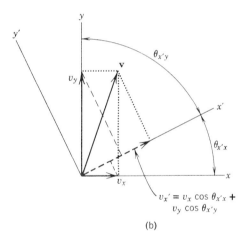

$$v_{x}' = v_x \cos \theta_{x'x} + v_y \cos \theta_{x'y}$$

(b)

FIGURE 1.13. Rectangular coordinate transformation corresponding to a clockwise rotation ξ about the z axis.

Suppose that a given vector $\mathbf{v}$ is represented by components v_x, v_y, v_z relative to the old coordinate axes and by components v_x', v_y', v_z' relative to the new coordinate axes. Each new component can be evaluated in terms of components along the old axes by projecting all of the old components onto the relevant axes of the new system. Figure (1.13b) illustrates this procedure for a two-dimensional case. For example

$$v_x' = \cos \theta_{x'x} v_x + \cos \theta_{x'y} v_y$$
$$= a_{xx} v_x + a_{xy} v_y.$$

The general vector transformation law is therefore

$$v'_i = a_{ij}v_j$$

$$i, j = x, y, z, \tag{1.33}$$

where the coefficients a_{ij} define a transformation matrix $[a]$. Since this transformation simply gives a new description of the same vector, the magnitude of $\mathbf{v}$ remains invariant,

$$\mathbf{v} \cdot \mathbf{v} = v_x^2 + v_y^2 + v_z^2 = (v'_x)^2 + (v'_y)^2 + (v'_z)^2.$$

If the vector is represented by a column matrix

$$\mathbf{v} \to [v] = \begin{bmatrix} v_x \\ v_y \\ v_z \end{bmatrix}$$

this condition is expressed by the matrix equation

$$\widetilde{[v]}[v] = \widetilde{[v']}[v'] = \widetilde{[v]}(\widetilde{[a]}[a])[v],$$

where (1.33) has been written as

$$[v'] = [a][v]$$

and use has been made of the matrix identity

$$\widetilde{([a][v])} = \widetilde{[v]}\widetilde{[a]}.$$

In other words, the transformation matrix must satisfy the condition

$$\widetilde{[a]}[a] = [I],$$

where $[I]$ is the identity matrix; or

$$[a]^{-1} = \widetilde{[a]}. \tag{1.34}$$

A matrix satisfying this condition is called *orthogonal*.

The transformation law (1.33) applies to any vector quantity, including the differential particle displacement vector $[du]$ and the differential coordinate vector $[dr]$. According to (1.22), these are related by the matrix equation

$$[du] = [\mathscr{E}][dr]. \tag{1.35}$$

From this relation and the vector transformation law (1.33) it is a simple matter to deduce the transformation law for $[\mathscr{E}]$. Transformation of $[du]$ to the new coordinate system is accomplished by using (1.33),

$$[du'] = [a][du] = [a][\mathscr{E}][dr]. \tag{1.36}$$

Similarly

$$[dr'] = [a][dr],$$

or

$$[a]^{-1}[dr'] = [dr].$$

Substitution into (1.35) then gives

$$[du'] = [a][\mathscr{E}][a]^{-1}[dr'].$$

This shows that the displacement gradient matrix in the new coordinate system is

$$[\mathscr{E}'] = [a][\mathscr{E}][a]^{-1} = [a][\mathscr{E}][\widetilde{a}], \qquad (1.37)$$

where use has been made of (1.34).

According to (1.24), the strain matrix $[S]$ is the symmetric part of the displacement gradient matrix. It must, therefore, also transform in the same manner. That is,

$$[S'] = [a][S][\widetilde{a}]; \qquad (1.38)$$

or, in terms of components,

$$S'_{ij} = a_{ik}S_{kl}(\tilde{a})_{lj} = a_{ik}S_{kl}a_{jl}.$$

Since the order of terms is not significant in this equation, the transformation may be rearranged in the easily remembered form

$$S'_{ij} = a_{ik}a_{jl}S_{kl}. \qquad (1.39)$$

The left-hand subscripts (i, j) on the a's match the strain subscripts on the left, and the right-hand subscripts (k, l) on the a's match the strain subscripts on the right.

Physical quantities, such as the displacement gradient $[\mathscr{E}]$ or the strain $[S]$, that transform according to (1.39) are called *second rank tensors*. Vector quantities, like the particle displacement $\mathbf{u}$, transform according to (1.33) and are called *first rank tensors*.

EXAMPLE 6. For a clockwise rotation of the coordinate axes through an angle ξ about the z axis, defined in the rotation sense of a *right-hand* screw advancing along the *positive z* direction (Fig. 1.13),

$$\cos \theta_{x'x} = a_{xx} = \cos \xi$$
$$\cos \theta_{x'y} = a_{xy} = \sin \xi$$
$$\cos \theta_{y'y} = a_{yy} = \cos \xi$$
$$\cos \theta_{y'x} = a_{yx} = -\sin \xi$$
$$\cos \theta_{z'z} = a_{zz} = 1$$

and all other a_{ij}'s are zero. The coordinate transformation matrix is therefore

$$[a] = \begin{bmatrix} \cos \xi & \sin \xi & 0 \\ -\sin \xi & \cos \xi & 0 \\ 0 & 0 & 1 \end{bmatrix}$$

Transformation of a particle displacement vector **u** to the rotated coordinate system is performed with the matrix equation

$$\begin{bmatrix} u'_x \\ u'_y \\ u'_z \end{bmatrix} = \begin{bmatrix} \cos \xi & \sin \xi & 0 \\ -\sin \xi & \cos \xi & 0 \\ 0 & 0 & 1 \end{bmatrix} \begin{bmatrix} u_x \\ u_y \\ u_z \end{bmatrix}$$

That is,

$$u'_x = \cos \xi \, u_x + \sin \xi \, u_y$$
$$u'_y = -\sin \xi \, u_x + \cos \xi \, u_y$$
$$u'_z = u_z.$$

In Example 4 the strain matrix for a static *pure shear* deformation in the *xy* plane was found to be

$$[S] = \begin{bmatrix} 0 & C & 0 \\ C & 0 & 0 \\ 0 & 0 & 0 \end{bmatrix}.$$

Applying (1.38) transforms this to

$$[S'] = \begin{bmatrix} \cos \xi & \sin \xi & 0 \\ -\sin \xi & \cos \xi & 0 \\ 0 & 0 & 1 \end{bmatrix} \begin{bmatrix} 0 & C & 0 \\ C & 0 & 0 \\ 0 & 0 & 0 \end{bmatrix} \begin{bmatrix} \cos \xi & -\sin \xi & 0 \\ \sin \xi & \cos \xi & 0 \\ 0 & 0 & 1 \end{bmatrix}$$

$$= \begin{bmatrix} C \sin 2\xi & C \cos 2\xi & 0 \\ C \cos 2\xi & -C \sin 2\xi & 0 \\ 0 & 0 & 0 \end{bmatrix}$$

in the rotated coordinate system.

It is seen from this that the strain components change radically when referred to a new coordinate system. In the original coordinate system only the shear strain components $S_{xy} = S_{yx}$ were present. Rotation of the coordinates adds an extension along the x' axis and a compression along the y' axis. When $\xi = \pi/4$ the shear strain components vanish completely, leaving only an extension along x' and a compression along y'. This is illustrated very clearly by Fig. 1.9b, which shows the deformation of a cubic block of material subjected to pure shear strain in the xy plane. Stretching along the x' axis and compression along the y' axis is clearly visible in the figure.

EXAMPLE 7. Inversion of the coordinate axis

$$\hat{\mathbf{x}}' = -\hat{\mathbf{x}}$$
$$\hat{\mathbf{y}}' = -\hat{\mathbf{y}}$$
$$\hat{\mathbf{z}}' = -\hat{\mathbf{z}}$$

is described by the transformation matrix

$$[a] = \begin{bmatrix} -1 & 0 & 0 \\ 0 & -1 & 0 \\ 0 & 0 & -1 \end{bmatrix}$$

In this case all vector components change sign,

$$u_i = -u_i$$
$$i = x, y, z$$

according to (1.33), but the strain components remain unchanged,

$$[S'] = \begin{bmatrix} -1 & 0 & 0 \\ 0 & -1 & 0 \\ 0 & 0 & -1 \end{bmatrix} \begin{bmatrix} S_{xx} & S_{xy} & S_{xz} \\ S_{xy} & S_{yy} & S_{yz} \\ S_{xy} & S_{yz} & S_{zz} \end{bmatrix} \begin{bmatrix} -1 & 0 & 0 \\ 0 & -1 & 0 \\ 0 & 0 & -1 \end{bmatrix} = [S].$$

Since the partial derivatives in (1.21) are unaffected when the vector components u_i and r_j both reverse sign, this result is not unexpected. An important physical consequence of this property of the strain components will be noted in Example 3 of Chapter 8.

EXAMPLE 8. In transforming inhomogeneous strains it is necessary to change both the dependent strain variables and the independent spatial coordinate variables. Consider, for instance, the x-polarized y-propagating shear strain given by (1.28) in Example 5, and let the coordinate axis be rotated about z, as in Example 6. The strain components transform in exactly the same way as the static shear strain components of Example 6. According to (1.33) the coordinate position vector

$$\mathbf{r} = \hat{\mathbf{x}}x + \hat{\mathbf{y}}y + \hat{\mathbf{z}}z$$

transforms as

$$[r'] = [a][r].$$

Therefore,

$$[r] = [a]^{-1}[r'] = \widetilde{[a]}[r'].$$

Thus, the relation between old and new position coordinates is

$$x = x' \cos \xi - y' \sin \xi$$
$$y = x' \sin \xi + y' \cos \xi$$
$$z = z'.$$

Using this result and the strain transformation derived in Example 6, one finds that the strain field is described in the rotated coordinate system by

$$[S'] = \frac{k}{2} \sin (\omega t - k(x' \sin \xi + y' \cos \xi)) \begin{bmatrix} \sin 2\xi & \cos 2\xi & 0 \\ \cos 2\xi & -\sin 2\xi & 0 \\ 0 & 0 & 0 \end{bmatrix} \quad (1.40)$$

This rather complicated transformation law will be seen later to account for many of the difficulties encountered in acoustic vibration problems.

EXAMPLE 9.   An alternative method of arriving at the strain field (1.40) is to first transform the displacement field (1.25) and then to apply the strain-displacement relation in the rotated coordinate system. This is a useful shortcut when the strain field in the original coordinate system is not required. From Example 6,

$$\mathbf{u} = \hat{\mathbf{x}} \cos (\omega t - ky)$$

transforms into

$$\mathbf{u} = (\hat{\mathbf{x}}' \cos \xi - \hat{\mathbf{y}}' \sin \xi) \cos (\omega t - ky)$$

and transformation of the spatial coordinate, as in Example 8, gives

$$\mathbf{u} = (\hat{\mathbf{x}}' \cos \xi - \hat{\mathbf{y}}' \sin \xi) \cos (\omega t - k(x' \sin \xi + y' \cos \xi)).$$

If the strain-displacement relation

$$S_{i'j'} = \frac{1}{2} \left(\frac{\partial u_{i'}}{\partial r_{j'}} + \frac{\partial u_{j'}}{\partial r_{i'}} \right)$$

is now applied, the result is

$$S_{x'x'} = \frac{\partial u_{x'}}{\partial x'} = k \sin \xi \cos \xi \sin (\omega t - k(x' \sin \xi + y' \cos \xi))$$

$$= \frac{k}{2} \sin 2\xi \sin (\omega t - k(x' \sin \xi + y' \cos \xi))$$

$$S_{y'y'} = \frac{\partial u_{y'}}{\partial y'} = -\frac{k}{2} \sin 2\xi \sin (\omega t - k(x' \sin \xi + y' \cos \xi))$$

$$S_{x'y'} = S_{y'x'} = \frac{1}{2}\left(\frac{\partial u_{x'}}{\partial y'} + \frac{\partial u_{y'}}{\partial x'}\right)$$

$$= \frac{k}{2} \cos 2\xi \sin (\omega t - k(x' \sin \xi + y' \cos \xi)),$$

which agrees with (1.40).

F. SYMBOLIC NOTATION AND ABBREVIATED SUBSCRIPTS

It is standard practice in physics and engineering to represent vector quantities by bold face letter symbols rather than giving their components. Thus, the electric and magnetic fields in electromagnetic theory are written symbolically as $\mathbf{E}$, $\mathbf{H}$ rather than using the subscript notation E_i, H_i. Product and differentiation operations are also represented symbolically. For example, the scalar (or dot) product of two vectors is

$$\mathbf{E} \cdot \mathbf{D} = E_x D_x + E_y D_y + E_z D_z = E_i D_i$$

and the divergence of a vector is

$$\nabla \cdot \mathbf{D} = \frac{\partial}{\partial x} D_x + \frac{\partial}{\partial y} D_y + \frac{\partial}{\partial z} D_z = \frac{\partial}{\partial r_i} D_i.$$

Second rank tensor quantities encountered in electromagnetism are also represented symbolically rather than by their components. For example, the electric displacement vector $\mathbf{D}$ and the electric field vector $\mathbf{E}$ in a crystalline medium are not always parallel. In such cases the components of $\mathbf{D}$ are related to the components of $\mathbf{E}$ by three linear equations

$$D_x = \epsilon_{xx} E_x + \epsilon_{xy} E_y + \epsilon_{xz} E_z$$

$$D_y = \epsilon_{yx} E_x + \epsilon_{yy} E_y + \epsilon_{yz} E_z$$

$$D_z = \epsilon_{zx} E_x + \epsilon_{zy} E_y + \epsilon_{zz} E_z; \qquad (1.41)$$

or, in matrix form,

$$[D] = [\epsilon][E]. \qquad (1.42)$$

Since the permittivity matrix $[\epsilon]$ in (1.42) relates one vector quantity to another, it transforms in exactly the same way as the displacement gradient matrix $[\mathscr{E}]$ in (1.35) and is therefore a second rank tensor. It is usually written symbolically as $\boldsymbol{\epsilon}$. A symbolic representation of (1.41) is then

$$\mathbf{D} = \boldsymbol{\epsilon} \cdot \mathbf{E}, \qquad (1.43)$$

where the dot indicates summation over the second subscript on ϵ_{ij}.

In this notation, no attempt is made to distinguish explicitly between second rank tensors and vectors. Tensor rank is established by the nature of the physical quantity itself and does not require special labeling, other than the use of standard symbols for all quantities. This symbolic notation facilitates manipulation of the electromagnetic field equations by reducing the amount of detail which must be carried along, and also exposes more clearly the principal steps in a derivation or calculation. For the same reasons, this kind of symbolism is also useful in acoustics. In this case, bold face letters are used to symbolize the second rank displacement gradient and strain fields, $\mathscr{E}$ and $\mathbf{S}$, as well as the vector particle displacement field $\mathbf{u}$. The same convention will be used for other second and higher rank tensors, to be introduced later. As in electromagnetism, the rank of a tensor does not have to be indicated explicitly by the symbolism, because it is already specified by the nature of the physical variables; $\mathbf{u}$ is a first rank tensor, $\mathscr{E}$ and $\mathbf{S}$ are second rank tensors, etc. With the introduction of symbolic notation, the basic equations of this chapter may be written much more compactly. The standard symbol† for the gradient of a vector $\mathbf{u}$ is $\nabla \mathbf{u}$. The linearized definition of displacement gradient (1.22) is therefore written as

$$\mathscr{E} = \nabla \mathbf{u} \qquad (1.44)$$

in symbolic notation. Correspondingly, (1.35) takes the form

$$d\mathbf{u} = \mathscr{E} \cdot d\mathbf{r}. \qquad (1.45)$$

As in (1.43), the dot indicates summation over the second subscript on $\mathscr{E}_{ij}$.

According to (1.24) the strain matrix equals the symmetric part of the displacement gradient matrix. This correspondence is expressed symbolically as

$$\mathbf{S} = \tfrac{1}{2}(\mathscr{E} + \tilde{\mathscr{E}}). \qquad (1.46)$$

The relationship between strain and particle displacement can now be written down by substituting (1.44) into (1.46),

$$\mathbf{S} = \tfrac{1}{2}(\nabla \mathbf{u} + \widetilde{\nabla \mathbf{u}}).$$

---

† Symbols for tensors of rank higher than one are often called *dyadics* and a corresponding formalism has been developed for manipulating them. This formalism is not used explicitly in this book, but the symbols are treated in a manner consistent with dyadic concepts. Complete treatments of dyadic methods are given in References 1 and 3 at the end of the chapter.

Because this equation is used so frequently in acoustics theory, it is desirable to introduce a more compacted symbolism for the right-hand side. The operation described takes the symmetric part of the displacement gradient, and it will therefore be represented by appending a subscript s (for symmetric) to the gradient symbol. Thus,

$$\tfrac{1}{2}(\nabla \mathbf{u} + \widetilde{\nabla \mathbf{u}}) = \nabla_s \mathbf{u}. \tag{1.47}$$

The strain-displacement relation is therefore

$$\mathbf{S} = \nabla_s \mathbf{u}. \tag{1.48}$$

In solving specific problems it is, of course, eventually necessary to select a coordinate system and resolve the fields into components with respect to these coordinates. In acoustics this step is commonly simplified by introducing a system of abbreviated subscripts for the strain components. Since the strain tensor (1.21) is symmetric, each component can be specified by one subscript rather than two. These are defined according to the scheme

$$\mathbf{S} = \begin{bmatrix} S_{xx} & S_{xy} & S_{xz} \\ S_{xy} & S_{yy} & S_{yz} \\ S_{xz} & S_{yz} & S_{zz} \end{bmatrix} = \begin{bmatrix} S_1 & \tfrac{1}{2}S_6 & \tfrac{1}{2}S_5 \\ \tfrac{1}{2}S_6 & S_2 & \tfrac{1}{2}S_4 \\ \tfrac{1}{2}S_5 & \tfrac{1}{2}S_4 & S_3 \end{bmatrix}, \tag{1.49}$$

where the order of numbering in the abbreviated system follows the cyclic pattern shown. The convention of introducing factors $\tfrac{1}{2}$ is standard practice in elasticity theory, the reason being that it simplifies some of the key equations in Chapters 3 and 5. In this abbreviated subscript notation, the strain may be written as a six-element column matrix rather than as a nine-element square matrix. That is

$$S = \begin{bmatrix} S_1 \\ S_2 \\ S_3 \\ S_4 \\ S_5 \\ S_6 \end{bmatrix} \tag{1.50}$$

EXAMPLE 10. For the x-polarized y-propagating strain wave of Example 5, the only nonzero strain components are

$$S_{yx} = S_{xy} = \frac{k}{2} \sin{(\omega t - ky)}$$

from (1.28). According to (1.49) this gives

$$S_6 = S_{xy} = 2\,k \sin(\omega t - ky),$$

and therefore

$$\mathbf{S} = k \sin(\omega t - ky) \begin{bmatrix} 0 \\ 0 \\ 0 \\ 0 \\ 0 \\ 1 \end{bmatrix} \qquad (1.51)$$

In example 8 this same strain field was referred to a set of coordinate axes rotated by an angle ξ about the z axis. This resulted in a transformed strain matrix given by (1.40), where

$$S_{x'x'} = S_{1'} = \frac{k}{2} \sin 2\xi \sin(\omega t - k(x' \sin \xi + y' \cos \xi))$$

$$S_{y'y'} = S_{2'} = -\frac{k}{2} \sin 2\xi \sin(\omega t - k(x' \sin \xi + y' \cos \xi))$$

$$S_{x'y'} = S_{yx'} = \frac{S_{6'}}{2} = \frac{k}{2} \cos 2\xi \sin(\omega t - k(x' \sin \xi + y' \cos \xi)).$$

Therefore

$$\mathbf{S'} = \frac{k}{2} \sin(\omega t - k(x' \sin \xi + y' \cos \xi)) \begin{bmatrix} \sin 2\xi \\ -\sin 2\xi \\ 0 \\ 0 \\ 0 \\ 2 \cos 2\xi \end{bmatrix}$$

in abbreviated subscript notation.

It is clear from Example 10 that the strain transformation law (1.38) cannot be applied directly in abbreviated subscripts. This, however, is not a significant disadvantage for the abbreviated notation. Conversion back to full subscripts can always be made before transformation or, alternatively, the abbreviated subscript transformation law derived in Section D of Chapter 3 may be used. To avoid confusion in this respect, it is important to have a clear notational indication of the type of subscript used. The convention used in this book will be to denote abbreviated subscripts by upper case letters (S_I) and full subscripts by lower case letters (S_{ij}).

One immediate advantage of the factors $\frac{1}{2}$ introduced in (1.49) is that the strain components in abbreviated subscripts are related in a simple way to the particle displacement components. From (1.21) and (1.49)

$$
\begin{bmatrix} S_1 \\ S_2 \\ S_3 \\ S_4 \\ S_5 \\ S_6 \end{bmatrix} = \begin{bmatrix} \dfrac{\partial u_x}{\partial x} \\[6pt] \dfrac{\partial u_y}{\partial y} \\[6pt] \dfrac{\partial u_z}{\partial z} \\[6pt] \dfrac{\partial u_y}{\partial z} + \dfrac{\partial u_z}{\partial y} \\[6pt] \dfrac{\partial u_x}{\partial z} + \dfrac{\partial u_z}{\partial x} \\[6pt] \dfrac{\partial u_x}{\partial y} + \dfrac{\partial u_y}{\partial x} \end{bmatrix} = \begin{bmatrix} \dfrac{\partial}{\partial x} & 0 & 0 \\[6pt] 0 & \dfrac{\partial}{\partial y} & 0 \\[6pt] 0 & 0 & \dfrac{\partial}{\partial z} \\[6pt] 0 & \dfrac{\partial}{\partial z} & \dfrac{\partial}{\partial y} \\[6pt] \dfrac{\partial}{\partial z} & 0 & \dfrac{\partial}{\partial x} \\[6pt] \dfrac{\partial}{\partial y} & \dfrac{\partial}{\partial x} & 0 \end{bmatrix} \begin{bmatrix} u_x \\ u_y \\ u_z \end{bmatrix}
$$

or

$$ S_I = \nabla_{Ij} u_j. \tag{1.52} $$

This is a matrix representation in rectangular coordinates of the strain-displacement relation (1.48). It may be used, for example, to calculate the strain (1.51) in Example 9 directly from the displacement field (1.25) of Example 5. The symmetric gradient operator ∇_s in (1.48) thus has a matrix representation

$$
\nabla_s \to \nabla_{Ij} = \begin{bmatrix} \dfrac{\partial}{\partial x} & 0 & 0 \\[6pt] 0 & \dfrac{\partial}{\partial y} & 0 \\[6pt] 0 & 0 & \dfrac{\partial}{\partial z} \\[6pt] 0 & \dfrac{\partial}{\partial z} & \dfrac{\partial}{\partial y} \\[6pt] \dfrac{\partial}{\partial z} & 0 & \dfrac{\partial}{\partial x} \\[6pt] \dfrac{\partial}{\partial y} & \dfrac{\partial}{\partial x} & 0 \end{bmatrix} \tag{1.53}
$$

in rectangular coordinates. Representations of ∇_s in cylindrical and spherical coordinates are derived in Part B of Appendix 1.

PROBLEMS

1. Find the strain fields corresponding to the particle displacement fields

(a) $u_x = Kx$, $u_y = Ky$

(b) $u_x = Kx$, $u_y = -Ky$

(c) $u_x = Ky$, $u_y = Kx$.

2. Sketch grid and field line diagrams illustrating the field distributions of Problem 1.

3. Transform the strain fields of Problem 1 to a coordinate system rotated $45°$ clockwise about the z axis.

4. Obtain the transformed strain fields of Problem 3 by first transforming the displacement fields of Problem 1 and then calculating strain in the new coordinate system.

5. Express the strain fields of Problems 1 and 3 in abbreviated subscript notation.

6. It was demonstrated in Section C that the antisymmetric displacement gradient matrix

$$[\mathscr{E}] = \begin{bmatrix} 0 & \mathscr{E}_{xy} & 0 \\ -\mathscr{E}_{xy} & 0 & 0 \\ 0 & 0 & 0 \end{bmatrix}$$

represents a pure rotation through a small angle. Show that this corresponds to a counterclockwise rotation about the z axis and that the rotation angle ϕ is given by the equation

$$\mathscr{E}_{xy} = \sin \phi \approx \phi.$$

Consider a rotation axis defined by the unit vector

$$\hat{\alpha} = \hat{x}\alpha_x + \hat{y}\alpha_y + \hat{z}\alpha_z$$

and prove that the displacement gradient matrix representing pure rotation through a small *clockwise* angle χ about this axis is

$$[\mathscr{E}] = \chi \begin{bmatrix} 0 & -\alpha_z & \alpha_y \\ \alpha_z & 0 & -\alpha_x \\ -\alpha_y & \alpha_x & 0 \end{bmatrix}$$

where $\alpha_x^2 + \alpha_y^2 + \alpha_z^2 = 1$. Verify that the relation

$$[du] = [\mathscr{E}][dL]$$

is, in this case, equivalent to

$$d\mathbf{u} = \chi\hat{\boldsymbol{\alpha}} \times d\mathbf{L}$$

7. An x-polarized z-propagating shear wave has the displacement field

$$\mathbf{u} = \hat{\mathbf{x}} \cos{(\omega t - kz)}.$$

Calculate the strain field and transform to a coordinate system rotated 45° clockwise about the z axis.

8. Express the displacement and strain fields of part (a) Problem 1 in terms of cylindrical coordinates (Appendix 1).

9. Using a diagram similar to Fig. 1.4, explain the presence of a strain component $S_{\phi\phi}$ in Problem 8.

10. An arbitrary two-dimensional strain field in the xy plane is described by strain components S_{xx}, S_{yy}, $S_{xy} = S_{yx}$. Show that a clockwise rotation of coordinates through an angle ξ about the z axis transforms the strain components into

$$S'_{xx} = \frac{S_{xx} + S_{yy}}{2} + \frac{S_{xx} - S_{yy}}{2} \cos 2\xi + S_{xy} \sin 2\xi$$

$$S'_{yy} = \frac{S_{xx} + S_{yy}}{2} - \frac{S_{xx} - S_{yy}}{2} \cos 2\xi - S_{xy} \sin 2\xi$$

$$S'_{xy} = S'_{yx} = \frac{S_{yy} - S_{xx}}{2} \sin 2\xi + S_{xy} \cos 2\xi.$$

11. Use the equations derived in Problem 10, find the angles ξ that give maximum values of tensile, compressive, and shear strain. Express the maximum strains in terms of S_{xx}, S_{yy}, and S_{xy}.

REFERENCES

1. P. C. Chou and N. J. Pagano, *Elasticity—Tensor, Dyadic and Engineering Approaches*, Ch. 2, van Nostrand, New York, 1967.
2. L. D. Landau and E. M. Lifshitz, *Theory of Elasticity*, pp. 1–4, Pergamon, New York, 1970.
3. G. Nadeau, *Introduction to Elasticity*, pp. 33–40, Holt, Rinehart, and Winston, New York, 1964.
4. J. F. Nye, *Physical Properties of Crystals*, Ch. 1, 2, 6, Oxford, England, 1964.
5. M. J. P. Musgrave, *Crystal Acoustics*, pp. 9–21 Holden-Day, San Francisco, 1970.

CHAPTER 2

STRESS AND THE DYNAMICAL EQUATIONS

A. BODY FORCES AND BODY TORQUES

Chapter 1 introduced the acoustic field variables, particle displacement $\mathbf{u}(\mathbf{r}, t)$ and strain $\mathbf{S}(\mathbf{r}, t)$, that characterize particle motion and deformation in a vibrating material medium. When a body vibrates acoustically, elastic restoring forces (or *stresses*) develop between neighboring particles. In a freely vibrating body these are the only forces present. If the vibration is driven by an external agency, two kinds of excitation forces (*body forces* and *surface* or *traction forces*) must also be considered. To analyze vibration problems, all of these forces must first be defined in a quantitative manner and then related mathematically to the fields $\mathbf{u}(\mathbf{r}, t)$ and $\mathbf{S}(\mathbf{r}, t)$.

Body forces are long-range forces acting directly upon particles in the interior of a body. For example, the gravitational field produces a static body force

$$\mathbf{F} \, dV = \rho \mathbf{g} \, dV \qquad (2.1)$$

on each particle† of volume dV. In (2.1) ρ is the mass density of the medium

---

† In Section A of Chapter 1 a distinction was drawn between the material particles considered in acoustics theory and the atomic particles that comprise the medium. In acoustics, all media are considered to be continuous and the material particles are infinitesimal volume elements.

and $\mathbf{g}$ is the gravitational field vector. This kind of force may also be produced by applying electric and magnetic fields to media having *permanent* electric or magnetic polarizations. *Body torques* also appear in such cases. In a medium with a permanent magnetic moment $\mathbf{M}$ per unit volume, application of a magnetic field of strength $\mathbf{H}$ thus exerts a body torque

$$\mathbf{G}\,dV = -\mathbf{M} \times \mathbf{H}\,dV \tag{2.2}$$

on every material particle of volume dV. It is characteristic of body forces and body torques that they are proportional to the volume of the particle on which they act and they are therefore defined by volume densities of force and torque, $\mathbf{F}$ and $\mathbf{G}$ in (2.1) and (2.2). In the MKS system used in this book $\mathbf{F}$ is measured in units of newtons/meter^3 and $\mathbf{G}$ in units of newtons/meter^2.

B. TRACTION FORCES AND STRESSES

An alternative method of exciting acoustic vibrations in a material body is to apply surface forces at its boundary. In this case the applied excitation does not act directly on particles within the body but is transmitted to them by means of the elastic forces (or stresses) acting between neighboring particles (Fig. 2.1). Like the traction forces applied at the boundary, the stresses between particles act upon a surface rather than a volume and may therefore be described as internal traction forces. Stresses and external traction forces are both measured in units of newtons/meter^2.

Stresses within a vibrating medium are defined by taking the material particles to be volume elements of some orthogonal coordinate system. Figure 2.2 shows such a particle in rectangular Cartesian coordinates. Each face is subjected to forces applied by the contiguous material. To specify these forces, three force components are required for each face of the particle. The traction force, or force *per unit area*, acting on the area element facing the $+x$ direction in Fig. 2.2 is thus

$$\mathbf{T}_x = \hat{\mathbf{x}}T_{xx} + \hat{\mathbf{y}}T_{yx} + \hat{\mathbf{z}}T_{zx}. \tag{2.3}$$

Similarly, the traction forces on area elements facing the $+y$ and $+z$ directions are

$$\mathbf{T}_y = \hat{\mathbf{x}}T_{xy} + \hat{\mathbf{y}}T_{yy} + \hat{\mathbf{z}}T_{zy} \tag{2.4}$$

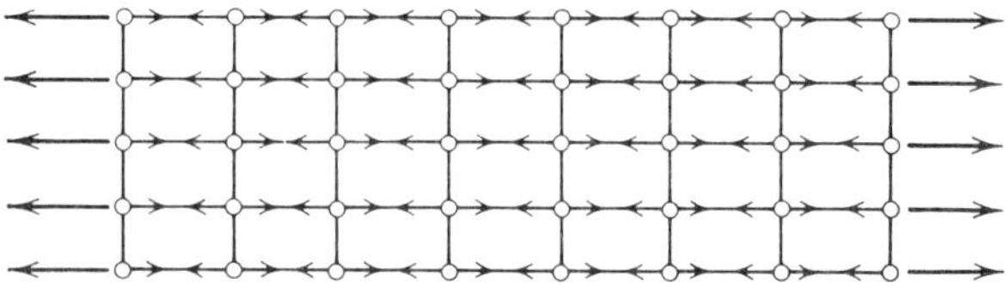

FIGURE 2.1. **Transmission of an applied traction force from particle to particle into the interior of a solid body.**

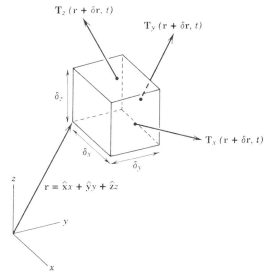

FIGURE 2.2.   Traction forces acting on a material particle. $\mathbf{T}_x$, $\mathbf{T}_y$, and $\mathbf{T}_z$ are forces *per unit area*.

and

$$\mathbf{T}_z = \hat{\mathbf{x}}T_{xz} + \hat{\mathbf{y}}T_{yz} + \hat{\mathbf{z}}T_{zz}. \qquad (2.5)$$

The components $T_{ij}(i, j = x, y, z)$ of these force densities are called *stress components*. In a vibrating medium these are always functions of spatial position. In Fig. 2.2 one must therefore take the limit δx, δy, $\delta z = 0$ in order to use (2.3)–(2.5) for defining the stress field at the point $\mathbf{r} = \hat{\mathbf{x}}x + \hat{\mathbf{y}}y + \hat{\mathbf{z}}z$. According to this definition, $T_{ij}(\mathbf{r}, t)$ is the ith component of force density acting on the $+j$ face of an infinitesimal volume element at position $\mathbf{r}$†.

EXAMPLE 1.   The static compressional displacement and strain of Examples 1 and 2 in Chapter 1 are produced by applying compressional forces F to the ends of the bar. Since a compressed body tends to expand laterally, it is also necessary to apply forces at the sides (Fig. 2.3a) in order to constrain the particle displacement to the x-direction, as assumed in the first example. If the cross sectional area of the bar is A, the right-hand end of the bar (facing the $+x$ direction) is subjected to a traction force

$$\mathbf{T}_x = -\hat{\mathbf{x}}\frac{F}{A}.$$

---

† This follows the convention used in References 2, 4, and 5, at the end of the chapter. In References 1 and 3 the roles of the subscripts are interchanged.

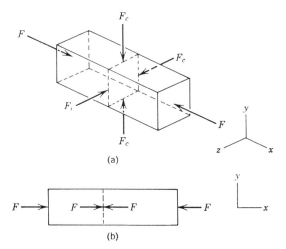

FIGURE 2.3.   **Traction forces acting on a uniformly compressed bar.**

The stress components acting on this surface are thus

$$T_{xx} = -\frac{F}{A}$$
$$T_{yx} = 0$$
$$T_{zx} = 0.$$

At the left-hand end of the bar (facing the $-x$ direction) the traction force is reversed,

$$\mathbf{T}_{-x} = \hat{\mathbf{x}}\frac{F}{A}.$$

The stresses T_{xx}, T_{yx}, T_{zx} are therefore *minus* the force components acting on a surface facing the $-x$ direction. Since the body is in static equilibrium, the forces are the same at every cross section of the bar (Fig. 2.3b) and the stress is uniform throughout,

$$T_{xx} = -\frac{F}{A}.$$

Two additional components of stress, T_{yy} and T_{zz}, are produced by the constraining forces F_c applied to the sides of the bar. If the bar is subjected to tension, all the forces in Fig. 2.3 are reversed and the stress component T_{xx} becomes

$$T_{xx} = +\frac{F}{A}.$$

Compressional stresses are negative and tensile stresses are positive, corresponding to the sign conventions for strains in Example 2 of Chapter 1.

EXAMPLE 2.   Static, pure shear deformation of a cube of material (Fig. 1.9b) is produced by applying forces F to its faces as shown in Fig. 2.4. If the area of each face is A, the traction forces on the $+x$ and $+y$ faces are

$$\mathbf{T}_x = \hat{\mathbf{y}}\,\frac{F}{A}$$

and

$$\mathbf{T}_y = \hat{\mathbf{x}}\,\frac{F}{A}\,,$$

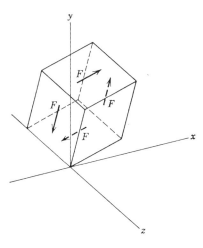

FIGURE   2.4.   **Traction   forces applied to a uniformly sheared block.**

respectively. As in Example 1, these traction forces are uniform throughout the block and the stress field is therefore described by the constant shear stress components

$$T_{yx} = \frac{F}{A}$$

$$T_{xy} = \frac{F}{A}\,.$$

B.1   Stress on an Arbitrarily Oriented Surface

Although the stresses in a deformed medium have been defined in terms of forces acting on surfaces parallel to the coordinate planes, they may also be calculated for an arbitrarily oriented surface. To see how this is done, consider the tetrahedral volume element in Fig. 2.5. Three of the faces are normal

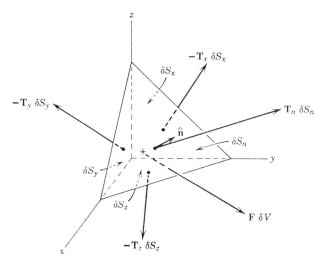

FIGURE 2.5.   Calculation of the traction force on a surface
normal to $\hat{n}$. The traction forces $T_x = \hat{x}T_{xx} + \hat{y}T_{yx} + \hat{z}T_{zx}$,
etc. are evaluated at $x, y, z = 0$. Higher order terms drop out
when $\delta V \to 0$.

to the rectangular coordinate axes and the fourth face has its outward-directed
normal $\hat{n}$ in an arbitrary direction. The fourth face, of area δS_n, may be an
arbitrarily oriented area within the body or it may lie in the bounding surface.

Forces acting on the volume element δV include traction forces on the
surfaces $\delta S_x, \delta S_y, \delta S_z, \delta S_n$ and a body force $\mathbf{F} \, \delta V$, which may include both
applied and inertial force terms. These forces must all be in balance. Pro-
jection of the traction forces onto the x axis gives

$$T_{xn} \, \delta S_n - T_{xx} \, \delta S_x - T_{xy} \, \delta S_y - T_{xz} \, \delta S_z + F_x \, \delta V = 0. \qquad (2.6)$$

If the volume element is allowed to approach zero, the body force term drops
out because δV goes to zero faster than the δS's. Since δS_x is the projection
of δS_n on the yz plane (Fig. 2.5)

$$\delta S_x = n_x \, \delta S_n,$$

where $\hat{n}$ is the unit vector normal to δS_n. Similarly,

$$\delta S_y = n_y \, \delta S_n$$

$$\delta S_z = n_z \, \delta S_n$$

and (2.6) becomes

$$T_{xn} = T_{xx}n_x + T_{xy}n_y + T_{xz}n_z. \qquad (2.7)$$

Repetition of this calculation for the y and z directions leads to

$$T_{yn} = T_{yx}n_x + T_{yy}n_y + T_{yz}n_z \tag{2.8}$$

$$T_{zn} = T_{zx}n_x + T_{zy}n_y + T_{zz}n_z. \tag{2.9}$$

The set of linear equations (2.7)–(2.9) is conveniently expressed in matrix form as

$$
\begin{bmatrix} T_{xn} \\ T_{yn} \\ T_{zn} \end{bmatrix}
=
\begin{bmatrix} T_{xx} & T_{xy} & T_{xz} \\ T_{yx} & T_{yy} & T_{yz} \\ T_{zx} & T_{zy} & T_{zz} \end{bmatrix}
\begin{bmatrix} n_x \\ n_y \\ n_z \end{bmatrix}
\tag{2.10}
$$

Traction force vector $\mathbf{T}_n$ Stress matrix   Normal vector $\hat{\mathbf{n}}$

or

$$[T_n] = [T][n]. \tag{2.11}$$

The components of $\mathbf{T}_n$ in (2.10) are the stresses acting on a surface normal to $\hat{\mathbf{n}}$.

EXAMPLE 3. In Example 2 there were only two nonzero stress components

$$T_{xy} = T_{yx} = \frac{F}{A}.$$

The stress matrix in (2.10) is therefore

$$
[T] = \begin{bmatrix} 0 & \dfrac{F}{A} & 0 \\ \dfrac{F}{A} & 0 & 0 \\ 0 & 0 & 0 \end{bmatrix}
$$

where the T_{yx} component represents a y-directed traction force on a $+x$-oriented surface and the T_{xy} component represents an x-directed traction force on a $+y$-oriented surface (Fig. 2.6a and 2.6b).

Now consider surfaces normal to the rotated coordinate axes x', y' in Fig. 2.6c

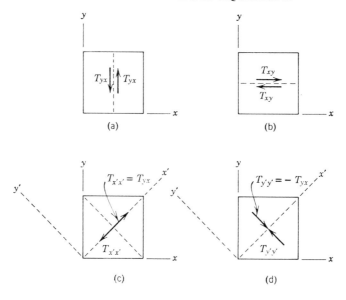

FIGURE 2.6.   Stress components in a uniformly sheared block.
The deformation is assumed to be infinitesimally small.

and 2.6d. If $\hat{\mathbf{n}}$ is taken parallel to the x' axis,

$$n_x = \frac{1}{\sqrt{2}}$$

$$n_y = \frac{1}{\sqrt{2}}$$

and

$$\begin{bmatrix} T_{xn} \\ T_{yn} \\ T_{zn} \end{bmatrix} = \frac{1}{\sqrt{2}} \begin{bmatrix} \dfrac{F}{A} \\ \dfrac{F}{A} \\ 0 \end{bmatrix}$$

from (2.10). This gives components of $\mathbf{T}_n$ referred to the x and y axes. To find components along the x' and y' axes one must perform the calculations

$$T_{x'x'} = \hat{\mathbf{x}}' \cdot \mathbf{T}_n = \frac{1}{\sqrt{2}} T_{nx} + \frac{1}{\sqrt{2}} T_{ny} = \frac{F}{A}$$

$$T_{y'x'} = \hat{\mathbf{y}}' \cdot \mathbf{T}_n = -\frac{1}{\sqrt{2}} T_{nx} + \frac{1}{\sqrt{2}} T_{ny} = 0.$$

The traction force on the $+x'$-oriented surface in Fig. 2.6c is thus directed along the x' axis and is positive. This is a pure tensile stress, corresponding to the stretching along x' illustrated in Fig. 1.9b.

When $\hat{n}$ is taken parallel to the y' axis,

$$n_x = -\frac{1}{\sqrt{2}}$$

$$n_y = \frac{1}{\sqrt{2}}$$

and

$$\begin{bmatrix} T_{xn} \\ \\ T_{yn} \\ \\ T_{zn} \end{bmatrix} = \frac{1}{\sqrt{2}} \begin{bmatrix} \dfrac{F}{A} \\ \\ -\dfrac{F}{A} \\ \\ 0 \end{bmatrix}.$$

In this case the traction force is directed along y' but is negative. This is a pure compression

$$T_{y'y'} = -\frac{F}{A},$$

again corresponding to the physical situation in Fig. 1.9b.

B.2 Transformation Properties

Comparison of (2.11) with (1.35) shows that the stress matrix, like the displacement gradient matrix $[\mathscr{E}]$, relates one vector to another. The argument used in deriving (1.37) therefore shows that the stress matrix transforms according to the law

$$[T'] = [a][T][\widetilde{a}], \tag{2.12}$$

or

$$T'_{ij} = a_{ik}a_{jl}T_{kl}. \tag{2.13}$$

The stress and strain matrices thus transform in exactly the same way and are both second rank tensors. Following the symbolism introduced in Section 1.F the stress tensor is represented by a bold face letter $\mathbf{T}$, and (2.10) may be written symbolically as

$$\mathbf{T}_n = \mathbf{T} \cdot \hat{n}, \tag{2.14}$$

where the dot indicates summation over the second subscript on T_{ij}.

EXAMPLE 4. The transformation law (2.12) provides another method for calculating the stresses $T_{x'x'}$ and $T_{y'y'}$ in Fig. 2.6c and 2.6d. According to Example

6 of Chapter 1 the transformation matrix for a 45° clockwise rotation of coordinate axes about z is

$$[a] = \begin{bmatrix} \dfrac{1}{\sqrt{2}} & \dfrac{1}{\sqrt{2}} & 0 \\[2mm] -\dfrac{1}{\sqrt{2}} & \dfrac{1}{\sqrt{2}} & 0 \\[2mm] 0 & 0 & 1 \end{bmatrix}$$

Application of the transformation (2.12) to the stress

$$[T] = \begin{bmatrix} 0 & \dfrac{F}{A} & 0 \\[2mm] \dfrac{F}{A} & 0 & 0 \\[2mm] 0 & 0 & 0 \end{bmatrix}$$

from Example 3, thus gives

$$[T'] = \begin{bmatrix} \dfrac{F}{A} & 0 & 0 \\[2mm] 0 & -\dfrac{F}{A} & 0 \\[2mm] 0 & 0 & 0 \end{bmatrix}$$

in agreement with the calculations of Example 3.

C. THE DYNAMICAL EQUATIONS OF ACOUSTICS

The deformation of an acoustically vibrating body has now been characterized by a particle displacement field $u(r, t)$ and a strain field $S(r, t)$; and the forces associated with this deformation have been characterized by a stress field $T(r, t)$, a body force field $F(r, t)$, and a body torque field $G(r, t)$. Traction forces applied at boundaries of the body are expressed in terms of the stress field by the relation (2.14), with $\hat{n}$ taken as the *outward-directed* unit vector at the boundary. To proceed with the analysis of acoustic vibrations, the particle displacement u must be related to the applied forces and torques F, G and to the elastic restoring forces T by applying the dynamical laws of mechanics. It was seen in Sections B and C of Chapter 1 that particles in a vibrating medium generally experience both translational and rotational motions. These two kinds of motion will be treated separately.

C.1   Translational Equation of Motion

Consider a vibrating material particle of arbitrary shape, with volume δV and surface area δS. The forces associated with its vibration are a body force $\mathbf{F}\,\delta V$ and traction forces applied to its surface by the neighboring particles. The applied surface forces are calculated from (2.14), which may be used for both the external traction forces at the boundary of a body and the internal traction forces between particles within the body. The integrated surface force acting on the particle is therefore

$$\int_{\delta S} \mathbf{T} \cdot \hat{\mathbf{n}}\, dS.$$

Newton's Law then states that

$$\int_{\delta S} \mathbf{T} \cdot \hat{\mathbf{n}}\, dS + \int_{\delta V} \mathbf{F}\, dV = \int_{\delta V} \rho \frac{\partial^2 \mathbf{u}}{\partial t^2}\, dV, \tag{2.15}$$

where ρ is the equilibrium mass density of the medium.†

If the particle volume is sufficiently small, the integrands of the volume integrals in (2.15) are essentially constant, and

$$\frac{\displaystyle\int_{\delta S} \mathbf{T} \cdot \hat{\mathbf{n}}\, dS}{\delta V} = \rho \frac{\partial^2 \mathbf{u}}{\partial t^2} - \mathbf{F}. \tag{2.16}$$

The limit of the left-hand side of this equation as $\delta V \to 0$ is defined as the divergence of the stress, represented symbolically as‡

$$\nabla \cdot \mathbf{T} = \lim_{\delta V \to 0} \frac{\displaystyle\int_{\delta S} \mathbf{T} \cdot \hat{\mathbf{n}}\, dS}{\delta V}. \tag{2.17}$$

In this limit (2.16) becomes

$$\nabla \cdot \mathbf{T} = \rho \frac{\partial^2 \mathbf{u}}{\partial t^2} - \mathbf{F}, \tag{2.18}$$

the *translational equation of motion* for a vibrating medium.

---

† Changes in the particle volume due to deformation lead to first order changes in the mass density. These changes must be considered in nonlinear analysis, but can be ignored in the linearized approximation.

‡ In dyadic notation divergence of the stress should, strictly speaking, be written as $\mathbf{T} \cdot \nabla$. However, it will be seen later in this section that $T_{ij} = T_{ji}$ for all problems of interest here. Under these conditions $\mathbf{T} \cdot \nabla = \nabla \cdot \mathbf{T}$.

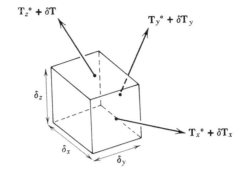

FIGURE 2.7.   Evaluation of $\nabla \cdot \mathbf{T}$ in rectangular Cartesian coordinates.

The symbolic form of the translational equation of motion (2.18) is independent of coordinates. To apply it to a specific problem the divergence of stress must be evaluated in some suitable coordinate system. This is accomplished by assuming that the material particles are volume elements in the coordinate system of interest and applying the fundamental definition (2.17). In rectangular Cartesian coordinates the particles thus become elementary cubes (Fig. 2.7). If the stresses at the center of the cube in the figure are T°_{xx}, T°_{xy}, etc., then the traction forces acting on the *outside* of the $+x$, $+y$, $+z$ cube faces are

$$\mathbf{T}^\circ_x + \delta\mathbf{T}_x = \mathbf{T}^\circ_x + \frac{\partial}{\partial x}(\hat{\mathbf{x}}T_{xx} + \hat{\mathbf{y}}T_{yx} + \hat{\mathbf{z}}T_{zx})\frac{\delta x}{2}$$

$$\mathbf{T}^\circ_y + \delta\mathbf{T}_y = \mathbf{T}^\circ_y + \frac{\partial}{\partial y}(\hat{\mathbf{x}}T_{xy} + \hat{\mathbf{y}}T_{yy} + \hat{\mathbf{z}}T_{zy})\frac{\delta y}{2}$$

$$\mathbf{T}^\circ_z + \delta\mathbf{T}_z = \mathbf{T}^\circ_z + \frac{\partial}{\partial z}(\hat{\mathbf{x}}T_{xz} + \hat{\mathbf{y}}T_{yz} + \hat{\mathbf{z}}T_{zz})\frac{\delta z}{2}. \qquad (2.19)$$

Similarly, the traction forces acting on the *outside* of the $-x$, $-y$, $-z$ cube faces are

$$\mathbf{T}^\circ_{-x} - \delta\mathbf{T}_{-x} = -\mathbf{T}^\circ_x + \delta\mathbf{T}_x$$

$$\mathbf{T}^\circ_{-y} - \delta\mathbf{T}_{-y} = -\mathbf{T}^\circ_y + \delta\mathbf{T}_y$$

$$\mathbf{T}^\circ_{-z} - \delta\mathbf{T}_{-z} = -\mathbf{T}^\circ_z + \delta\mathbf{T}_z, \qquad (2.20)$$

where $\delta\mathbf{T}_x$, $\delta\mathbf{T}_y$, $\delta\mathbf{T}_z$ are the same as in (2.19). The surface integral of

$$\mathbf{T} \cdot \hat{\mathbf{n}} = \mathbf{T}_x n_x + \mathbf{T}_y n_y + \mathbf{T}_z n_z$$

over the surface of the cube thus becomes

$$\int_{\delta S} \mathbf{T} \cdot \hat{\mathbf{n}} \, dS = \left(\frac{\partial}{\partial x} \mathbf{T}_x + \frac{\partial}{\partial y} \mathbf{T}_y + \frac{\partial}{\partial z} \mathbf{T}_z \right) \delta x \, \delta y \, \delta z, \qquad (2.21)$$

and

$$\nabla \cdot \mathbf{T} = \lim_{\delta V \to 0} \frac{\int_{\delta S} \mathbf{T} \cdot \hat{\mathbf{n}} \, dS}{\delta x \, \delta y \, \delta z} = \left(\frac{\partial}{\partial x} \mathbf{T}_x + \frac{\partial}{\partial y} \mathbf{T}_y + \frac{\partial}{\partial z} \mathbf{T}_z \right). \qquad (2.22)$$

In a rectangular Cartesian coordinate system the unit vectors in (2.19) are not functions of the coordinates, and the derivatives apply, therefore, only to the stress components.† One finds, then, that the divergence of stress in rectangular coordinates is

$$\nabla \cdot \mathbf{T} = \hat{\mathbf{x}} \left(\frac{\partial}{\partial x} T_{xx} + \frac{\partial}{\partial y} T_{xy} + \frac{\partial}{\partial z} T_{xz} \right)$$

$$+ \hat{\mathbf{y}} \left(\frac{\partial}{\partial x} T_{yx} + \frac{\partial}{\partial y} T_{yy} + \frac{\partial}{\partial z} T_{yz} \right)$$

$$+ \hat{\mathbf{z}} \left(\frac{\partial}{\partial x} T_{zx} + \frac{\partial}{\partial y} T_{zy} + \frac{\partial}{\partial z} T_{zz} \right).$$

This may be written compactly as

$$(\nabla \cdot \mathbf{T})_i = \frac{\partial}{\partial r_j} T_{ij}$$

$$i, j = x, y, z \qquad (2.23)$$

and the translational equation of motion in rectangular Cartesian coordinates is‡

$$\frac{\partial}{\partial r_j} T_{ij} = \rho \frac{\partial^2 u_i}{\partial t^2} - F_i$$

$$i, j = x, y, z \qquad (2.24)$$

from (2.18).

---

† In cylindrical and spherical coordinates, derivatives of the unit vectors and the elemental surface areas must also be considered (Part C of Appendix 1).

‡ This equation often appears in the literature as $T_{ij,j} = \rho \ddot{u}_i - F_i$, where the double dot over u_i denotes the second derivative with respect to time and the comma subscript on T denotes a spatial partial derivative with respect to the jth spatial coordinate. Summation over repeated subscripts is understood.

C.2   Rotational Equation of Motion

To examine the dynamics of particle *rotation*, one must consider the body torque acting on a particle, as well as the torques arising from traction forces acting on the particle surface. Figure 2.8 shows the z-components of these torques for a rectangular particle. The traction force torque is found by multiplying the appropriate stress components by the surface area of the particle face and by the moment arm. This gives a net traction force torque

$$(T_{yx} - T_{xy})\, \delta x\, \delta y\, \delta z,$$

and the rotational equation of motion is therefore

$$(T_{yx} - T_{xy} + G_z)\, \delta x\, \delta y\, \delta z = I_z \frac{\partial^2}{\partial t^2}\, \theta_z,$$

where G_z is the z component of body torque density, θ_z is the rotation angle about z, and

$$I_z = \rho\, \delta x\, \delta y\, \delta z (\delta x^2 + \delta y^2)_{\text{average}}$$

is the moment of inertia about the z axis. As δx, δy, $\delta z \to 0$ the moment of inertia goes to zero faster than the volume of the element, and the equation reduces to

$$T_{yx} - T_{xy} + G_z = 0.$$

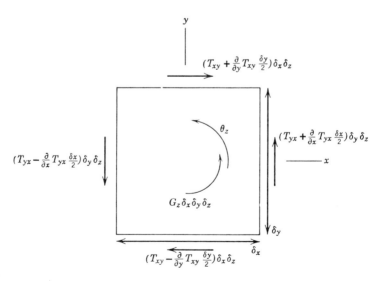

FIGURE 2.8.   **Body torque and traction force components that produce particle rotation about the z axis.**

Analogous results are obtained for rotation about the x and y axes, giving the general *rotational equation of motion*

$$T_{ji} - T_{ij} + G_k = 0, \tag{2.25}$$

where subscripts occur in the cyclic patterns

$$
\begin{array}{ccc}
i & j & k \\
x & y & z \\
y & z & x \\
z & x & y
\end{array}
$$

The rotational equation of motion (2.25) shows, first of all, that there are no inertial effects associated with particle rotation. This apparently paradoxical result comes about because the inertial forces decrease faster than the torques when the elementary volume approaches zero. Another consequence of (2.25) is that the stress matrix is unsymmetric ($T_{ij} \neq T_{ji}$) only when body torques are present. It has been seen in Section A of this chapter that such torques occur in media with permanent electric or magnetic polarization (ferroelectric or ferromagnetic materials). However, even in strongly polarized materials the body torques $\mathbf{G}$ are found to be of negligible importance in linearized vibration theory, and they will therefore be neglected. Under these conditions the stress matrix is always symmetric, and particle rotation plays no part in the dynamics of the vibration.

EXAMPLE 5. Consider a uniform plane wave of xy shear stress propagating along the y axis,

$$T_{xy} = T_{yx} = \sin{(\omega t - ky)}. \tag{2.26}$$

From (2.23) the divergence of this stress field is

$$
\begin{aligned}
\nabla \cdot \mathbf{T} &= \hat{\mathbf{x}} \, \frac{\partial}{\partial r_j} \, T_{xj} + \hat{\mathbf{y}} \, \frac{\partial}{\partial r_j} \, T_{yj} + \hat{\mathbf{z}} \, \frac{\partial}{\partial r_j} \, T_{zj} \\
&= \hat{\mathbf{x}} \, \frac{\partial}{\partial y} \, T_{xy} = -\hat{\mathbf{x}} k \cos{(\omega t - ky)},
\end{aligned}
$$

since the stress components are functions of only one spatial coordinate (y). If this is a freely propagating wave with no body force sources ($\mathbf{F} = 0$), the translational equation of motion (2.24) requires that the particle displacement field associated with (2.26) satisfy the conditions

$$-k \cos{(\omega t - ky)} = \rho \, \frac{\partial^2}{\partial t^2} u_x$$

$$0 = \frac{\partial^2}{\partial t^2} u_y$$

$$0 = \frac{\partial^2}{\partial t^2} u_z. \tag{2.27}$$

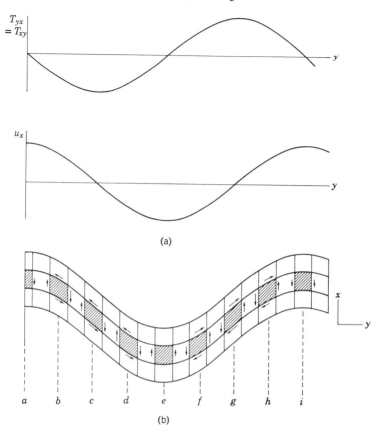

FIGURE 2.9. **Relationship between the stress and particle displacement fields in an x-polarized y-propagating uniform plane wave. Because of mechanical inertia the net forces on the particles in (b) are $180°$ out of phase with the displacements.**

That is,†

$$\mathbf{u} = \hat{\mathbf{x}}\,\frac{k}{\rho\omega^2}\,\cos\,(\omega t - ky). \tag{2.28}$$

Except for the multiplying factor $k/\rho\omega^2$, this is the same as the particle displacement field (1.25) for the x-*polarized y-propagating shear wave* of Example 5 in Chapter 1. The strain field accompanying (2.26) and (2.28) is therefore

$$S_{xy} = S_{yx} = \frac{1}{2}\,\frac{k^2}{\rho\omega^2}\,\sin\,(\omega t - ky). \tag{2.29}$$

† The particle displacement field

$$\mathbf{u} = \mathbf{c}t,$$

where **c** is arbitrary, also satisfies the differential equations (2.27). This, however, is a *rigid* motion of the medium and is of no interest in acoustics.

An analogous development can be carried through for the z-polarized y-propagating shear wave and the y-propagating compressional wave of Example 5 in Chapter 1.

Figure 2.9a illustrates the phase relationship of stress and particle displacement fields for an x-polarized y-propagating shear wave. A physical explanation for the 90° spatial phase shift between stress and displacement is given by Fig. 2.9b. Particle displacements are shown by a grid diagram similar to that of Fig. 1.11a. For selected grid squares, or "particles", the stresses applied by neighboring particles are indicated with small arrows. Consider first a point of maximum positive stress (plane g in the figure). At this point the traction force on the $+y$ face of a particle is in the $+x$ direction and an identical force acts in the $-x$ direction on the $-y$ face. Since $T_{yx} = T_{xy}$, there is also a $+y$ force of the same magnitude acting on the $+x$ face and the negative of this force acting on the $-x$ face. Since all forces are in balance, particles at this point are not displaced laterally. They simply deform and rotate. Note, however, that the rotation is produced without application of a net torque to the particle. This reflects the absence of an inertial term in the rotational equation of motion (2.25). At plane c, where the stress has its maximum negative value, all forces are reversed in sign.

Next, consider a point where the stress passes through zero from negative to positive (plane e in the figure). The grid has been drawn with the center of a particle at the point of zero stress. Owing to the finite size of a particle, the $+y$ particle face thus experiences a positive stress and the $-y$ particle face experiences a negative stress. This means that there are small $+x$-directed forces applied to both the $+y$ and $-y$ faces of a particle. It is these unbalanced forces that produce a maximum particle displacement at this plane. For the $+x$ and $-x$ faces of a particle the *average* stress is zero, and there is nothing to produce a displacement along the y axis. If the stress at the center of a particle passes through zero from positive to negative (plane a or plane i), the net force on the particle is in the $-x$ direction and the displacement is reversed in sign. At the planes b, d, f, h between points of zero and maximum stress there are again unbalanced forces on the $+y$ and $-y$ faces. These produce particle displacement along x. Forces at the $+x$ and $-x$ particle faces, on the other hand, are always in perfect balance because the stress field (2.26) is independent of the x coordinate. There is therefore no particle motion along y.

D. ABBREVIATED SUBSCRIPTS

When the stress matrix is symmetric, the abbreviated subscript notation introduced in Section 1.F can also be used to describe stress components. In this case the convention is to omit the factors $\frac{1}{2}$ that appeared in (1.49), and

$$\mathbf{T} = \begin{bmatrix} T_{xx} & T_{xy} & T_{xz} \\ T_{xy} & T_{yy} & T_{yz} \\ T_{xz} & T_{yz} & T_{zz} \end{bmatrix} = \begin{bmatrix} T_1 & T_6 & T_5 \\ T_6 & T_2 & T_4 \\ T_5 & T_4 & T_3 \end{bmatrix} \tag{2.30}$$

The stress can now be written as a six-element column matrix†

$$
\mathbf{T} =
\begin{bmatrix}
T_1 \\
T_2 \\
T_3 \\
T_4 \\
T_5 \\
T_6
\end{bmatrix}
\tag{2.31}
$$

EXAMPLE 6. In Example 5 the stress field for an x-polarized y-propagating shear wave was

$$
T_{xy} = T_{yx} = \sin(\omega t - ky)
$$

and the stress matrix is therefore

$$
\mathbf{T} = \sin(\omega t - ky)
\begin{bmatrix}
0 & 1 & 0 \\
1 & 0 & 0 \\
0 & 0 & 0
\end{bmatrix}
$$

According to (2.30), this becomes

$$
\mathbf{T} = \sin(\omega t - ky)
\begin{bmatrix}
0 \\
0 \\
0 \\
0 \\
0 \\
1
\end{bmatrix}
\tag{2.32}
$$

in abbreviated subscript notation.

Since both stress and strain are second rank tensor quantities the xy shear stress in this example transforms in the same way as the xy shear strain in Example 8 of Chapter 1. A clockwise rotation of coordinate axes through an angle ξ about the z axis therefore converts (2.32) to

$$
\mathbf{T}' = \sin(\omega t - k(x' \sin \xi + y' \cos \xi))
\begin{bmatrix}
\sin 2\xi \\
-\sin 2\xi \\
0 \\
0 \\
0 \\
\cos 2\xi
\end{bmatrix}
$$

† This is possible only when the stress matrix is symmetric.

Note that the factor 2 that appeared in the sixth element of the transformed strain $\mathbf{S}'$ in Example 10 of Chapter 1 is not present in the stress. This is simply a consequence of the different definitions of abbreviated subscript components in (1.49) and (2.30).

The cautionary remarks regarding transformation of the strain field (Section 1.F) also apply here. To transform the stress, components must either be converted to full subscript notation before applying (2.13), or the stress transformation laws derived in Section D of Chapter 3 may be used directly.

The divergence of $\mathbf{T}$ in the translational equation of motion (2.18) can be written in terms of abbreviated subscripts as

$$
\nabla \cdot \mathbf{T} =
\begin{bmatrix}
\dfrac{\partial}{\partial x} \overset{T_1}{\underset{}{\left(T_{xx}\right)}} + \dfrac{\partial}{\partial y} \overset{T_6}{\underset{}{\left(T_{xy}\right)}} + \dfrac{\partial}{\partial z} \overset{T_5}{\underset{}{\left(T_{xz}\right)}} \\[2em]
\dfrac{\partial}{\partial x} \overset{T_6}{\underset{}{\left(T_{xy}\right)}} + \dfrac{\partial}{\partial y} \overset{T_2}{\underset{}{\left(T_{yy}\right)}} + \dfrac{\partial}{\partial z} \overset{T_4}{\underset{}{\left(T_{yz}\right)}} \\[2em]
\dfrac{\partial}{\partial x} \overset{T_5}{\underset{}{\left(T_{xz}\right)}} + \dfrac{\partial}{\partial y} \overset{T_4}{\underset{}{\left(T_{yz}\right)}} + \dfrac{\partial}{\partial z} \overset{T_3}{\underset{}{\left(T_{zz}\right)}}
\end{bmatrix}
\tag{2.33}
$$

and this may be expressed as a matrix operator multiplying the stress column matrix; that is,

$$
\nabla \cdot \mathbf{T} =
\begin{bmatrix}
\dfrac{\partial}{\partial x} & 0 & 0 & 0 & \dfrac{\partial}{\partial z} & \dfrac{\partial}{\partial y} \\[1.5em]
0 & \dfrac{\partial}{\partial y} & 0 & \dfrac{\partial}{\partial z} & 0 & \dfrac{\partial}{\partial x} \\[1.5em]
0 & 0 & \dfrac{\partial}{\partial z} & \dfrac{\partial}{\partial y} & \dfrac{\partial}{\partial x} & 0
\end{bmatrix}
\begin{bmatrix}
T_1 \\ T_2 \\ T_3 \\ T_4 \\ T_5 \\ T_6
\end{bmatrix}
\tag{2.34}
$$

The translational equation of motion (2.18) then has a matrix representation

$$\nabla_{iJ} T_J = \rho \frac{\partial^2 u_i}{\partial t^2} - F_i$$

$$i = x, y, z$$

$$J = 1, 2, 3, 4, 5, 6 \tag{2.35}$$

in abbreviated subscript notation. The divergence of stress operation thus has a matrix representation

$$\nabla \cdot \rightarrow \nabla_{iJ} = \begin{bmatrix} \dfrac{\partial}{\partial x} & 0 & 0 & 0 & \dfrac{\partial}{\partial z} & \dfrac{\partial}{\partial y} \\[2mm] 0 & \dfrac{\partial}{\partial y} & 0 & \dfrac{\partial}{\partial z} & 0 & \dfrac{\partial}{\partial x} \\[2mm] 0 & 0 & \dfrac{\partial}{\partial z} & \dfrac{\partial}{\partial y} & \dfrac{\partial}{\partial x} & 0 \end{bmatrix} \tag{2.36}$$

in rectangular Cartesian coordinates. In this case the divergence matrix operator is just the transpose of the symmetric gradient matrix operator (1.53). For cylindrical and spherical coordinates this transpose relationship no longer applies (Part C of Appendix 1).

In any coordinate system, the divergence of stress may be calculated directly from the abbreviated subscript stress components by using the appropriate matrix ∇_{iJ}.

EXAMPLE 7. When ∇_{iJ} is applied to the y-propagating stress field (2.32) in Example 6, only the derivative $\partial/\partial y$ is significant. The divergence of **T** is therefore

$$\nabla \cdot \mathbf{T} = \begin{bmatrix} 0 & 0 & 0 & 0 & 0 & \dfrac{\partial}{\partial y} \\[2mm] 0 & \dfrac{\partial}{\partial y} & 0 & 0 & 0 & 0 \\[2mm] 0 & 0 & 0 & \dfrac{\partial}{\partial y} & 0 & 0 \end{bmatrix} \begin{bmatrix} 0 \\ 0 \\ 0 \\ 0 \\ 0 \\ \sin(\omega t - ky) \end{bmatrix}$$

$$= \begin{bmatrix} -k \cos (\omega t - ky) \\ 0 \\ 0 \end{bmatrix}$$

in agreement with Example 5.

PROBLEMS

1. Which of the time-independent stress fields

(a) $T_{xx} = A,$ $\quad T_{yy} = A$

(b) $T_{xx} = A,$ $\quad T_{yy} = -A$

(c) $T_{xx} = A\,x,$ $\quad T_{yy} = A\,y$

(d) $T_{xx} = A\,y,$ $\quad T_{yy} = A\,x$

(e) $T_{xx} = A\,x,$ $\quad T_{yy} = A\,y,$ $\quad T_{yx} = T_{xy} = -A\,xy$

satisfy the (zero body force) equilibrium condition $\nabla \cdot \mathbf{T} = 0$?

2. Express the stress fields of Problem 1 in abbreviated subscript notation.

3. Using the stress fields given in Problem 1, calculate the traction force on a surface normal to the unit vector

$$\hat{\mathbf{n}} = \hat{\mathbf{y}}\,1/\sqrt{2} + \hat{\mathbf{z}}\,1/\sqrt{2}.$$

Show that transformation of the stress field to a coordinate system rotated $45°$ about the x axis leads to the same answer.

4. Assuming that the stress fields in Problem 1 vary with time as $\sin \omega t$, use the equation of motion (with body force $\mathbf{F} = 0$) to find the corresponding particle displacement fields.

5. Using the translational and rotational equations of motion, verify that

$$\mathbf{T}_n = \mathbf{T} \cdot \hat{\mathbf{n}}$$

for time-varying stress fields.

6. The two-dimensional stress field T_{xx}, T_{yy}, $T_{xy} = T_{yx}$ is transformed by a *clockwise* rotation of coordinates about z. Show that the transformation law is the same as the one obtained for strain in Problem 10 of Chapter 1.

7. Show that the transformed stress components T'_{xx}, T'_{yy}, $T'_{xy} = T'_{yx}$ in Problem 6 satisfy the equation

$$\left(\sigma - \frac{T_{xx} + T_{yy}}{2} \right)^2 + \tau^2 = \left(\frac{T_{xx} - T_{yy}}{2} \right)^2 + T_{xy}^2$$

$$\sigma = T'_{xx} \text{ or } T'_{yy}$$

$$\tau = -T'_{xy} \text{ or } T'_{yx},$$

which defines *Mohr's circle of stress*. (The definition of variable τ is in accord with the usual convention for drawing and interpreting Mohr's circle.) Prove that the strain transformation in Problem 10 Chapter 1 can also be represented by a circle.

8. Prove that points with coordinates $(T_{xx}, -T_{xy})$ and (T_{yy}, T_{yx}) satisfy the equation defining Mohr's circle and lie at opposite ends of a diameter, as illustrated for the case $T_{yy} < T_{xx}$. Show that the transformed points $(T'_{xx}, -T'_{xy})$ and (T'_{yy}, T'_{yx}), corresponding to a clockwise coordinate rotation angle ξ, lie at the ends of a diameter that is rotated through an angle 2ξ, as shown in the figure.

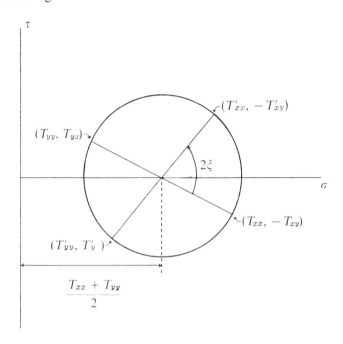

9. The Mohr's circle construction of Problems 7 and 8 is a useful method for visualizing the effect of coordinate transformations on a stress field. It can also be used for determining the maximum values of normal (i.e., compressive or tensile) and shear stress, and orientations of the surfaces subject to these maximum stresses. Use the Mohr's circle construction to find the maximum normal stress and shear stress in the following fields

(a) $T_{xx} = A,\qquad T_{yy} = A$

(b) $T_{xx} = A,\qquad T_{yy} = -A$

(c) $T_{xx} = A,\qquad T_{yy} = B,\qquad T_{xy} = T_{yx} = C$

(d) $T_{xy} = T_{yx} = \sin(\omega t - ky).$

10. Transform the stress fields of Problem 9 into cylindrical coordinates (Part D of Appendix 1).

REFERENCES

1. P. C. Chou and N. J. Pagano, *Elasticity-Tensor, Dyadic and Engineering Approaches*, Ch. 1, van Nostrand, New York, 1967.

2. L. D. Landau and E. M. Lifshitz, *Theory of Elasticity*, pp. 4–7, Pergamon, New York, 1970.

3. G. Nadeau, *Introduction to Elasticity*, pp. 40–48, Holt, Rinehart and Winston, New York, 1964.

4. J. F. Nye, *Physical Properties of Crystals*, Ch. 5, Oxford, England, 1964.

5. M. J. P. Musgrave, *Crystal Acoustics*, Ch. 3, Holden-Day, San Francisco, 1970.

CHAPTER 3

ELASTIC PROPERTIES OF SOLIDS

A. ELASTIC STIFFNESS AND COMPLIANCE

In Chapter 1 deformation in an acoustically vibrating body was described by the *strain field* $S(\mathbf{r}, t)$, which is related to the particle displacement field $\mathbf{u}(\mathbf{r}, t)$ through the stress-displacement equation

$$S(\mathbf{r}, t) = \nabla_s \mathbf{u}(\mathbf{r}, t), \tag{3.1}$$

and in Chapter 2 the elastic restoring forces were defined in terms of the *stress field* $T(\mathbf{r}, t)$. Within a freely vibrating medium both inertial and elastic restoring forces act upon each particle, and it is the interplay of these forces that produces oscillatory motion in a manner analogous to the free vibrations of a macroscopic system of masses and springs. Accordingly, the elastic restoring forces in a medium may be described as microscopic "spring" forces. In Section 2.C a relation between the inertial and elastic restoring forces (or stresses) was derived by applying the dynamical equations of mechanics to a single particle, and it was found that inertia influences only the translational part of the particle motion. Inertial and elastic restoring forces in a freely vibrating medium are thus related through the translational

57

equation of motion†

$$\nabla \cdot \mathbf{T} = \rho \frac{\partial^2 \mathbf{u}}{\partial t^2} . \qquad (3.2)$$

It is now necessary to establish a connection between the elastic restoring forces and the material deformation, that is, to define the microscopic "spring constants" of the medium. This is the purpose of the present chapter.

For small deformations it is an experimentally observed fact that the strain in a deformed body is linearly proportional to the applied stress (Hooke's Law). As increasing deformations are imposed, the relationship between strain and stress becomes increasingly nonlinear, but the body still returns to its original state when the stress is removed. This is called the region of linear and nonlinear *elastic deformation* (Fig. 3.1). If, however, the strain is increased past a certain limit, typically in the range 10^{-4} to 10^{-3} for relatively rigid materials, the deformation is no longer elastic. Beyond this *elastic limit* the medium deforms permanently (*plastic deformation*) and ultimately fractures. Ordinarily the plastic deformation region is not of interest in the study of acoustics, and one uses the elastic strain-stress relationship to determine the microscopic "spring constants." The *linear* "spring constants"

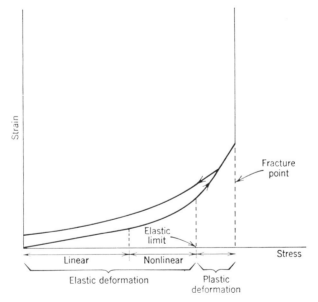

FIGURE 3.1. **Typical stress-strain relation for a solid material.**

† For a freely vibrating medium the body force field $\mathbf{F}$ in (2.18) is zero.

relevant to small amplitude vibration theory are therefore defined by Hooke's Law.

Hooke's Law states that the strain is linearly proportional to the stress, or conversely, that the stress is linearly proportional to the strain. The second form is stated mathematically by writing each component of stress (elastic restoring force) as a general linear function of all the strain components. For example,

$$
\begin{aligned}
T_{xx} = {} & c_{xxxx}S_{xx} + c_{xxxy}S_{xy} + c_{xxxz}S_{xz} \\
& + c_{xxyx}S_{yx} + c_{xxyy}S_{yy} + c_{xxyz}S_{yz} \\
& + c_{xxzx}S_{zx} + c_{xxzy}S_{zy} + c_{xxzz}S_{zz}.
\end{aligned}
\tag{3.3}
$$

In general, then,

$$
T_{ij} = c_{ijkl}S_{kl}
$$

$$
i, j, k, l = x, y, z
\tag{3.4}
$$

with summation over the repeated subscripts k and l. The "microscopic spring constants" c_{ijkl} in (3.4) are called *elastic stiffness constants*. Like macroscopic spring constants, they have small values for easily deformed materials and large values for very rigid materials. Since (3.4) contains nine equations (corresponding to all possible combinations of the subscripts ij) and each equation contains nine strain variables, there are 81 elastic stiffness constants. These are not all independent however. It will be seen in Section 3.C that

$$
c_{ijkl} = c_{jikl} = c_{ijlk} = c_{jilk},
$$

which reduces the number of independent constants to 36. Also, Section 5.C shows that

$$
c_{ijkl} = c_{klij},
$$

and this means that the constants are further reduced to 21. This is the maximum number of constants for *any* medium. Usually the number is much less than this, because of additional restrictions imposed by the microscopic nature of the medium (Sections 6.B and 7.A).

Alternatively, the strains may be expressed as general linear functions of all the stresses,

$$
S_{ij} = s_{ijkl}T_{kl}
$$

$$
i, j, k, l = x, y, z
\tag{3.5}
$$

In this case the constants s_{ijkl}, called *compliance constants*, are measures of the deformability of the medium and have large values for easily deformed materials, small values for rigid materials. Relation (3.5) and its converse

(3.4) are called *elastic constitutive relations*.† The compliance constants s_{ijkl} describe the elastic properties of a medium in a manner analogous to the description of its electrical properties by the permittivity matrix elements ϵ_{ij} in Section 1.F, where the *electrical* constitutive relation corresponding to (3.5) is

$$D_i = \epsilon_{ij} E_j$$

$$i, j = x, y, z. \tag{3.6}$$

Since strain is dimensionless, it follows from (3.4) that stiffness has the same dimension as stress (newtons/(meter)²). Compliance is, conversely, measured in (meter)²/newton. Experimentally observed stiffnesses range from approximately 0.1×10^{10} newtons/m^2 for rubberlike materials to more than 10×10^{10} newtons/m^2 for relatively rigid materials like metals and single crystal insulators. Corresponding compliances range from 1000×10^{-12} for rubberlike materials to 10×10^{-12} for rigid materials. For rigid materials the stress at the elastic limit (strain $\approx 10^{-4}$ to 10^{-3}) is in the range of 10^7 to 10^8 newtons/m^2.

EXAMPLE 1.   Hooke's Law for a Single Crystal Medium of Cubic Symmetry. To provide some illustrative examples for this chapter, it will be necessary to anticipate some of the results of Chapters 6 and 7, where the independent stiffness and compliance constants are determined for various types of materials. Consider, for example, a single crystal material in which the atoms are arranged in a cubic array or lattice. The microscopic structure of such a material, which is said to belong to the cubic class of crystals, may be represented by the 3-dimensional geometric lattice shown in Fig. 3.2. Axes aligned along the edges of the basic cube elements, or cells, of the lattice are called crystal axes. In this book *crystal axes* will be designated by upper case letters (X, Y, Z) in order to distinguish them from rectangular *coordinate axes* x, y, z, which may be arbitrarily oriented with respect to the crystal lattice.

If the coordinate axes x, y, z are chosen to coincide with the cubic crystal axes, as in Fig. 3.2, the nonzero elastic stiffness constants are

$$c_{xxxx} = c_{yyyy} = c_{zzzz}$$

$$c_{xxyy} = c_{yyxx} = c_{xxzz} = c_{zzxx} = c_{yyzz} = c_{zzyy}$$

$$c_{yxyx} = c_{yxxy} = c_{zyyz} = c_{zyzy}$$

$$= c_{xzxz} = c_{xzzx} = c_{zxxz} = c_{zxzx}$$

$$= c_{xyxy} = c_{xyyx} = c_{yxxy} = c_{yxyx}.$$

---

† These have been defined in rectangular Cartesian coordinates. Stiffness and compliance constants in cylindrical and spherical coordinates are discussed in Part D of Appendix 1.

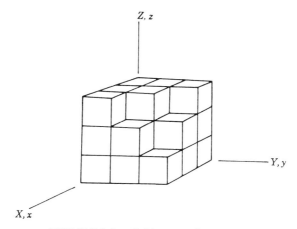

FIGURE 3.2.   **Cubic crystal structure.**

The stiffness equations (3.4) therefore reduce to

$$T_{xx} = c_{xxxx}S_{xx} + c_{xxyy}S_{yy} + c_{xxyy}S_{zz}$$
$$T_{yy} = c_{xxyy}S_{xx} + c_{xxxx}S_{yy} + c_{xxyy}S_{zz}$$
$$T_{zz} = c_{xxyy}S_{xx} + c_{xxyy}S_{yy} + c_{xxxx}S_{zz}$$
$$T_{yz} = T_{zy} = 2c_{yzyz}S_{yz}$$
$$T_{xz} = T_{zx} = 2c_{xzxz}S_{xz}$$
$$T_{xy} = T_{yx} = 2c_{xyxy}S_{xy}, \tag{3.7}$$

since $T_{ij} = T_{ji}$ and $S_{ij} = S_{ji}$. These stiffness equations reflect the structural symmetries of the crystal lattice. For instance, it is clear from the lattice geometry in Fig. 3.2 that a uniaxial compression along the y-axes ($S_{yy} = -S$, $S_{xx} = S_{zz} = 0$) should require the same amount of applied pressure (or stress) as the same degree of uniaxial compression along the x-axes ($S_{xx} = -S$, $S_{yy} = S_{zz} = 0$). That is, T_{yy} in the first case should equal T_{xx} in the second case. This is, indeed, what the elastic constitutive relation (3.7) states; namely,

$$T_{xx} = -c_{xxxx}S$$
$$T_{yy} = -c_{xxxx}S.$$

Constraints are thus imposed on the stiffness constants by the principle that symmetrically equivalent directions in a crystal must have equivalent elastic properties, and it is this principle that is used in Chapter 7 to determine the number of independent elastic constants for all classes of crystalline materials.

B.   TRANSFORMATION PROPERTIES

As in Example 1 above, the stiffness and compliance constants for crystalline materials are normally given with respect to crystal axes. This, however, may not always be the most convenient choice of axes for solving specific problems, and it is therefore necessary to consider how the stiffness and compliance constants may be transformed into other coordinate systems. Since Hooke's Law applies in all coordinate systems, the required transformation laws may be deduced from (3.4), (3.5) and the already-derived transformation properties of stress and strain,

$$T'_{mn} = a_{mi}T_{ij}a_{nj} \tag{2.13}$$

and

$$S'_{op} = a_{ok}S_{kl}a_{pl}. \tag{1.39}$$

Application of (2.13) to (3.4) gives

$$T'_{mn} = a_{mi}c_{ijkl}S_{kl}a_{nj} = a_{mi}a_{nj}c_{ijkl}S_{kl}, \tag{3.8}$$

since the order of terms is not significant. Transformation of Hooke's Law is then completed by expressing S_{kl} in terms of the new coordinates. Inversion of (1.39) gives

$$S_{kl} = (a^{-1})_{ko}S'_{op}(a^{-1})_{lp} = a_{ok}S'_{op}a_{pl} = a_{ok}a_{pl}S'_{op},$$

where $(a^{-1})_{ij} = (\tilde{a})_{ij} = (a)_{ji}$ from (1.34), and substitution into (3.8) then gives

$$T'_{mn} = a_{mi}a_{nj}c_{ijkl}a_{ok}a_{pl}S'_{op}. \tag{3.9}$$

Comparison of this result with (3.4) shows that the stiffnesses in the new coordinate system are[†]

$$c'_{mnop} = a_{mi}a_{nj}a_{ok}a_{pl}c_{ijkl}. \tag{3.10}$$

Similarly, for the compliances

$$s'_{mnop} = a_{mi}a_{nj}a_{ok}a_{pl}s_{ijkl}. \tag{3.11}$$

Physical quantities that transform according to (3.10) and (3.11) are called fourth rank tensors. The general pattern of tensor transformation laws should now be clear. In Section 1.E, a first rank tensor (or vector) was seen to transform according to the law

$$v'_i = a_{ij}v_j$$

and a second rank tensor, like the strain, was shown to transform according to

$$S'_{ij} = a_{ik}a_{jl}S_{kl}.$$

The fourth rank stiffness tensor has just been seen to transform as

$$c'_{mnop} = a_{mi}a_{nj}a_{ok}a_{pl}c_{ijkl}.$$

[†] Transformations to cylindrical and spherical coordinate systems are discussed in Part D of Appendix 1.

In each case the number of "a" factors is equal to the rank of the tensor. The left subscripts on the "a"s always correspond to the subscripts on the transformed quantity, while the right subscripts correspond to the subscripts on the untransformed quantity.

Following the convention established in Section 1.F, the compliance and stiffness tensors will be denoted symbolically by boldface letters s and c. Hooke's Law is then written symbolically as

$$\mathbf{T} = \mathbf{c}:\mathbf{S} \tag{3.12}$$

and

$$\mathbf{S} = \mathbf{s}:\mathbf{T}. \tag{3.13}$$

The double scalar (or double dot) product of a fourth rank and a second rank tensor is defined by the summation over pairs of subscripts in (3.4) and (3.5). As before, it is unnecessary for the symbolism to designate explicitly the tensor rank. Standard notation will be used for all physical quantities and, as explained in Section 1.F, the nature of a physical variable implicitly specifies its tensor rank. It has just been shown, for example, that stiffness and compliance are fourth rank tensors and no further identification of the rank of c and s is necessary. With this information, and definitions of the various tensor product and differentiation operations, symbolic equations can be converted unambiguously into component form.

EXAMPLE 2. In Example 6 of Chapter 1 the transformation matrix describing a clockwise rotation through an angle ξ about the z-axis was shown to be

$$[a] = \begin{bmatrix} \cos\xi & \sin\xi & 0 \\ -\sin\xi & \cos\xi & 0 \\ 0 & 0 & 1 \end{bmatrix}$$

The effect of this transformation (Fig. 3.3) on the stiffness constants given in Example 1 of this chapter is evaluated by substituting into (3.10) the stiffness constants and the elements of $[a]$. Taking the stiffness constants in their listed order in Example 1, one finds that their transformed values are

$$\begin{aligned}
c'_{xxxx} &= a_{xx}^4 c_{xxxx} + a_{xy}^4 c_{yyyy} + a_{xy}^4 c_{zzzz} \\
&\quad + a_{xx}^2 a_{xy}^2 c_{xxyy} + a_{xy}^2 a_{xx}^2 c_{yyxx} + \cdots \\
&= (\cos^4\xi)c_{xxxx} + (\sin^4\xi)c_{xxxx} + (2\cos^2\xi\sin^2\xi)c_{xxyy} + \cdots,
\end{aligned}$$

and so on.

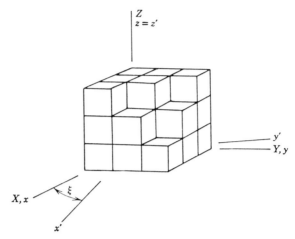

FIGURE 3.3.   Rotation of coordinate axes about the Z crystal axis.

C.   ABBREVIATED SUBSCRIPTS

Examples 1 and 2 clearly illustrate the inconvenience of using full subscript notation for writing and transforming Hooke's Law. This difficulty can be avoided by using the abbreviated subscripts introduced in Sections 1.F and 2.D, which is always possible when the stress components are symmetric $(T_{ij} = T_{ji})$. When the stress symmetry condition is satisfied, terms such as

$$c_{xyxy}S_{xy}$$

and

$$c_{yxxy}S_{xy}$$

in (3.4) are always equal. Therefore

$$c_{ijkl} = c_{jikl}. \tag{3.14}$$

Furthermore, since

$$S_{ij} = S_{ji}$$

there is no way to distinguish experimentally between terms such as

$$c_{xyxy}S_{xy}$$

and

$$c_{xyyx}S_{yx}.$$

Because of this, no purpose is served by distinguishing c_{ijkl} from c_{ijlk}; and the condition

$$c_{ijkl} = c_{ijlk} \tag{3.15}$$

is therefore imposed. Similar arguments show that

$$S_{ijkl} = S_{jikl} \tag{3.16}$$

$$S_{ijkl} = S_{ijlk}. \tag{3.17}$$

With these constraints on the stiffness and compliance constants, the four subscripts may be reduced to two by using abbreviated subscript notation as in Section F of Chapter 1 and Section D of Chapter 2, where

I	ij
1	xx
2	yy
3	zz
4	yz, zy
5	xz, zx
6	$xy, yx.$

Relationships between the stiffness with full subscripts and those with abbreviated subscripts are established by considering individual terms in (3.4). For example,

$$T_{xx} = c_{xxyy}S_{yy}$$

is replaced in abbreviated subscript notation by

$$T_1 = c_{12}S_2,$$

since $T_1 = T_{xx}$ and $S_2 = S_{yy}$. This gives

$$c_{12} = c_{xxyy}.$$

Similarly, a term

$$T_{xy} = c_{xyyy}S_{yy}$$

is equivalent to

$$T_6 = c_{xyyy}S_2 = c_{62}S_2,$$

and therefore

$$c_{62} = c_{xyyy}.$$

In a term

$$T_{xy} = c_{xyxy}S_{xy} + c_{xyyx}S_{yx} = 2c_{xyxy}S_{xy}$$

conversion of the stress and strain to abbreviated subscripts gives

$$T_6 = 2c_{xyxy}\left(\frac{S_6}{2}\right) = c_{xyxy}S_6.$$

Consequently,

$$c_{66} = c_{xyxy}.$$

The general relationship is therefore

$$c_{IJ} = c_{ijkl}. \tag{3.18}$$

By similar arguments it is found that

$$s_{IJ} = s_{ijkl} \times \begin{cases} 1 \text{ for } I \text{ and } J = 1, 2, 3 \\ 2 \text{ for } I \text{ or } J = 4, 5, 6 \\ 4 \text{ for } I \text{ and } J = 4, 5, 6. \end{cases} \tag{3.19}$$

The differences between (3.18) and (3.19) result from the manner in which the factors of two are introduced into the definition of strain in the abbreviated notation (1.49). By using a different definition for the strains, the 2's and 4's could be removed from the compliances and put into the stiffnesses, but common convention specifies the forms given here.

According to the equivalence relation (3.18), Hooke's Law (3.4) may be written as a matrix equation. Consider, for example, the equation (3.3) for T_{xx}. With the introduction of abbreviated subscripts, this becomes

$$T_1 = c_{11}S_1 + c_{16}\frac{S_6}{2} + c_{15}\frac{S_5}{2}$$

$$+ c_{16}\frac{S_6}{2} + c_{12}S_2 + c_{14}\frac{S_4}{2}$$

$$+ c_{15}\frac{S_5}{2} + c_{14}\frac{S_4}{2} + c_{13}S_3$$

$$= c_{1J}S_J, \qquad J = 1, 2, 3, 4, 5, 6.$$

A complete expression for Hooke's Law is therefore

$$T_I = c_{IJ}S_J$$
$$I, J = 1, 2, 3, 4, 5, 6. \tag{3.20}$$

Similarly, it follows from (3.19) that (3.5) reduces to

$$S_I = s_{IJ}T_J$$
$$I, J = 1, 2, 3, 4, 5, 6 \tag{3.21}$$

when abbreviated subscripts are introduced. The individual terms again combine in just the right way to remove factors of 2 and 4 in the final result.

The matrix formulation of Hooke's Law has many advantages. It is economical of space and it also clarifies certain physical relationships. For example, one can see immediately from (3.20) and (3.21) that the compliance matrix $[s]$ in (3.21) is simply the inverse of the stiffness matrix $[c]$ in (3.20). That is

$$[s] = [c]^{-1} \tag{3.22}$$

and therefore

$$[s][c] = [c][s] = [I], \tag{3.23}$$

the identity matrix. This relationship is very useful for converting stiffness constants to compliance constants, and vice versa; it is also a valuable aid in performing algebraic manipulations of the acoustic field equations.

EXAMPLE 3. In abbreviated subscript notation, the nonzero compliances for the cubic crystal material considered in Example 1 are

$$c_{11} = c_{22} = c_{33} = c_{xxxx}$$

$$c_{12} = c_{21} = c_{13} = c_{31} = c_{23} = c_{32} = c_{xxyy}$$

$$c_{44} = c_{55} = c_{66} = c_{yzyz}$$

and the stiffness matrix is therefore

$$[c] = \begin{bmatrix} c_{11} & c_{12} & c_{12} & 0 & 0 & 0 \\ c_{12} & c_{11} & c_{12} & 0 & 0 & 0 \\ c_{12} & c_{12} & c_{11} & 0 & 0 & 0 \\ 0 & 0 & 0 & c_{44} & 0 & 0 \\ 0 & 0 & 0 & 0 & c_{44} & 0 \\ 0 & 0 & 0 & 0 & 0 & c_{44} \end{bmatrix}$$

According to (3.22) the compliance constants are obtained by inverting $[c]$. This gives

$$[s] = \begin{bmatrix} s_{11} & s_{12} & s_{12} & 0 & 0 & 0 \\ s_{12} & s_{11} & s_{12} & 0 & 0 & 0 \\ s_{12} & s_{12} & s_{11} & 0 & 0 & 0 \\ 0 & 0 & 0 & s_{44} & 0 & 0 \\ 0 & 0 & 0 & 0 & s_{44} & 0 \\ 0 & 0 & 0 & 0 & 0 & s_{44} \end{bmatrix}$$

with

$$s_{11} = \frac{c_{11} + c_{12}}{(c_{11} - c_{12})(c_{11} + 2c_{12})}$$

$$s_{12} = \frac{-c_{12}}{(c_{11} - c_{12})(c_{11} + 2c_{12})}$$

$$s_{44} = \frac{1}{c_{44}}.$$

EXAMPLE 4. Shear Wave Propagation Along a Cubic Crystal Axis. In Example 5 of Chapter 2 it was seen that an x-polarized y-propagating shear wave has a stress field

$$T_6 = T_{xy} = \sin(\omega t - ky)$$

and a strain field

$$S_6 = 2S_{xy} = \frac{k^2}{\rho\omega^2}\sin(\omega t - ky).$$

These fields must be related to each other through the elastic constitutive equation of the propagation medium. If the propagation medium is a cubic crystal and the coordinate axes x, y, z coincide with the crystal axes X, Y, Z as in Fig. 3.2, one has from Example 3 that

$$
\begin{bmatrix} 0 \\ 0 \\ 0 \\ 0 \\ 0 \\ T_6 \end{bmatrix}
=
\begin{bmatrix}
c_{11} & c_{12} & c_{12} & 0 & 0 & 0 \\
c_{12} & c_{11} & c_{12} & 0 & 0 & 0 \\
c_{12} & c_{12} & c_{11} & 0 & 0 & 0 \\
0 & 0 & 0 & c_{44} & 0 & 0 \\
0 & 0 & 0 & 0 & c_{44} & 0 \\
0 & 0 & 0 & 0 & 0 & c_{44}
\end{bmatrix}
\begin{bmatrix} 0 \\ 0 \\ 0 \\ 0 \\ 0 \\ S_6 \end{bmatrix}
$$

or

$$T_6 = c_{44}S_6.$$

This means that the field T_6, S_6 constitutes a valid acoustic wave solution only if

$$c_{44}k^2 = \rho\omega^2. \qquad (3.24)$$

This relationship between the wave vector k and the angular frequency ω is called the *dispersion relation* of the wave. The phase velocity for an X-polarized Y-propagating shear wave in a cubic crystal is therefore

$$V_s = \frac{\omega}{k} = (c_{44}/\rho)^{1/2}. \qquad (3.25)$$

It was seen in Section A of this chapter that the stiffness constants for relatively rigid materials are of the order of 10^{11} newtons/m^2. For a typical mass density ρ in the neighborhood of 4000 kg/m^3, the shear wave velocity is therefore of order

$$V_s = \left(\frac{10^{11}}{4 \times 10^3}\right)^{1/2} = 5 \times 10^3 \text{ m/s}$$

and the wavelength

$$\lambda_s = \frac{2\pi V_s}{\omega}$$

is in the neighborhood of 5×10^{-5} meters, or 50 microns, at a frequency of 100 MHz (100 Mc/s). These values are approximately 5 orders of magnitude smaller than the equivalent quantities for electromagnetic waves, and it is just this property of acoustic waves that makes them so attractive for many technological applications.

For a Z-polarized shear wave propagating along the Y cubic crystal axis the same analysis may be followed, and the phase velocity is found to be the same as for the X-polarized wave, that is

$$(V_s)_{Z\text{-polarization}} = (V_s)_{X\text{-polarization}} = (c_{44}/\rho)^{1/2}.$$

Shear waves that propagate along the same direction with the same velocity are called *degenerate*. This is an important characteristic because, in any linear system,

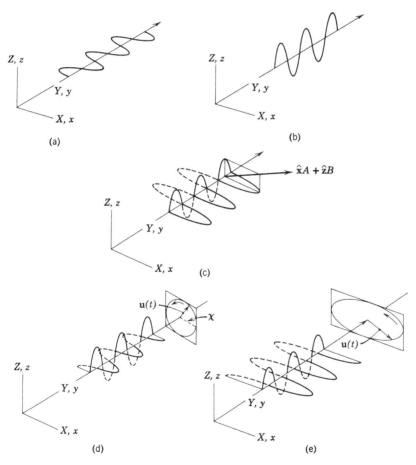

FIGURE 3.4. Combination of degenerate shear waves to form arbitrary particle displacement polarizations.

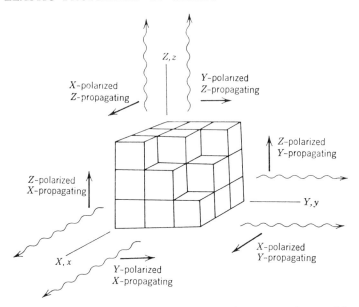

FIGURE 3.5. **Symmetrically-equivalent shear waves in a cubic crystal medium.**

degenerate waves may be combined in any arbitrary manner to produce wave solutions with a wide variety of polarizations. This is a physical phenomenon of fundamental importance.

Suppose that the particle displacement fields for X- and Z-polarized shear waves propagating along the Y cubic crystal axes are

$$\mathbf{u}_x = \hat{\mathbf{x}} \cos(\omega t - ky)$$

and

$$\mathbf{u}_z = \hat{\mathbf{z}} \cos(\omega t - ky),$$

respectively (Fig. 3.4a and 3.4b). If these are combined with arbitrary amplitudes and the same time phase, one obtains

$$\mathbf{u} = (\hat{\mathbf{x}}A + \hat{\mathbf{z}}B) \cos(\omega t - ky),$$

which is a shear wave traveling with the same phase velocity $(\rho/c_{44})^{1/2}$ but with *linearly polarized* particle displacement along the direction $\hat{\mathbf{x}}A + \hat{\mathbf{z}}B$ (Fig. 3.4c). If, on the other hand, the X- and Z-polarized waves are combined with equal amplitudes but $90°$ phase shift,

$$\mathbf{u} = \hat{\mathbf{x}} \cos(\omega t - ky) + \hat{\mathbf{z}} \sin(\omega t - ky),$$

the particle displacement is now *circularly polarized*,

$$u = (u_x^2 + u_z^2)^{1/2}$$

$$\tan \chi = \frac{u_z}{u_x} = \tan (\omega t - ky)$$

as illustrated in Fig. 3.4d. By taking arbitrary combinations of amplitude and phase, *elliptically polarized* particle displacement patterns of various orientations may be obtained (Fig. 3.4e).

It seems intuitively obvious from the symmetry of the cubic crystal lattice in Fig. 3.2 that shear waves traveling along the X and Z cubic crystal axes should have the same properties as for Y-propagating waves. That is, all these waves should have the same velocity $(c_{44}/\rho)^{1/2}$, regardless of the particle displacement polarization and propagation direction. Repetition of the above analysis shows that this is, indeed, the case. All six shear waves in Fig. 3.5 therefore have identical phase velocities.

EXAMPLE 5. Compressional Wave Propagation Along a Cubic Crystal Axis. It was shown in Example 5 of Chapter 1 that the particle displacement and strain fields of a y-propagating compressional wave are

$$\mathbf{u} = \hat{\mathbf{y}} \cos (\omega t - ky) \tag{3.25}$$

and

$$S_2 = S_{yy} = k \sin (\omega t - ky). \tag{3.26}$$

If the medium is a cubic single crystal and coordinate axes coincide with the crystal axes as in Fig. 3.2, the stress field is calculated from (3.26) by means of the constitutive equation

$$
\begin{bmatrix} T_1 \\ T_2 \\ T_3 \\ T_4 \\ T_5 \\ T_6 \end{bmatrix}
=
\begin{bmatrix}
c_{11} & c_{12} & c_{12} & 0 & 0 & 0 \\
c_{12} & c_{11} & c_{12} & 0 & 0 & 0 \\
c_{12} & c_{12} & c_{11} & 0 & 0 & 0 \\
0 & 0 & 0 & c_{44} & 0 & 0 \\
0 & 0 & 0 & 0 & c_{44} & 0 \\
0 & 0 & 0 & 0 & 0 & c_{44}
\end{bmatrix}
\begin{bmatrix} 0 \\ S_2 \\ 0 \\ 0 \\ 0 \\ 0 \end{bmatrix}
$$

That is

$$T_1 = c_{12}S_2 = c_{12}k \sin (\omega t - ky)$$

$$T_2 = c_{11}S_2 = c_{11}k \sin (\omega t - ky)$$

$$T_3 = c_{12}S_2 = c_{12}k \sin (\omega t - ky). \tag{3.27}$$

From (2.24) the equation of motion for a free wave propagating along the y axis is

$$\frac{\partial}{\partial y} T_{xy} = -\rho\omega^2 u_x$$

$$\frac{\partial}{\partial y} T_{yy} = -\rho\omega^2 u_y$$

$$\frac{\partial}{\partial y} T_{zy} = -\rho\omega^2 u_z, \tag{3.28}$$

since $\mathbf{F} = 0$ and the stress field varies only with y. Among the stress components (3.27) for a y-propagating compressional wave, only $T_2 = T_{yy}$ enters into these equations. Consequently,

$$-\rho\omega^2 u_x = 0$$

$$-\rho\omega^2 u_y = \frac{\partial}{\partial y} T_{yy} = -c_{11}k^2 \cos(\omega t - ky)$$

$$-\rho\omega^2 u_z = 0,$$

and substitution of the originally assumed displacement field (3.25) for u_y shows that the dispersion relation

$$c_{11}k^2 = \rho\omega^2 \tag{3.29}$$

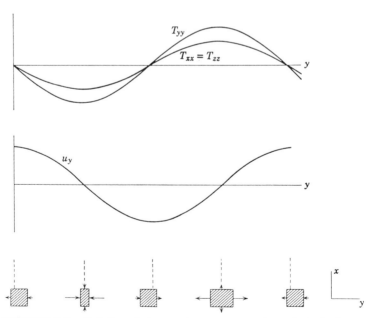

FIGURE 3.6.   **Relations between stress components and strain in a y-propagating compressional wave. The net force on a particle is 180° out of phase with the displacement because of mechanical inertia. Compare with Fig. 2.9.**

must be satisfied. The phase velocity for a compressional wave propagating along the Y cubic crystal axis is therefore[†]

$$V_l = \frac{\omega}{k} = (c_{11}/\rho)^{1/2}. \qquad (3.30)$$

As in Example 4, the X and Z axes are symmetrically equivalent to the Y axis, and (3.30) is therefore the compressional wave velocity for propagation along *any* of the cubic crystal axes. Compressional stiffness constants are typically several times larger than shear stiffness constants, and compressional waves are generally faster than shear waves.[‡]

This analysis has shown that the stress components $T_1 = T_{xx}$ and $T_3 = T_{zz}$ in the compressional wave do not take any part in the dynamics of the vibration. They, so to speak, simply go along for the ride. The reason for this is illustrated by Fig. 3.6. As in Fig. 2.9 for the shear wave case, this figure shows the stresses acting on selected material particles. The T_{xx} and T_{zz} stresses acting on a particle are always in balance because the stress field varies only with y.

D. TRANSFORMATIONS WITH ABBREVIATED SUBSCRIPTS

The preceding examples clearly demonstrate the economy of space and the ease of algebraic manipulation provided by abbreviated subscript notation. It is therefore of considerable importance to have a method for performing coordinate transformations directly in this notation without the additional effort of converting to full subscripts, applying the awkward calculation illustrated by Example 2, and then reconverting to the abbreviated notation. A very efficient matrix technique has been developed for this purpose by W. L. Bond.[§] In essence, it involves construction of 6×6 matrices that may be used to transform stress or strain by means of a single matrix multiplication.

Consider first the stress field **T**. In full subscript notation this transforms according to

$$T'_{ij} = a_{ik}a_{jl}T_{kl}$$
$$i, j, k, l = x, y, z. \qquad (2.13)$$

To convert to abbreviated subscripts, each stress component must be examined

---

[†] Compressional waves are often called *longitudinal* waves because the particle displacement is along the propagation direction, and the phase velocity is usually labeled with subscript l. Shear waves are, correspondingly, often called *transverse*.
[‡] There are, however, exceptions. See Fig. 3.11 in part B.5a of Appendix 3.
[§] Reference 5 at the end of the chapter.

individually. From (2.13) the transformed stress T'_{xx}, for instance, is

$$
\begin{aligned}
T'_{xx} &= a^2_{xx}T_{xx} + a_{xx}a_{xy}T_{xy} + a_{xx}a_{xz}T_{xz} \\
&\quad + a_{xy}a_{xx}T_{yx} + a^2_{xy}T_{yy} + a_{xy}a_{xz}T_{yz} \\
&\quad + a_{xz}a_{xx}T_{zx} + a_{xz}a_{xy}T_{zy} + a^2_{xz}T_{zz} \\
&= a^2_{xx}T_{xx} + a^2_{xy}T_{yy} + a^2_{xz}T_{zz} + 2a_{xy}a_{xz}T_{yz} + 2a_{xx}a_{xz}T_{xz} + 2a_{xx}a_{xy}T_{xy},
\end{aligned}
$$

since stress is symmetric ($T_{ij} = T_{ji}$). After conversion of the stress components to abbreviate subscript notation, this becomes

$$
T'_1 = a^2_{xx}T_1 + a^2_{xy}T_2 + a^2_{xz}T_3 + 2a_{xy}a_{xz}T_4 + 2a_{xx}a_{xz}T_5 + 2a_{xx}a_{xy}T_6.
$$

Repetition of the same procedure for each component of $\mathbf{T}'$ gives the matrix transformation law

$$
\begin{aligned}
T'_H &= M_{HI}T_I \\
H, I &= 1, 2, 3, 4, 5, 6,
\end{aligned}
\tag{3.31}
$$

where the coefficients M_{HI} define a 6×6 transformation matrix

$$
[M] =
\begin{bmatrix}
a^2_{xx} & a^2_{xy} & a^2_{xz} & 2a_{xy}a_{xz} & 2a_{xz}a_{xx} & 2a_{xx}a_{xy} \\
a^2_{yx} & a^2_{yy} & a^2_{yz} & 2a_{yy}a_{yz} & 2a_{yz}a_{yx} & 2a_{yx}a_{yy} \\
a^2_{zx} & a^2_{zy} & a^2_{zz} & 2a_{zy}a_{zz} & 2a_{zz}a_{zx} & 2a_{zx}a_{zy} \\
a_{yx}a_{zx} & a_{yy}a_{zy} & a_{yz}a_{zz} & a_{yy}a_{zz}+a_{yz}a_{zy} & a_{yx}a_{zz}+a_{yz}a_{zx} & a_{yy}a_{zx}+a_{yx}a_{zy} \\
a_{zx}a_{xx} & a_{zy}a_{xy} & a_{zz}a_{xz} & a_{xy}a_{zz}+a_{xz}a_{zy} & a_{xz}a_{zx}+a_{xx}a_{zz} & a_{xx}a_{zy}+a_{xy}a_{zx} \\
a_{xx}a_{yx} & a_{xy}a_{yy} & a_{xz}a_{yz} & a_{xy}a_{yz}+a_{xz}a_{yy} & a_{xz}a_{yx}+a_{xx}a_{yz} & a_{xx}a_{yy}+a_{xy}a_{yx}
\end{bmatrix}
\tag{3.32}
$$

The pattern of coefficients in this matrix is most easily remembered by partitioning the matrix array as shown, and applying the following rules:

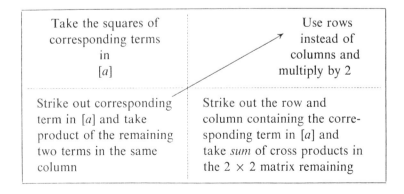

Take the squares of corresponding terms in [a]	Use rows instead of columns and multiply by 2
Strike out corresponding term in [a] and take product of the remaining two terms in the same column	Strike out the row and column containing the corresponding term in [a] and take *sum* of cross products in the 2 × 2 matrix remaining

In the lower left-hand submatrix of (3.32) consider, for example, the 2nd element of the 1st column. The corresponding 2nd element of the 1st column in the $[a]$ matrix is a_{yx}, and the remaining two elements of the 1st column in $[a]$ are a_{zx} and a_{xx}. One obtains the required matrix element in (3.32) by taking the product $a_{zx}a_{xx}$.

Starting from the strain transformation law

$$S'_{ij} = a_{ik}a_{jl}S_{kl} \tag{1.39}$$

and following the same line of argument, one finds that the matrix transformation law for strain is

$$S'_K = N_{KJ}S_J$$
$$K, J = 1, 2, 3, 4, 5, 6, \tag{3.33}$$

where the coefficients N_{KJ} define a 6×6 transformation matrix

$$[N] =
\begin{bmatrix}
a_{xx}^2 & a_{xy}^2 & a_{xz}^2 & a_{xy}a_{xz} & a_{xz}a_{xx} & a_{xx}a_{xy} \\
a_{yx}^2 & a_{yy}^2 & a_{yz}^2 & a_{yy}a_{yz} & a_{yz}a_{yx} & a_{yx}a_{yy} \\
a_{zx}^2 & a_{zy}^2 & a_{zz}^2 & a_{zy}a_{zz} & a_{zz}a_{zx} & a_{zx}a_{zy} \\
2a_{yx}a_{zx} & 2a_{yy}a_{zy} & 2a_{yz}a_{zz} & a_{yy}a_{zz}+a_{yz}a_{zy} & a_{yx}a_{zz}+a_{yz}a_{zx} & a_{yy}a_{zx}+a_{yx}a_{zy} \\
2a_{zx}a_{xx} & 2a_{zy}a_{xy} & 2a_{zz}a_{xz} & a_{xy}a_{zz}+a_{xz}a_{zy} & a_{xz}a_{zx}+a_{xx}a_{zz} & a_{xx}a_{zy}+a_{xy}a_{zx} \\
2a_{xx}a_{yx} & 2a_{xy}a_{yy} & 2a_{xz}a_{yz} & a_{xy}a_{yz}+a_{xz}a_{yy} & a_{xz}a_{yx}+a_{xx}a_{yz} & a_{xx}a_{yy}+a_{xy}a_{yx}
\end{bmatrix}$$

$$\tag{3.34}$$

Comparison with (3.32) shows that $[N]$ is the same as $[M]$, except for a shift of the factor 2 from the upper right-hand corner to the lower left-hand corner.

Now that abbreviated subscript laws have been established for T_I and S_J, transformation of Hooke's Law

$$[T] = [c][S] \tag{3.35}$$

can be carried out by the same simple method used to derive (3.10) and (3.11). Application of the Bond *stress* transformation matrix (3.32) to (3.35) leads to

$$[T'] = [M][c][S]. \tag{3.36}$$

The inverse of (3.33) is

$$[S] = [N]^{-1}[S'],$$

where $[N]$ is the Bond *strain* transformation matrix (3.34), and substitution for $[S]$ in (3.36) gives

$$[T'] = [M][c][N]^{-1}[S'].$$

Comparison with (3.35) now shows that the transformed stiffness matrix is simply

$$[c'] = [M][c][N]^{-1}. \tag{3.37}$$

In a similar manner one finds that the transformed compliance matrix is

$$[s'] = [N][s][M]^{-1}. \tag{3.38}$$

In equations (3.37) and (3.38), the stiffness and compliance transformation laws require inversion of the 6×6 matrices $[N]$ and $[M]$. Even in the case of 3×3 matrices this is not the most trivial task, and the prospect of having to invert a 6×6 matrix might lead one to question the usefulness of Bond's transformation method. Fortunately, these matrix inversions prove to be unnecessary. It was seen in (1.34) that the inverse of $[a]$ is simply $[\widetilde{a}]$, which is obtained by transposing the subscripts on all matrix elements. Thus, if one has found the matrix $[N]$ corresponding to a particular $[a]$, the matrix $[N]^{-1}$ corresponding to $[a]^{-1} = [\widetilde{a}]$ is obtained by transposing all subscripts in (3.34). Comparison with (3.32) shows that the result is simply $[\widetilde{M}]$. That is,

$$[N]^{-1} = [\widetilde{M}];$$

and substitution into (3.37) gives the easily applied stiffness transformation law

$$[c'] = [M][c][\widetilde{M}] \tag{3.39}$$

or

$$c'_{HK} = M_{HI}M_{KJ}c_{IJ}. \tag{3.40}$$

In a completely parallel way, the compliance transformation (3.38) is shown to be equivalent to

$$[s'] = [N][s][\widetilde{N}]. \tag{3.41}$$

EXAMPLE 6. The basic advantage of the Bond method for transforming stiffness and compliance is that it can be applied directly to elastic constants given in abbreviated subscript notation, as they always are in tables of elastic constants. It also involves shorter and less complicated algebra, and provides a more effective bookkeeping system as a guard against error. To illustrate these features of the method, it will be applied to the problem considered in Example 2 of this chapter; that is, the transformation of stiffness constants for a cubic crystal by clockwise rotation of coordinates through an angle ξ about the Z crystal axis. According to Example 2, the coordinate transformation matrix is

$$[a] = \begin{bmatrix} \cos \xi & \sin \xi & 0 \\ -\sin \xi & \cos \xi & 0 \\ 0 & 0 & 1 \end{bmatrix}$$

and, from (3.32), the corresponding Bond *stress* transformation matrix is

$$[M] = \begin{bmatrix} \cos^2 \xi & \sin^2 \xi & 0 & 0 & 0 & \sin 2\xi \\ \sin^2 \xi & \cos^2 \xi & 0 & 0 & 0 & -\sin 2\xi \\ 0 & 0 & 1 & 0 & 0 & 0 \\ 0 & 0 & 0 & \cos \xi & -\sin \xi & 0 \\ 0 & 0 & 0 & \sin \xi & \cos \xi & 0 \\ -\dfrac{\sin 2\xi}{2} & \dfrac{\sin 2\xi}{2} & 0 & 0 & 0 & \cos 2\xi \end{bmatrix} \tag{3.42}$$

To perform the transformation, the stiffness matrix $[c]$ in Example 3 is, according to (3.39), simply multiplied on the left by $[M]$ and on the right by $[\widetilde{M}]$. This gives

$$[c'] = \begin{bmatrix} c'_{11} & c'_{12} & c'_{13} & 0 & 0 & c'_{16} \\ c'_{12} & c'_{11} & c'_{13} & 0 & 0 & -c'_{16} \\ c'_{13} & c'_{13} & c'_{33} & 0 & 0 & 0 \\ 0 & 0 & 0 & c'_{44} & 0 & 0 \\ 0 & 0 & 0 & 0 & c'_{44} & 0 \\ c'_{16} & -c'_{16} & 0 & 0 & 0 & c'_{66} \end{bmatrix} \tag{3.43}$$

where

$$c'_{11} = c_{11} - \left(\frac{c_{11} - c_{12}}{2} - c_{44} \right) \sin^2 2\xi$$

$$c'_{12} = c_{12} + \left(\frac{c_{11} - c_{12}}{2} - c_{44} \right) \sin^2 2\xi$$

$$c'_{13} = c_{12}$$

$$c'_{16} = - \left(\frac{c_{11} - c_{12}}{2} - c_{44} \right) \sin 2\xi \cos 2\xi$$

$$c'_{33} = c_{11}$$

$$c'_{44} = c_{44}$$

$$c'_{66} = c_{44} + \left(\frac{c_{11} - c_{12}}{2} - c_{44} \right) \sin^2 2\xi.$$

Note that matrix elements that were equal before the transformation are now unequal and that some previously zero elements have now become nonzero.

EXAMPLE 7. Any general rotation of rectangular coordinates can be performed by applying successive rotations about different coordinate axes. A standard method of doing this is the following. The coordinates are first rotated clockwise through

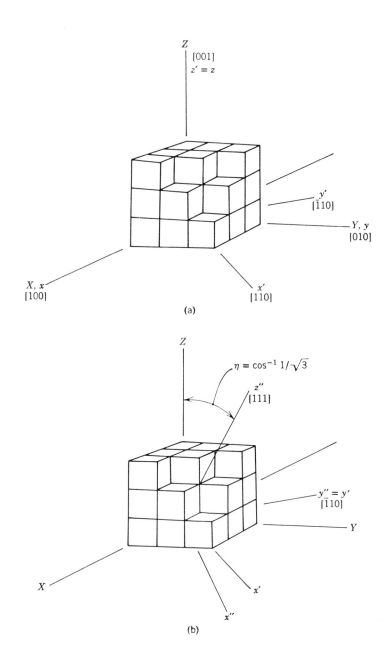

FIGURE 3.7. **Commonly used rotated coordinate systems for cubic crystals.**

an angle ξ about the z axis, then through a clockwise angle η about the *transformed* y axis,[†] and finally rotated clockwise through an angle ξ' about the *transformed* z axis. All coordinate rotations can be represented in terms of these angles ξ, η, ξ'. In the transformation of Example 6, for instance, the rotation angles are

$$\xi, \text{ arbitrary}$$
$$\eta = 0$$
$$\xi' = 0.$$

To illustrate the technique of calculating successive rotations, consider first the transformation of Example 6, with $\xi = 45°$. This puts the x' and y' axes along face diagonals of the basic cube elements of the lattice (Fig. 3.7a) and, from (3.43), the transformed stiffness matrix becomes

$$[c'] = \begin{bmatrix} c'_{11} & c'_{12} & c_{12} & 0 & 0 & 0 \\ c'_{12} & c'_{11} & c_{12} & 0 & 0 & 0 \\ c_{12} & c_{12} & c_{11} & 0 & 0 & 0 \\ 0 & 0 & 0 & c_{44} & 0 & 0 \\ 0 & 0 & 0 & 0 & c_{44} & 0 \\ 0 & 0 & 0 & 0 & 0 & \dfrac{c_{11} - c_{12}}{2} \end{bmatrix} \tag{3.44}$$

with

$$c'_{11} = \frac{c_{11} + c_{12} + 2c_{44}}{2}$$

$$c'_{12} = \frac{c_{11} + c_{12} - 2c_{44}}{2}.$$

Next, a clockwise rotation is performed about the y' axis in order to bring the z' axis in line with the body diagonal of the cube elements (Fig. 3.7b). This requires $\eta = \cos^{-1} 1/\sqrt{3}$, and a coordinate transformation matrix

$$[a'] = \begin{bmatrix} \cos \eta & 0 & -\sin \eta \\ 0 & 1 & 0 \\ \sin \eta & 0 & \cos \eta \end{bmatrix}$$

[†] As noted previously, lower case letters (x, y, z) refer to *coordinate* axes and upper case letters (X, Y, Z) refer to *crystal* axes. Coordinate transformations described by the matrices (3.32) and (3.34) are always referenced to the *coordinate* axes. The second transformation in a sequence is therefore performed with respect to the coordinate axes obtained *after* the first transformation.

From (3.32) the Bond *stress* transformation matrix for a clockwise rotation η about y' is

$$[M'] = \begin{bmatrix} \cos^2 \eta & 0 & \sin^2 \eta & 0 & -\sin 2\eta & 0 \\ 0 & 1 & 0 & 0 & 0 & 0 \\ \sin^2 \eta & 0 & \cos^2 \eta & 0 & \sin 2\eta & 0 \\ 0 & 0 & 0 & \cos \eta & 0 & \sin \eta \\ \dfrac{\sin 2\eta}{2} & 0 & -\dfrac{\sin 2\eta}{2} & 0 & \cos 2\eta & 0 \\ 0 & 0 & 0 & -\sin \eta & 0 & \cos \eta \end{bmatrix} \quad (3.45)$$

This matrix (with $\cos \eta = 1/\sqrt{3}$ and $\sin \eta = \sqrt{2/3}$) is therefore used to transform (3.44) according to (3.39), giving

$$[c''] = \begin{bmatrix} c''_{11} & c''_{12} & c''_{13} & 0 & c''_{15} & 0 \\ c''_{12} & c''_{11} & c''_{13} & 0 & -c''_{15} & 0 \\ c''_{13} & c''_{13} & c''_{33} & 0 & 0 & 0 \\ 0 & 0 & 0 & c''_{44} & 0 & -c''_{15} \\ c''_{15} & -c''_{15} & 0 & 0 & c''_{44} & 0 \\ 0 & 0 & 0 & -c''_{15} & 0 & c''_{66} \end{bmatrix} \quad (3.46)$$

$$c''_{11} = \frac{c_{11} + c_{12} + 2c_{44}}{2}$$

$$c''_{33} = \frac{c_{11} + 2c_{12} + 4c_{44}}{3}$$

$$c''_{12} = \frac{c_{11} + 5c_{12} - 2c_{44}}{6}$$

$$c''_{13} = \frac{c_{11} + 2c_{12} - 2c_{44}}{3}$$

$$c''_{15} = \frac{c_{12} - c_{11} + 2c_{44}}{3\sqrt{2}}$$

$$c''_{44} = \frac{c_{11} - c_{12} + c_{44}}{3}$$

$$c''_{66} = \frac{c_{11} - c_{12} + 4c_{44}}{6}$$

for the stiffness matrix referred to the rotated axes in Fig. 3.7b. In this case the coordinate rotation angles are

$$\xi = 45°$$
$$\eta = \cos^{-1} 1/\sqrt{3}$$
$$\xi' = 0.$$

Example 7 shows that only the two stress transformation matrices (3.42) and (3.45) are required for the most general rotation of coordinates. Rotation through the angle ξ is described by (3.42). This is followed by a rotation η, according to (3.45). Finally, (3.42) is used again with ξ' substituted for ξ. If a compliance matrix is to be transformed, (3.42) and (3.45) may be converted to $[N]$ matrices by comparing (3.34) with (3.32).

In experimental practice, several different methods are used to specify the orientation of a crystal. The IRE Standards on Piezoelectric Crystals, *Proceedings of the IRE*, **37**, pp. 1378–1395 (1949)† considers rotations of the coordinates about *all three axes*. The coordinate axes are labeled as t, l, w to correspond with the thickness, length and width of the usual rectangular plate sample geometry. A crystal rotation is described by first aligning the t, l, w coordinate axes along the crystal axes and then specifying one, two, or three rotations about particular coordinate axes. This system requires use of the z-rotation matrix (3.42), the y-rotation matrix (3.45), and, in addition, the x-rotation matrix

$$[M] = \begin{bmatrix} 1 & 0 & 0 & 0 & 0 & 0 \\ 0 & \cos^2 \xi_x & \sin^2 \xi_x & \sin 2\xi_x & 0 & 0 \\ 0 & \sin^2 \xi_x & \cos^2 \xi_x & -\sin 2\xi_x & 0 & 0 \\ 0 & -\dfrac{\sin 2\xi_x}{2} & \dfrac{\sin 2\xi_x}{2} & \cos 2\xi_x & 0 & 0 \\ 0 & 0 & 0 & 0 & \cos \xi_x & -\sin \xi_x \\ 0 & 0 & 0 & 0 & \sin \xi_x & \cos \xi_x \end{bmatrix}$$

(3.47)

where the rotation angle ξ_x is clockwise about the x coordinate axis.

Another procedure is to consider that the crystal itself is rotated with respect to a *fixed* set of coordinate axes. From this point of view the

† Updated standards are to appear in 1973.

transformation in Fig. 3.7a is equivalent to a clockwise crystal rotation through $-45°$ about z; and Fig. 3.7b is equivalent to a clockwise crystal rotation through $-45°$ about z, followed by a clockwise crystal rotation through $-\cos^{-1} 1/\sqrt{3}$ about y. A general *coordinate rotation* described by the angles ξ, η, ξ' thus corresponds to the *crystal rotation angles*

$$\phi = -\xi \text{ about the } z \text{ axis}$$

and

$$\theta = -\eta \text{ about the } y \text{ axis,}$$

followed by

$$\psi = -\xi' \text{ about the } z \text{ axis.}$$

These are called the *Euler Angles* of the rotation. Euler Angles for Fig. 3.7a are therefore

$$\phi = -45°, \qquad \theta = 0, \qquad \psi = 0,$$

and for Fig. 3.7b, they are

$$\phi = -45°, \qquad \theta = -\cos^{-1} 1/\sqrt{3}, \qquad \psi = 0.$$

When crystal rotation is specified in terms of Euler Angles, only the transformation matrix (3.42), with $\xi = -\phi$ and $-\psi$, and the transformation matrix (3.45), with $\eta = -\theta$, are needed for transforming the stiffness constants.

EXAMPLE 8. *Compressional Wave Propagation Along the [110] Direction in a Cubic Crystal.* In problem solving, coordinate transformations are used to obtain the coordinate system most suited to a problem. As an illustration, consider propagation of a compressional (or longitudinal) wave along the positive x' axis in Fig. 3.7a. In crystallographic terminology this is called the [110] direction. The three numbers in the square bracket give the relative components of the directional vector, referred to the crystal axes X, Y, Z. Positive crystal axis directions are thus designated by [100], [010], [001] respectively, and the z'' axis in Fig. 3.7b is [111]. Negative direction components are indicated by means of a bar over the number. The y' axis in Fig. 3.7a and the y' axis in Fig. 3.7b are therefore [$\bar{1}$10].

A uniform plane wave propagating along the x' axis has field components that are functions of x' only and, for a compressional wave, the particle displacement is along the x' direction. If coordinate axes (x, y, z) parallel to the crystal axes are used to describe this particle displacement field, two coordinate variables (x, y) and two particle displacement components (u_x, u_y) are required. This leads to three strain components

$$S_{xx} = \frac{\partial}{\partial x} u_x$$

$$S_{yy} = \frac{\partial}{\partial y} u_y$$

$$S_{xy} = \frac{1}{2}\left(\frac{\partial}{\partial x} u_y + \frac{\partial}{\partial y} u_x\right).$$

It is obviously simpler to use the rotated (x', y', z') coordinate system, where

$$\mathbf{u} = \hat{\mathbf{x}}' \cos (\omega t - kx') \tag{3.48}$$

and there is only one strain component

$$S_{x'x'} = \frac{\partial u_{x'}}{\partial x'} = k \sin (\omega t - kx'),$$

or

$$[S] = \begin{bmatrix} k \sin (\omega t - kx') \\ 0 \\ 0 \\ 0 \\ 0 \\ 0 \end{bmatrix} \tag{3.49}$$

The stress field is computed from (3.49) by multiplying with the transformed stiffness matrix $[c']$ of (3.44), giving

$$\begin{aligned}
T_{1'} &= T_{x'x'} = c'_{11} k \sin (\omega t - kx') \\
T_{2'} &= T_{y'y'} = c'_{12} k \sin (\omega t - kx') \\
T_{3'} &= T_{z'z'} = c_{12} k \sin (\omega t - kx').
\end{aligned} \tag{3.50}$$

Since the equation of motion (2.24) has the same form in any rectangular coordinate system, it reduces to

$$\frac{\partial}{\partial x'} T_{x'x'} = -\rho \omega^2 u_{x'}$$

$$\frac{\partial}{\partial x'} T_{y'x'} = -\rho \omega^2 u_{y'} \tag{3.51}$$

$$\frac{\partial}{\partial x'} T_{z'x'} = -\rho \omega^2 u_{z'}$$

for a uniform plane wave freely propagating along the x' axis. Of the three stress components in (3.50), only $T_{x'x'}$ enters into these equations and therefore

$$\frac{\partial}{\partial x'} T_{x'x'} = -c'_{11} k^2 \cos (\omega t - kx') = -\rho \omega^2 u_{x'}. \tag{3.52}$$

This is consistent with the assumed particle displacement field (3.48) only if the dispersion relation

$$c'_{11} k^2 = \rho \omega^2$$

is satisfied. The velocity of a compressional wave propagating along the [110] direction in a cubic crystal is thus

$$(V_l)_{[110]} = \left(\frac{c'_{11}}{\rho} \right)^{1/2} = \left(\frac{c_{11} + c_{12} + 2c_{44}}{2\rho} \right)^{1/2}. \tag{3.53}$$

The compressional wave velocity (3.53) for propagation along the [110] direction is not the same as the velocity for propagation along the crystal axis directions [100], [010], and [001], given by (3.30) in Example 5. This anisotropy of the wave velocity is simply a consequence of the elastic anisotropy in a crystalline medium. It is intuitively obvious from Fig. 3.7a that a compression along the cube diagonal direction [110] and a compression along the cube edge direction [010] will be subject to different stiffnesses and this gives different propagation velocities for these two directions. Just as the three cube edge directions ([100], [010], [001]) were seen to be elastically equivalent in Examples 1 and 5, so are the six face diagonal directions ([110], [$\bar{1}$10], [011], [0$\bar{1}$1], [101], [10$\bar{1}$]). The compressional wave velocity (3.53) therefore applies to propagation along *any* of the face diagonal directions in the crystal.

EXAMPLE 9. *Shear Wave Propagation Along the [110] Direction in a Cubic Crystal.* If a Z- (or [001]-) polarized particle displacement is assumed in Fig. 3.7a,

$$\mathbf{u} = \hat{\mathbf{z}}' \cos (\omega t - kx')$$

and the strain field is

$$S_{5'} = 2S_{z'x'} = k \sin (\omega t - kx').$$

From (3.35) and (3.44), the stress field corresponding to this strain is

$$T_{5'} = T_{z'x'} = c_{44}k \sin (\omega t - kx'),$$

and the equation of motion (3.51) requires that

$$\frac{\partial}{\partial x'} T_{5'} = -c_{44}k^2 \cos (\omega t - kx') = -\rho\omega^2 u_z.$$

This shows that the assumed displacement is a solution to the acoustic equations if the dispersion relation

$$c_{44}k^2 = \rho\omega^2$$

is satisfied. The velocity of an [001]-polarized shear wave propagating along the [110] direction is therefore

$$(V_s)_{[001]} = \left(\frac{c_{44}}{\rho}\right)^{1/2}. \tag{3.54}$$

Repetition of this calculation for a [110]-propagating shear wave with [$\bar{1}$10]-polarized particle displacement gives a velocity

$$(V_s)_{[\bar{1}10]} = \left(\frac{c_{11} - c_{12}}{2\rho}\right)^{1/2}. \tag{3.55}$$

The same results apply for propagation in all other face diagonal directions. Shear waves propagating along a face diagonal and polarized along a cube edge travel with velocity (3.54), while shear waves propagating along a face diagonal and polarized along a face diagonal travel with velocity (3.55).

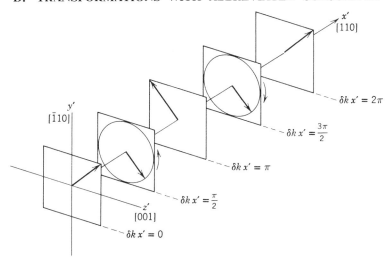

FIGURE 3.8. **Polarization transformations produced by combining non-degenerate shear waves in a *birefringent* medium.**

As in Example 8, the propagation velocity is not the same for a face diagonal direction as it is for a cube edge direction. In this case, however, there is an additional important change in propagation characteristics. Example 4 showed that both shear wave polarizations travel along a cube edge with the same velocity. For propagation along a face diagonal there are *two* shear wave velocities, one for cube edge polarization and the other for face diagonal polarization. Shear waves with this property are called *nondegenerate* and the medium is called *birefringent*. This means that shear waves traveling along a face diagonal cannot be combined to form an unchanging polarization pattern, as in Fig. 3.4. If two [110]-propagating shear waves are combined with equal amplitude and phase, the total particle displacement field is

$$\mathbf{u} = \hat{\mathbf{y}}' \cos\left(\omega t - \frac{2\pi}{\lambda_{[\bar{1}10]}} x'\right) + \hat{\mathbf{z}}' \cos\left(\omega t - \frac{2\pi}{\lambda_{[001]}} x'\right), \tag{3.56}$$

where $\lambda_{[\bar{1}10]} = (2\pi/\omega)V_{[110]}$ is the wavelength of the [$\bar{1}$10]-polarized wave and $\lambda_{[001]}$ is the wavelength of the [001]-polarized wave. At $x' = 0$ the y' and z' components of polarization are in time phase and the particle displacement field is linearly polarized (Fig. 3.8). As the waves propagate along x' the two components gradually shift phase because of their different velocities, and the polarization changes. If $\lambda_{[\bar{1}10]} < \lambda_{[001]}$, the y'–polarization has a 90° phase lag with respect to the z'–polarization when

$$\delta k\, x' = 2\pi\left(\frac{1}{\lambda_{[\bar{1}10]}} - \frac{1}{\lambda_{[100]}}\right)x' = \pi/2.$$

At this point the particle displacement is circularly polarized in the counterclockwise sense about $\hat{x}'$. When $\delta k\ x' = \pi$, the y' component of polarization is reversed in polarity with respect to the z' component, and the displacement is linearly polarized at right angles to the initial polarization. At $\delta k\ x' = 3\pi/2$, there is circular polarization in the clockwise sense, and at $\delta k\ x' = 2\pi$ the original polarization is reproduced. These changes in particle displacement repeat again and again as the observation point moves along x'.

E.   DAMPING AND ATTENUATION

It will be seen in Chapter 5 that material media with constitutive relations having the form of (3.12) or (3.13) do not have internal energy losses. Acoustic vibrations in such media are, as a consequence, completely undamped. This means that wavelike vibrations propagate without any decrease, or *attenuation*, of amplitude and that resonant oscillations persist indefinitely. Ideal materials of this kind do not exist in nature, although weakly damped materials are often approximated in this manner, and it is therefore necessary to look for a way of introducing damping into the elastic constitutive relation. Elastic damping usually depends, in a rather complicated manner, upon temperature, as well as the frequency and type of vibration. This is due to the rather impressive number of physical mechanisms that contribute to the phenomenon, and it is not possible to represent them all by a single modification of the constitutive relation. At room temperature, however, acoustic losses in many materials may be adequately described by a viscous damping term.

The form of the viscous damping term in the constitutive relation may be easily deduced by considering a simple physical analogue. In an ideal lossless medium Hooke's Law corresponds to the force-displacement relation for a spring (Fig. 3.9a); that is,

$$f = K x,$$

where the *applied* force f corresponds to stress and the spring displacement x to strain. Damping is added to this system by placing a viscous element, with response

$$f_d = K' \frac{\partial x}{\partial t},$$

in parallel with the spring (Fig. 3.9b). The response of the damped system

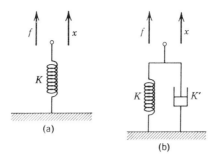

FIGURE 3.9. Mechanical analogues of ideal and damped elastic media.

is then specified by the differential equation

$$f = K x + K' \frac{\partial x}{\partial t}.$$

Following this analogy, the ideal Hooke's Law relation

$$T_I = c_{IJ} S_J$$

is modified to include damping by adding terms containing time derivatives of the strains;† that is,

$$T_I = c_{IJ} S_J + \eta_{IJ} \frac{\partial S_J}{\partial t}, \tag{3.57}$$

or

$$T_{ij} = c_{ijkl} S_{kl} + \eta_{ijkl} \frac{\partial S_{kl}}{\partial t} \tag{3.58}$$

in full subscript notation. Since the viscosity constants η_{ijkl} relate one second rank tensor ($\partial S_{kl}/\partial t$) to another second rank tensor (T_{ij}), it follows from the arguments of Section B in this chapter that they transform according to the law

$$\eta'_{mnop} = a_{mi} a_{nj} a_{ok} a_{pl} \eta_{ijkl}. \tag{3.59}$$

They are therefore components of a fourth rank tensor, the viscosity tensor

---

† Viscoelastic damping results from deformation rather than rigid displacement and must therefore depend upon the time derivatives of the strains rather than the time derivatives of the particle displacements.

η. Symbolically,

$$T = c:S + \eta:\frac{\partial S}{\partial t}, \qquad (3.60)$$

where the double dot indicates summation over full subscripts in (3.58) or summation over abbreviated subscripts in (3.57). In abbreviated subscripts η is transformed by the $[M]$ matrix (3.32), in the same way as c.

It is seen from (3.60) that the viscosity constants have units of newton seconds/(meter)^2 in the MKS system. Numerical values are, however, very frequently given in the CGS unit

$$1 \text{ centipoise} = 0.01 \frac{\text{dyne second}}{\text{cm}^2}.$$

A table of conversion ratios for MKS and cgs units will be found inside the back cover. Typical numerical values for the viscosity constants range from

$$0.1 \frac{\text{newton second}}{\text{m}^2} = 100 \text{ centipoises}$$

for metals and noncrystalline insulators, to approximately

$$0.0001 \frac{\text{newton second}}{\text{m}^2} = 0.1 \text{ centipoises},$$

for the lowest loss single crystal insulators.

EXAMPLE 10. It will be shown in Chapter 7 that the 6×6 viscosity matrix $[\eta]$ in (3.57) always has the same general form as the stiffness matrix $[c]$. For the single crystal cubic medium of Example 3 and Fig. 3.2,

$$[\eta] = \begin{bmatrix} \eta_{11} & \eta_{12} & \eta_{12} & 0 & 0 & 0 \\ \eta_{12} & \eta_{11} & \eta_{12} & 0 & 0 & 0 \\ \eta_{12} & \eta_{12} & \eta_{11} & 0 & 0 & 0 \\ 0 & 0 & 0 & \eta_{44} & 0 & 0 \\ 0 & 0 & 0 & 0 & \eta_{44} & 0 \\ 0 & 0 & 0 & 0 & 0 & \eta_{44} \end{bmatrix}$$

and the lossy constitutive equations are

$$T_1 = \left(c_{11} + \eta_{11}\frac{\partial}{\partial t}\right)S_1 + \left(c_{12} + \eta_{12}\frac{\partial}{\partial t}\right)S_2 + \left(c_{12} + \eta_{12}\frac{\partial}{\partial t}\right)S_3$$

$$T_2 = \left(c_{12} + \eta_{12}\frac{\partial}{\partial t}\right)S_1 + \left(c_{11} + \eta_{11}\frac{\partial}{\partial t}\right)S_2 + \left(c_{12} + \eta_{12}\frac{\partial}{\partial t}\right)S_3$$

$$T_3 = \left(c_{12} + \eta_{12}\frac{\partial}{\partial t}\right)S_1 + \left(c_{12} + \eta_{12}\frac{\partial}{\partial t}\right)S_2 + \left(c_{11} + \eta_{11}\frac{\partial}{\partial t}\right)S_3$$

$$T_4 = \left(c_{44} + \eta_{44}\frac{\partial}{\partial t}\right)S_4$$

$$T_5 = \left(c_{44} + \eta_{44}\frac{\partial}{\partial t}\right)S_5$$

$$T_6 = \left(c_{44} + \eta_{44}\frac{\partial}{\partial t}\right)S_6. \tag{3.61}$$

EXAMPLE 11. ***Attenuation of Acoustic Plane Waves Propagating Along the Cube Edge and the Face Diagonal Directions in a Cubic Crystal.*** A number of the previous examples have dealt with propagation along the cube edge and the face diagonal directions in a lossless cubic crystal. The effect of viscous damping on these waves will now be considered. In the earlier examples of *lossless* wave propagation, the starting point was an assumed particle displacement wave function. For example,

$$\mathbf{u} = \hat{\mathbf{x}}\cos\left(\omega t - ky\right), \tag{3.62}$$

was used in the case of an x-polarized y-propagating shear wave. According to the well-known trigonometric relation

$$e^{i(\omega t - ky)} = \cos\left(\omega t - ky\right) + i\sin\left(\omega t - ky\right),$$

(3.62) may also be written as

$$\mathbf{u} = \text{real part of } \hat{\mathbf{x}}e^{i(\omega t - ky)} = \mathscr{R}e\,\hat{\mathbf{x}}e^{i(\omega t - ky)}.$$

The advantage of this formulation is that the operation of taking the real part can be postponed to the end of the calculation, since the acoustic equations are both *real* and *linear*. That is, the entire calculation is first carried out using a particle displacement field

$$\mathbf{u} = \hat{\mathbf{x}}e^{i(\omega t - ky)}, \tag{3.63}$$

and the real part of the final result is then taken so as to find a solution related to the actual particle displacement field (3.62). This technique greatly simplifies the analysis, because every differentiation operation becomes a multiplication,

$$\frac{\partial}{\partial t}\mathbf{u} = i\omega\mathbf{u}$$

$$\frac{\partial}{\partial y}\mathbf{u} = -ik\mathbf{u}, \tag{3.64}$$

and every integration becomes a division.

Use of exponential wave functions is always a useful technique but it becomes almost essential in wave attenuation calculations, where the trigonometric formulation is extremely unwieldy. The amplitude of a wave propagating in a lossy medium steadily decreases as its energy is dissipated, or absorbed, by the medium. In this case the particle displacement field of an x-polarized y-propagating shear wave takes the form

$$\mathbf{u} = \hat{\mathbf{x}}e^{-\alpha y} \cos(\omega t - ky)$$

instead of (3.62). Alternatively, one may take the displacement field to be the real part of

$$\mathbf{u} = \hat{\mathbf{x}}e^{-\alpha y}e^{i(\omega t - ky)} \tag{3.65}$$

and the corresponding strain field is

$$S_{xy} = \frac{1}{2}\frac{\partial u_x}{\partial y} = -i\frac{(k - i\alpha)}{2}e^{i\omega t}e^{-i(k - i\alpha)y}.$$

If the medium is cubic and coordinates are aligned with the crystal axes (Fig. 3.2), the stress field is

$$T_6 = -i(k - i\alpha)(c_{44} + i\omega\eta_{44})e^{i\omega t}e^{-i(k - i\alpha)y}. \tag{3.66}$$

from (3.61). For fields varying only with the y coordinate, the equation of motion reduces to

$$\frac{\partial}{\partial y}T_{iy} = \rho\frac{\partial^2}{\partial t^2}u_i$$

$$i = x, y, z$$

and in this case one has

$$-i(k - i\alpha)T_{xy} = -i(k - i\alpha)T_6 = -\omega^2\rho u_x. \tag{3.67}$$

Substitution of (3.65) and (3.66) into (3.67) then gives the dispersion relation

$$(k - i\alpha)^2(c_{44} + i\omega\eta_{44}) = \rho\omega^2. \tag{3.68}$$

From (3.68),

$$(k^2 - \alpha^2 - i2\alpha k)(c_{44} + i\omega\eta_{44}) = \rho\omega^2.$$

Separation into real and imaginary parts gives

$$c_{44}(k^2 - \alpha^2) + 2\alpha k\omega\eta_{44} = \rho\omega^2$$

$$i((k^2 - \alpha^2)\omega\eta_{44} - 2\alpha kc_{44}) = 0,$$

and these equations may be further rearranged into the form

$$k^2 - \alpha^2 = \frac{\rho}{c_{44}}\frac{\omega^2}{\left(1 + \left(\frac{\omega\eta_{44}}{c_{44}}\right)^2\right)} \tag{3.69}$$

$$2\alpha k = \frac{\rho}{c_{44}}\frac{\omega^3\eta_{44}/c_{44}}{\left(1 + \left(\frac{\omega\eta_{44}}{c_{44}}\right)^2\right)}. \tag{3.70}$$

By eliminating k from (3.69) and (3.70) one obtains

$$\alpha^2 = \frac{\rho \omega^2}{2c_{44}} \left(\frac{1}{\left(1 + \left(\frac{\omega \eta_{44}}{c_{44}} \right)^2 \right)^{1/2}} - \frac{1}{\left(1 + \left(\frac{\omega \eta_{44}}{c_{44}} \right)^2 \right)} \right), \tag{3.71}$$

which can then be substituted into (3.69) to find k^2. For most solids the viscosity coefficient is sufficiently small that

$$\left(\frac{\omega \eta_{44}}{c_{44}} \right)^2 \ll 1$$

even at frequencies as high as 1 GHz (1000 Mc/sec). When this condition is satisfied, the rather complicated expression (3.71) reduces to

$$\alpha^2 = \frac{\omega^4}{4} \frac{\rho}{c_{44}} \left(\frac{\eta_{44}}{c_{44}} \right)^2 \tag{3.72}$$

and the corresponding approximation for k^2 is

$$k^2 = \frac{\rho \omega^2}{c_{44}} \left(1 + \frac{3}{4} \left(\frac{\omega \eta_{44}}{c_{44}} \right)^2 \right)^{-1}. \tag{3.73}$$

To examine the physical significance of these results one takes the real part of (3.65),

$$\mathscr{R}e \, \mathbf{u} = \hat{\mathbf{x}} e^{-\alpha y} \cos (\omega t - ky). \tag{3.74}$$

The *wave number* k again has its usual interpretation. That is, the wavelength is

$$\lambda = 2\pi/k$$

and the phase velocity is

$$V_p = \omega/k.$$

According to (3.73), then, the effect of viscous damping is to introduce a second order decrease in k and, consequently, a small increase in the phase velocity. A more important effect is that the wave amplitude varies with y according to the real exponential function $e^{-\alpha y}$, where α is called the *attenuation factor*. Since the wave amplitude in a lossy medium must decay in the direction of propagation, the signs of α and k in (3.72) and (3.73) must both be positive for a positive-traveling wave and negative for a negative-traveling wave (Fig. 3.10). The wave number k for a positive-traveling wave is therefore

$$k = \omega(\rho/c_{44})^{1/2} \left(1 + \frac{3}{8} \left(\frac{\omega \eta_{44}}{c_{44}} \right)^2 \right)^{-1/2} \tag{3.75}$$

and the corresponding attenuation factor for the low-loss case is

$$\alpha = \frac{\omega}{2} \left(\frac{\rho}{c_{44}} \right)^{1/2} \left(\frac{\omega \eta_{44}}{c_{44}} \right) = \frac{\omega^2}{2} \left(\frac{\rho}{c_{44}} \right)^{1/2} \left(\frac{\eta_{44}}{c_{44}} \right). \tag{3.76}$$

From (3.76), the attenuation factor is proportional to the square of the frequency. This dependence on frequency is typical of viscously damped materials and has the

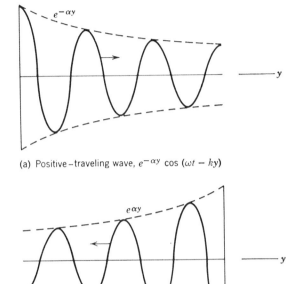

(a) Positive-traveling wave, $e^{-\alpha y} \cos(\omega t - ky)$

(b) Negative-traveling wave, $e^{\alpha y} \cos(\omega t + ky)$

FIGURE 3.10.   **Attenuated wave functions in a lossy medium.**

very important practical consequence that higher and higher quality materials must be used as the operating frequency is increased. Single crystal materials always have the lowest damping constants and must therefore be used in most applications above 100 MHz, even though their anisotropic wave velocity characteristics are often undesirable.

In (3.76) one sees that the attenuation *per wavelength*

$$\alpha\lambda = \frac{2\pi\alpha}{k} \qquad (3.77)$$

is a function only of the parameter

$$Q = \frac{c_{44}}{\omega\eta_{44}}, \qquad (3.78)$$

which is called the *acoustic quality factor* (or acoustic Q)†. From the symmetry equivalence of the different coordinate directions in Fig. 3.5 it is clear that this quality factor also applies to the z-polarized y-propagating wave and to shear waves propagating along any of the cubic crystal axes. Calculations for longitudinal

† A more general definition of Q is given in Section E of Chapter 11 in Volume 2.

TABLE 3.1. Acoustic Q's for Propagation Along the Cube Edge and Face Diagonal Directions in a Cubic Crystal

Propagation Direction	Polarization	$Q \approx \dfrac{\alpha}{2k}$
Cube edge	Longitudinal	$\dfrac{c_{11}}{\omega \eta_{11}}$
	Shear	$\dfrac{c_{44}}{\omega \eta_{44}}$
Face diagonal	Longitudinal	$\dfrac{c_{11} + c_{12} + 2c_{44}}{\omega(\eta_{11} + \eta_{12} + 2\eta_{44})}$
	Cube edge shear	$\dfrac{c_{44}}{\omega \eta_{44}}$
	Face diagonal shear	$\dfrac{c_{11} - c_{12}}{\omega(\eta_{11} - \eta_{12})}$

waves propagating along a cube edge direction and for shear and longitudinal waves propagating along a face diagonal direction give the acoustic Q's listed in Table 3.1. Values of k and α for these waves may be obtained from (3.75) and (3.76) by substituting the lossless wave vectors from previous calculations and the acoustic Q's from the table. Table 3.1 shows that measurement of the viscosity constants for a cubic material requires observations of the attenuation for at least three independent kinds of acoustic waves.†

Example 11 has shown that the field amplitude of a wave propagating along the $+y$ axis in a lossy medium varies as $e^{-\alpha y}$. The ratio of the field amplitudes at two points y_1 and $y_2 > y_1$ is therefore

$$\frac{e^{-\alpha y_1}}{e^{-\alpha y_2}} = e^{-\alpha(y_1 - y_2)}. \tag{3.79}$$

This is called the attenuation from y_1 to y_2. In engineering and experimental work it is usual to measure this quantity on a logarithmic scale, the main reason being that this compresses the very wide range of amplitude ratios encountered in practice. Use of a logarithmic scale also allows addition of successive attenuations. One method of converting to a logarithmic scale is to simply take the natural logarithm of the amplitude ratio (3.79). This gives the attenuation measured in *nepers*; that is

$$\text{Attenuation} = \alpha(y_2 - y_1) \text{ nepers.} \tag{3.80}$$

† See, for example, Reference 9 at the end of the chapter.

A neper is actually a dimensionless quantity and the name is used simply to indicate that the attenuation is measured on a natural logarithmic scale. The attenuation factor α in (3.76) is therefore said to have units of nepers/meter. A more commonly used logarithmic scale is based on common logarithms, and the basic unit is called the *decibel* (dB). In decibels, the attenuation from y_2 to y_1 is defined as

$$\text{Attenuation} = 10 \log{(e^{-\alpha(y_1 - y_2)})^2} = 20(\log e)\alpha(y_2 - y_1)\, dB \quad (3.81)$$

and is, consequently, a measure of (amplitude ratio)^2. According to (3.80) and (3.81) the attenuation factor α is converted from nepers/meter to dB/meter by multiplying with $20(\log e)$. That is,

$$\alpha(\text{dB/meter}) = 20(\log e)\alpha(\text{nepers/meter}) \approx 8.686\,\alpha(\text{nepers/meter}). \quad (3.82)$$

At room temperature, most single crystal materials have acoustic attenuations that increase with the square of the frequency. The principal physical mechanisms contributing to this viscous damping are the *thermoelastic mechanism* and the *Akhieser mechanism*. Thermoelastic attenuation is due to irreversible heat conduction from compression regions to rarefaction regions in a compressional (or longitudinal) wave. It occurs only in longitudinal waves, since shear waves do not produce changes in size of the elementary volume elements. At finite temperatures there exists in all materials an equilibrium distribution of thermally-excited acoustic waves (called phonons). Passage of a coherently-excited acoustic wave disturbs this phonon equilibrium, with a resulting energy absorption or damping. This damping mechanism is known as Akhieser (or phonon) damping. Damping may also occur at room temperature in polycrystalline materials, which consist of small, randomly oriented single crystal grains. The mechanism in this case is scattering at the grain boundaries. This, however, does not always lead to a viscous effect (attenuation $\sim \omega^2$) and a wide variety of attenuation-frequency curves is met experimentally.

The lowest room temperature acoustic attenuations are found in single crystal insulators, where the attenuation can be accounted for almost entirely by the Akhieser mechanism. Single crystal semiconductors come next in order of lossiness, which can again be attributed to the Akhieser process. In metals the predominant damping mechanism depends upon the type of wave. Thermoelastic effects contribute about one-half of the observed compressional wave attenuation and the remainder is explained by Akhieser damping. For shear waves there is no thermoelastic damping and the Akhieser mechanism is believed to be the major contributor. Polycrystalline materials usually have the largest values of attenuation. Table 3.2 gives measured values of acoustic attenuation for a number of single crystal cubic materials. These

TABLE 3.2.   Acoustic Attenuation in Single Crystal Cubic Materials at 1 GHZ (After Wauk)

Material	Propagation Direction	Polarization	Attenuation Factor α (dB/meter)
INSULATORS			
Magnesium oxide	Cube edge	Longitudinal	330
	Cube edge	Shear	40
Strontium titanate	Cube edge	Longitudinal	600
Yttrium iron garnet	Cube edge	Longitudinal	200
	Cube edge	Shear	34
Yttrium aluminum garnet	Cube edge	Longitudinal	20–32
	Cube edge	Shear	110
SEMICONDUCTORS			
Germanium	Cube edge	Longitudinal	2300
	Cube edge	Shear	1000
Silicon	Cube edge	Longitudinal	1000
	Body diagonal	Longitudinal	650
METALS			
Aluminum	Face diagonal	Longitudinal	7500
Copper	Cube edge	Longitudinal	27,000
Gold	Face diagonal	Longitudinal	20,000

values have not all been measured at the same frequency but are extrapolated to 1 GHz according to a (frequency)^2 law. The results show that attenuation generally increases in the order INSULATOR $\rightarrow$ SEMICONDUCTOR $\rightarrow$ METAL. Yttrium aluminum garnet has the lowest observed attenuation at 1 GHz, but a number of crystals in other symmetry classes have α's in the same general range. Figure 3.11 shows attenuation versus frequency curves for a number of materials that are commonly used in acoustic delay line devices. Attenuation is given for a length l_μ equivalent to one microsecond of delay time. That is,

$$l_\mu(\text{meters}) = 10^{-6} \, V_p \, (\text{meters/s})$$

From (3.76) and (3.82), the attenuation factor for a shear wave propagating along a cube edge direction is

$$\alpha = 4.343 \left(\frac{\rho}{c_{44}}\right)^{1/2} \left(\frac{\eta_{44}}{c_{44}}\right) \omega^2 \qquad (3.83)$$

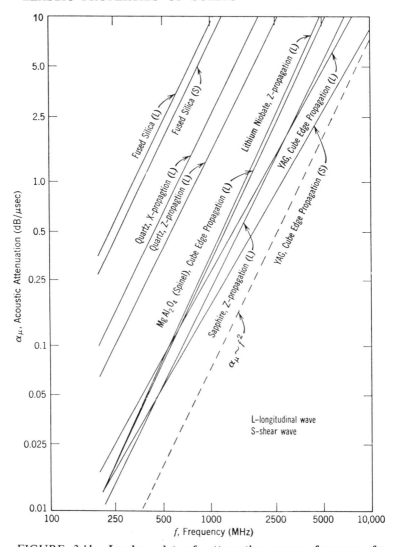

FIGURE 3.11. Log-log plot of attenuation versus frequency for selected acoustic materials. The dashed curve shows the quadratic frequency dependence typical of a viscously damped medium. (After Olson)

in dB/meter. If one substitutes the material parameters for yttrium aluminum garnet, (from Part A of Appendix 2)

$$\rho = 4550 \text{ kg/m}^2$$

$$c_{44} = 11.5 \times 10^{10} \text{ newtons/m}^2,$$

this becomes

$$\alpha = 7.50 \times 10^{-15} \ \eta_{44}\omega^2 \text{ dB/m},$$

where η_{44} is in units of newton seconds/meter^2. Taking $\alpha = 110$ dB/m from Table 3.2 for the attenuation factor of shear waves propagating along the cube edge direction in yttrium aluminum garnet, one obtains a viscosity coefficient

$$\eta_{44} = 37.2 \times 10^{-5} \frac{\text{newton seconds}}{\text{meter}^2},$$

or 0.372 centipoises. For longitudinal waves propagating along the cube edge of yttrium aluminum garnet (3.83) is modified by substituting c_{11} and η_{11}, and the viscosity constant calculated from Table 3.2 is

$$\eta_{11} = 33.1 \times 10^{-5} \frac{\text{newton seconds}}{(\text{meter})^2},$$

or 0.331 centipoises.

PROBLEMS

1. The elastic stiffness constants for sapphire are (Parts A-1 and A-2 of Appendix 2)

$$c_{11} = c_{22} \qquad\qquad c_{44} = c_{55}$$

$$c_{12} = c_{21} \qquad\qquad c_{56} = c_{65} = c_{14}$$

$$c_{13} = c_{31} = c_{23} = c_{32} \qquad\qquad c_{66} = \tfrac{1}{2}(c_{11} - c_{12})$$

$$c_{14} = c_{41} = -c_{24} = -c_{42}$$

in abbreviated subscript notation. Write down all of the constants in full subscript notation.

2. Using the constants given in Problem 1, illustrate the stresses produced in sapphire by strains $S_1, S_2, \ldots, S_6$. The strain may be represented by a deformed cube as in Fig. 2.4, and the stress by traction forces on the cube surfaces.

3. The form of the elastic compliance matrix $[s]$ and the elastic stiffness matrix $[c]$ for a hexagonal crystal is given in Part A.2 of Appendix 2. Express the elements of $[s]$ in terms of the elements of $[c]$, and vice versa.

4. Show that sapphire has an X-polarized X-propagating plane wave solution and an X-polarized Y-propagating solution. Find three solutions for propagation along the Z axis. Use the stiffness constants given in Problem 1.

5. The Z-propagating shear wave solutions in Example 4 are superposed, using amplitude A with phase angle zero for the X-polarized wave and amplitude B with phase angle θ for the Y-polarized wave. Find the shape and orientation of the elliptical path traced out by the tip of the particle displacement vector.

6. Using the Bond transformation matrix $[N]$, transform the strain fields of Problem 1 Chapter 1 to a coordinate system rotated 45° clockwise about the z axis. Compare with Problem 5 Chapter 1.

7. Derive (3.44) and (3.46).

8. Show that the stiffness matrix for trigonal classes *3, 3* (Part A.2 of Appendix 2) assumes the same form as the matrix for trigonal classes *32, 3m, 3m* when the coordinate system is rotated clockwise through an angle ξ, with

$$\tan 3\xi = \frac{c_{25}}{c_{14}},$$

about the Z axis.

9. Transform the cubic stiffness matrix of Example 3 and the hexagonal stiffness matrix of Problem 3 into cylindrical coordinates r, ϕ, z (Appendix 1).

10. Using Example 9 and the material constants given in Appendix 2, calculate the distance required to reverse the rotation direction of a circularly polarized shear wave propagating along the [110] direction in yttrium aluminum garnet. Repeat for magnesium oxide and diamond.

11. Show that the stress-strain relation (3.60) is represented by the complex stiffness matrix

$$[c_{IJ} + i\omega\eta_{IJ}]$$

when the fields have $e^{i\omega t}$ time-dependence. Find the corresponding complex compliance matrix.

12. The viscosity constants for sapphire satisfy the same relations as the stiffness constants given in Problem 1. Derive expressions giving attenuation constants for the plane wave solutions in Problem 4.

REFERENCES

Stress-strain Relations

1. P. C. Chou and N. J. Pagano, *Elasticity—Tensor, Dyadic and Engineering Approaches*, Ch. 3, van Nostrand, New York, 1967.

2. L. D. Landau and E. M. Lifshitz, *Theory of Elasticity*, pp. 10–13, Pergamon, New York, 1970.

3. G. Nadeau, *Introduction to Elasticity*, pp. 48–54, Holt, Rinehart and Winston, New York, 1964.

4. J. F. Nye, *Physical Properties of Crystals*, pp. 131–135, Oxford, England, 1964.

5. W. Bond, "The Mathematics of the Physical Properties of Crystals," pp. 1–72, *BSTJ*, **22** (1943).

Attenuation and Damping

6. L. D. Landau and E. M. Lifshitz, *Theory of Elasticity*, pp. 122–124, Pergamon, New York, 1970.

7. W. P. Mason, *Physical Acoustics and the Properties of Solids*, pp. 28–30, Ch. 7–11, van Nostrand, New York, 1958.

8. L. R. Kinsler and A. R. Frey, *Fundamentals of Acoustics*, pp. 221–225, Wiley, New York, 1962.

9. J. Lamb and J. Richter, "Anisotropic Acoustic Attenuation, with New Measurements for Quartz at Room Temperature," *Proc. Royal Soc. (London)*, **A293**, pp. 479–492 (1966).

10. P. G. Klemens, "Effect of Thermal and Phonon Processes on Ultrasonic Attenuation," pp. 201–234, in *Physical Acoustics*, vol. 3B, W. P. Mason, Ed., Academic Press, New York, 1965.

11. M. F. Lewis, "Attenuation of Longitudinal Ultrasonic Waves in Insulators at Room Temperature," *J. Acous. Soc. Am.*, **43**, pp. 852–858 (1968).

12. M. T. Wauk, *Attenuation in Microwave Acoustic Transducers and Resonators*, Ph.D. Dissertation, Department of Applied Physics, Stanford University, 1969.

13. F. A. Olson, "Microwave Acoustic Devices," *Space Age News*, **12**, pp 47–50 March 1969.

14. H. J. Maris, "Interaction of Sound Waves with Thermal Phonons in Dielectric Crystals," pp. 279–345, in *Physical Acoustics*, vol. 7, W. P. Mason and R. N. Thurston, Eds., Academic Press, New York, 1971.

15. E. P. Papadakis, "Ultrasonic Attenuation Caused by Scattering in Poly-crystalline Media," pp. 269–328 in *Physical Acoustics*, vol. 4B, W. P. Mason, Ed., Academic Press, New York, 1968.

CHAPTER 4

ACOUSTICS AND ELECTROMAGNETISM

A. THE ELECTROMAGNETIC-ACOUSTIC ANALOGY

The foundations of acoustic field theory in solids have now been laid. There are two basic field equations, the *strain-displacement relation*

$$S = \nabla_s u \tag{4.1}$$

and the *equation of motion*,

$$\nabla \cdot T = \rho \frac{\partial^2 u}{\partial t^2} - F. \tag{4.2}$$

Since there are three field variables (u, S, T) and only two equations, one additional condition is required and this is provided by the *elastic constitutive equation*

$$T = c:S + \eta:\frac{\partial S}{\partial t}. \tag{4.3}$$

101

In some cases this Hooke's Law relation does not fully describe the response of a solid to acoustic strain. Certain materials become electrically polarized when they are strained. This effect, called the *direct piezoelectric effect*, manifests itself experimentally by the appearance of bound electrical charges at the surfaces of a strained medium. It is a linear phenomenon, and the polarization changes sign when the sign of the strain is reversed. Piezoelectricity is intimately related to the microscopic structure of solids and, although a complicated subject, can be explained qualitatively in terms of a rather simple atomic model. Briefly, the atoms of a solid (and also the electrons within the atoms themselves) are displaced when the material is deformed. This displacement produces microscopic electrical dipoles within the medium, and in certain crystal structures these dipole moments combine to give an average macroscopic moment (or *electrical polarization*). A more detailed discussion of this model will be given in Chapter 8, where the lattice symmetry conditions required for existence of the effect are also derived.

The direct piezoelectric effect is always accompanied by the *converse piezoelectric effect*, whereby a solid becomes strained when placed in an electric field. Like the direct effect, this is also linear and the piezoelectric strain reverses sign with reversal of the applied electric field. Since the piezoelectric strain produced by an electric field will always generate internal stresses, the converse piezoelectric effect must be included in (4.3) by adding a stress term that is *linearly proportional to the electric field*. This *linear* electrically induced stress will be present only in materials with microscopic structures appropriate to the existence of piezoelectricity. There is, however, another kind of electrically induced stress that occurs in all materials. This stress, called *electrostrictive stress*, is a *quadratic function of the electric field*. It is produced by the same microscopic mechanism that causes the converse piezoelectric effect, namely, by electrical forces acting on the ionized atoms that form the crystal lattice; but, by contrast with the piezoelectric stress, it produces a macroscopic effect in all materials.

Since electrostriction is a second-order phenomenon, its role will be negligible in the small signal approximation of a linear theory. Piezoelectricity, on the other hand, introduces linear coupling between the acoustic field equations and Maxwell's electromagnetic field equations. This coupling will be studied at length in Chapter 8. In magnetic materials two analogous effects are observed. *Magnetostriction* is a quadratic magnetically induced stress present in materials of all symmetry classes, and *piezomagnetism* is a linear magnetoacoustic coupling that occurs only when certain lattice symmetry conditions are satisfied. *Intrinsic* piezomagnetic effects are not, at the present time, of any practical significance. Technologically useful linear magnetoacoustic coupling can, however, be realized by applying both a dc bias field $(\mathbf{H})_{dc}$ and a time-varying signal field $(\mathbf{H})_{signal}$ to certain kinds of magnetic materials. In this case the quadratic (or magnetostrictive) effect produces stress terms, proportional to terms such as $(H_i)_{dc}(H_j)_{signal}$, that are linearly dependent upon the applied signal field. This *biased piezomagnetism* is a strong effect† and has many important engineering applications. Magnetic media are, however, beyond the scope of this book and the subject will not be pursued further.‡

In one way or another, piezoelectricity (and biased piezomagnetism) provide the physical basis for almost all practical applications of acoustic fields. This is because they provide an effective means for electrically generating and detecting acoustic vibrations. In order to design the electroacoustic converters, or *transducers*, used for this purpose, it is necessary to establish a mathematical formalism relating the coupled electromagnetic and acoustic fields. This formalism will be developed in Chapter 8. It would hardly seem necessary to consider at this point the familiar equations governing uncoupled electromagnetic fields in nonpiezoelectric media. There is, however, an important benefit to be gained from doing so. By making some very simple notational changes, one may cast the acoustic field equations (4.1) and (4.2) into a form that very closely parallels the well-known Maxwell's equations of electromagnetism. This procedure provides much more than a satisfying mathematical symmetry. The field problems of greatest interest in acoustics are of the same general nature as problems, such as uniform plane wave propagation, guided waves, periodic waveguides, coupled modes, resonators, and filters, that have received much attention in electromagnetism, particularly in the area of microwave theory. Presentation of the acoustic field equations in a form analogous to Maxwell's equations simplifies the task of transferring to acoustics the analytical methodology and techniques that have been applied to problems of this kind in electromagnetism. In this chapter, the electromagnetic-acoustic analogy is first established and then illustrated by comparing the basic characteristics of electromagnetic and acoustic uniform plane waves. Later chapters, especially Chapters 10 to 13, will apply electromagnetic techniques and concepts to a number of more sophisticated acoustic problems.

---

† Biased piezoelectricity, which may be produced from electrostriction in an entirely analogous way, is generally very weak.

‡ For a treatment in detail see: B. A. Auld, "Magnetostatic and Magnetoelastic Wave Propagation in Solids," pp. 2–103 in *Applied Solid State Science*, vol. 2, R. Wolfe, Ed., Academic Press, New York, 1971; R. Birss, *Symmetry and Magnetism*, Ch. 4 and Ch. 5, North-Holland, Amsterdam, 1966; and D. A. Berlincourt, D. R. Curran and H. Jaffe, "Piezoelectric and Piezomagnetic Materials and Their Function in Transducers," Ch. 3 in *Physical Acoustics*, vol. 1A, W. P. Mason, Ed., Academic Press, New York, 1964.

B. THE ELECTROMAGNETIC FIELD EQUATIONS

The standard symbolic form for Maxwell's equations in rationalized units is

$$-\nabla \times \mathbf{E} = \frac{\partial \mathbf{B}}{\partial t} \tag{4.4}$$

and

$$\nabla \times \mathbf{H} = \frac{\partial \mathbf{D}}{\partial t} + \mathbf{J}_c + \mathbf{J}_s , \tag{4.5}$$

where $\mathbf{J}_c$ is the conduction current density and $\mathbf{J}_s$ is the source current density. Four basic field quantities $\mathbf{E}, \mathbf{H}, \mathbf{D}, \mathbf{B}$ are involved and are related to each other by the constitutive equations for the medium. It is customary to take $\mathbf{E}, \mathbf{H}$ as the basic field quantities. The linear constitutive relations for media with no piezoelectric or piezomagnetic effects are then

$$\mathbf{D} = \boldsymbol{\epsilon} \cdot \mathbf{E} \tag{4.6}$$

$$\mathbf{B} = \boldsymbol{\mu} \cdot \mathbf{H} \tag{4.7}$$

for the electric displacement and magnetic flux density, and

$$\mathbf{J}_c = \boldsymbol{\sigma} \cdot \mathbf{E} \tag{4.8}$$

for the conduction current density. Since these equations relate one vector quantity to another, the permittivity, permeability, and conductivity tensors $\boldsymbol{\epsilon}, \boldsymbol{\mu}$, and $\boldsymbol{\sigma}$ are of second rank. In the MKS system the electromagnetic field variables and constitutive parameters have the following units,

E	volts/m
H	amp/m
D	coulombs/m²
B	webers/m² $=$ volt sec/m²
$\mathbf{J}_c, \mathbf{J}_s$	amp/m²
$\boldsymbol{\epsilon}$	farads/m
$\boldsymbol{\mu}$	henries/m
$\boldsymbol{\sigma}$	mhos/m

In free space the permittivity and permeability reduce to scalar multipliers

$$\boldsymbol{\epsilon} \to \epsilon_0 \approx (\tfrac{1}{36\pi}) \times 10^{-9} \text{ farads/m}$$

$$\boldsymbol{\mu} \to \mu_0 = 4\pi \times 10^{-7} \text{ henries/m.}$$

Expressed in terms of $\mathbf{E}$ and $\mathbf{H}$ by means of the constitutive relations, Maxwell's equations (4.4) and (4.5) become

$$-\nabla \times \mathbf{E} = \mu \cdot \frac{\partial \mathbf{H}}{\partial t} \tag{4.9}$$

and

$$\nabla \times \mathbf{H} = \epsilon \cdot \frac{\partial \mathbf{E}}{\partial t} + \sigma \cdot \mathbf{E} + \mathbf{J}_s . \tag{4.10}$$

These symbolic equations are frequently displayed in component form by means of a matrix representation, with fields described by the column matrices

$$\mathbf{E} = \begin{bmatrix} E_x \\ E_y \\ E_z \end{bmatrix} \qquad \mathbf{H} = \begin{bmatrix} H_x \\ H_y \\ H_z \end{bmatrix}$$

and the curl operator described by the matrix-differential operator

$$[\nabla \times] = \begin{bmatrix} 0 & -\dfrac{\partial}{\partial z} & \dfrac{\partial}{\partial y} \\[2ex] \dfrac{\partial}{\partial z} & 0 & -\dfrac{\partial}{\partial x} \\[2ex] -\dfrac{\partial}{\partial y} & \dfrac{\partial}{\partial x} & 0 \end{bmatrix} \tag{4.11}$$

in rectangular Cartesian coordinates. Since

$$-[\nabla \times] = \widetilde{[\nabla \times]}$$

the electromagnetic field operators in (4.4) and (4.5) are transposes of each other in rectangular coordinates. This will prove to be a strong point of similarity with the acoustic field equations.

The equation of charge conservation

$$\nabla \cdot (\sigma \cdot \mathbf{E} + \mathbf{J}_s) = -\frac{\partial \rho_e}{\partial t} , \tag{4.12}$$

where ρ_e is the electrical charge density in coulombs/m^3, must also be satisfied by the electromagnetic field. In addition, useful divergence relations for $\mathbf{B}$ and $\mathbf{D}$,

$$\nabla \cdot \mathbf{B} = 0 \tag{4.13}$$

and

$$\nabla \cdot \mathbf{D} = \rho_e , \tag{4.14}$$

can be derived from (4.4), (4.5), and (4.12) (Problem 4.2).

C. THE ACOUSTIC FIELD EQUATIONS

By contrast with the usual form of Maxwell's equations, the basic acoustic field equations given by (4.1) and (4.2) in Section A involve second order time derivatives. To help establish the analogy with Maxwell's electromagnetic equations, first order time derivatives are introduced by using the particle velocity variable

$$\mathbf{v} = \frac{\partial \mathbf{u}}{\partial t}$$

rather than the displacement $\mathbf{u}$. The equation of motion (4.2) then becomes

$$\nabla \cdot \mathbf{T} = \frac{\partial \mathbf{p}}{\partial t} - \mathbf{F}, \tag{4.15}$$

where

$$\mathbf{p} = \rho \mathbf{v} \tag{4.16}$$

is defined as the *momentum density* (kg/m² s); and the strain-displacement relation (4.1) becomes

$$\nabla_s \mathbf{v} = \frac{\partial \mathbf{S}}{\partial t}. \tag{4.17}$$

Equations (4.15) and (4.17) now completely parallel the Maxwell equations (4.4) and (4.5). This strong analogy extends even to the transpose relationship between the acoustic field operators $\nabla \cdot$ and ∇_s in rectangular coordinates ((1.53) and (2.36)),† and it will also appear in the field theorems to be used in Chapter 5. There are four basic acoustic field quantities $\mathbf{T}$, $\mathbf{v}$, $\mathbf{S}$, $\mathbf{p}$, which are interrelated by constitutive relations for the medium. The strongest analogy with the electromagnetic equations is obtained by taking $\mathbf{T}$ as equivalent to $\mathbf{E}$ and $\mathbf{v}$ as equivalent to $\mathbf{H}$. Momentum density $\mathbf{p}$ then corresponds to $\mathbf{B}$, and the strain $\mathbf{S}$ corresponds to $\mathbf{D}$. One of the required constitutive relations is given by (4.16). To obtain the other, it is necessary to rearrange (4.3) so that $\mathbf{S}$ appears as the dependent variable. This is accomplished by multiplying with the compliance. In abbreviated subscript notation

$$s_{JI}T_I = S_J + \tau_{JK} \frac{\partial}{\partial t} S_K, \tag{4.18}$$

where

$$\tau_{JK} = s_{JI}\eta_{IK}$$

† In cylindrical and spherical coordinates neither the acoustic or the electromagnetic operators satisfy a transpose relationship. See Parts A, B, and C of Appendix 1.

is defined as the relaxation matrix. The time derivative of strain is then eliminated from (4.18) by using (4.17); and the elastic constitutive relation becomes

$$\mathbf{S} = \mathbf{s} : \mathbf{T} - \boldsymbol{\tau} : \nabla_s \mathbf{v}. \tag{4.19}$$

Using the constitutive relations, (4.15) and (4.17) can now be expressed entirely in terms of the stress field $\mathbf{T}$ and the particle velocity field $\mathbf{v}$,

$$\nabla \cdot \mathbf{T} = \rho \frac{\partial \mathbf{v}}{\partial t} - \mathbf{F} \tag{4.20}$$

and

$$\left(1 + \boldsymbol{\tau} : \frac{\partial}{\partial t}\right) \nabla_s \mathbf{v} = \mathbf{s} : \frac{\partial \mathbf{T}}{\partial t}. \tag{4.21}$$

These are the acoustic field equations corresponding to (4.9) and (4.10).

D. COMPARISON OF ELECTROMAGNETIC AND ACOUSTIC PLANE WAVES

Several examples of uniform acoustic plane waves were given in Chapter 3. Solutions of this kind, which have fields that are completely uniform over planes normal to the propagation direction, are easy to obtain mathematically and provide useful approximations to real waves in many practical problems. For example, the acoustic field at large distances from a piezoelectric transducer is, for many purposes, quite accurately represented by a uniform plane wave. Uniform plane waves also provide a basis for studying wave interactions at the boundaries of a finite body. An important practical example is acoustic wave propagation on structures that are relatively narrow in directions normal to the propagation direction. In this situation rather complicated wave solutions may be treated as superpositions of uniform plane waves. Problems of this kind will be discussed in Chapter 10.

For the same general reasons, uniform plane waves have always played an important role in electromagnetic theory. They therefore provide a good starting point for exploring the analogy between electromagnetic and acoustic fields. The cubic crystal media considered in the examples of Chapter 3 are electromagnetically isotropic and they do not therefore exhibit all of the interesting properties of electromagnetic plane wave propagation. A more general case—hexagonal crystal media—will be assumed here.[†] If coordinate axes x, y, z are chosen to coincide with the crystal axes X, Y, Z

[†] As in the cubic case, certain classes of hexagonal crystals are nonpiezoelectric and exhibit no linear interaction between electromagnetic and acoustic fields.

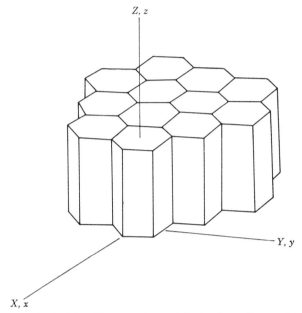

FIGURE 4.1.   Hexagonal crystal structure. In crystallography the unit cells for this case are usually taken to be
rhombic prisms, but the arrangement chosen here shows
more clearly the sixfold rotation symmetry about Z.

(Fig. 4.1), the dielectric permittivity matrix is

$$\boldsymbol{\epsilon} \rightarrow [\epsilon] = \begin{bmatrix} \epsilon_{xx} & 0 & 0 \\ 0 & \epsilon_{xx} & 0 \\ 0 & 0 & \epsilon_{zz} \end{bmatrix} \tag{4.22}$$

from Part C.1 of Appendix 2, and the elastic compliance matrix is

$$[s] = \begin{bmatrix} s_{11} & s_{12} & s_{13} & 0 & 0 & 0 \\ s_{12} & s_{11} & s_{13} & 0 & 0 & 0 \\ s_{13} & s_{13} & s_{33} & 0 & 0 & 0 \\ 0 & 0 & 0 & s_{44} & 0 & 0 \\ 0 & 0 & 0 & 0 & s_{44} & 0 \\ 0 & 0 & 0 & 0 & 0 & s_{66} \end{bmatrix} \tag{4.23}$$

with

$$S_{66} = 2(s_{11} - s_{12}),$$

from Part A.2 of Appendix 2. The stiffness matrix has the same form as (4.23), but with

$$c_{66} = \frac{(c_{11} - c_{12})}{2}.$$

EXAMPLE 1.   Uniform Plane Wave Propagation Along the X-axis of a Hexagonal Crystal.

(a) Electromagnetic Waves.   Since the fields vary only with the x coordinate, only the partial derivative $\partial/\partial x$ in (4.11) is nonzero,

$$[\nabla \times] = \begin{bmatrix} 0 & 0 & 0 \\ 0 & 0 & -\dfrac{\partial}{\partial x} \\ 0 & \dfrac{\partial}{\partial x} & 0 \end{bmatrix}$$

For a nonmagnetic medium ($\mu \to \mu_0$) with no conductive losses ($\sigma = 0$) and no current sources ($\mathbf{J}_s = 0$), Maxwell's equations (4.9) and (4.10) may be reduced to

$$-\nabla \times \mathbf{E} = \mu_0 \frac{\partial \mathbf{H}}{\partial t} \qquad\qquad \nabla \times \mathbf{H} = \boldsymbol{\epsilon} \cdot \frac{\partial \mathbf{E}}{\partial t}$$

$$0 = \mu_0 \frac{\partial H_x}{\partial t} \quad \text{(a)} \qquad\qquad 0 = \epsilon_{xx} \frac{\partial E_x}{\partial t} \quad \text{(d)}$$

$$\frac{\partial E_z}{\partial x} = \mu_0 \frac{\partial H_y}{\partial t} \quad \text{(b)} \qquad\qquad -\frac{\partial H_z}{\partial x} = \epsilon_{xx} \frac{\partial E_y}{\partial t} \quad \text{(e)}$$

$$-\frac{\partial E_y}{\partial x} = \mu_0 \frac{\partial H_z}{\partial t} \quad \text{(c)} \qquad\qquad \frac{\partial H_y}{\partial x} = \epsilon_{zz} \frac{\partial E_z}{\partial t} \quad \text{(f)}$$

$$(4.24)$$

by substituting the permittivity matrix (4.22). From (4.24(a)) and (4.24(d)) it follows that H_x and E_x are both zero, except for time-independent fields that are not of interest in the study of wave propagation. The magnetic and electric fields are, therefore, both transverse to the propagation direction, and the solutions are called *transverse waves*.

Examination of the remaining equations shows that they may be grouped in pairs, with one pair containing H_y, E_z and the other containing H_z, E_y. Consider the first pair. If (4.24(b)) is differentiated with respect to t and (4.24(f)) is differentiated with respect to x, combination of the two results gives an equation

$$\frac{\partial^2 H_y}{\partial x^2} = \mu_0 \epsilon_{zz} \frac{\partial^2 H_y}{\partial t^2}$$

for H_y. This is the well-known one-dimensional wave equation, which has a general solution of the form

$$H_y = f_1(t - x/V_p) + f_2(t + x/V_p),$$

where

$$V_p = (\mu_0 \epsilon_{zz})^{-1/2}$$

and $f_1(t - x/V_p)$, $f_2(t + x/V_p)$ are arbitrary differentiable functions of the arguments $(t - x/V_p)$ and $(t + x/V_p)$, respectively. These include, among others, the trigonometric and exponential forms of the time-harmonic wave functions

$$\cos \omega(t - x/V_p), \qquad e^{i\omega(t-x/V_p)},$$

etc., that were considered in Chapter 3. The first function f_1 is a wave traveling with phase velocity V_p in the $+x$ direction and the second function f_2 is a wave traveling with velocity V_p in the $-x$ direction. Since $\mu_0 = 4\pi \times 10^{-7}$ henries/meter and ϵ_{xx} is of the order of 10^{-11} farads/meter, these electromagnetic velocities are in the range of 10^8 meters/s,† approximately five orders of magnitude larger than acoustic velocities (Example 4 in Chapter 3).

For the positive-traveling wave

$$H_{y+} = f_1(t - x/V_p), \tag{4.25}$$

it is found from (4.24(f)) that

$$\frac{\partial E_{z+}}{\partial t} = \frac{1}{\epsilon_{zz}} \frac{\partial}{\partial x} H_{y+} = -\left(\frac{\mu_0}{\epsilon_{zz}}\right)^{1/2} f_1'(t - x/V_p),$$

where the prime denotes differentiation with respect to the argument of the function f_1. If this is integrated with respect to t, one obtains

$$E_{z+} = -\left(\frac{\mu_0}{\epsilon_{zz}}\right)^{1/2} f_1(t - x/V_p) = -\left(\frac{\mu_0}{\epsilon_{zz}}\right)^{1/2} H_{y+}, \tag{4.26}$$

since the time-independent constant of integration is not of interest in a wave propagation problem. Repeating the same steps for a negative-traveling wave, one finds that

$$E_{z-} = \left(\frac{\mu_0}{\epsilon_{zz}}\right)^{1/2} H_{y-}. \tag{4.27}$$

These results show that the electric and magnetic fields are perpendicular to one another and that one field reverses sign with respect to the other when the propagation direction is reversed. Figure (4.2a) illustrates this behavior for the positive-traveling time-harmonic wave

$$f_1\left(t - \frac{x}{V_p}\right) = \cos \omega\left(t - \frac{x}{V_p}\right).$$

† The dimension of henries $\times$ farads is (seconds)^2.

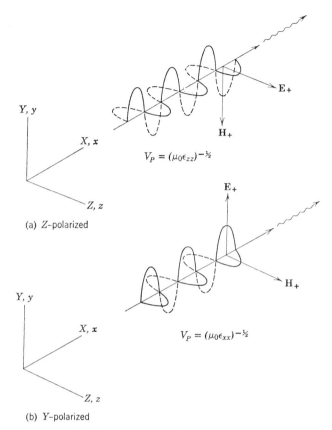

(a) Z-polarized

(b) Y-polarized

FIGURE 4.2.   **Electromagnetic wave propagation in the positive X axis direction of a hexagonal medium.**

A second plane wave solution is obtained by considering (4.24(c)) and (4.24(e)). In this case, a repetition of the above steps leads to

$$V_p = (\mu_0 \epsilon_{xx})^{-1/2}$$

and

$$E_{y\pm} = \pm \left(\frac{\mu_0}{\epsilon_{xx}}\right)^{1/2} H_{z\pm} . \tag{4.28}$$

where the upper signs refer to a positive-traveling wave and the lower signs to a negative-traveling wave. For the time-harmonic case, the field patterns of the positive-traveling wave are as illustrated in Fig. 4.2b.

To summarize, there are *two* kinds of electromagnetic uniform plane waves that can propagate along the X crystal axis of a hexagonal crystal. In both waves the electric and magnetic fields are polarized *transverse* to the propagation direction and at right angles to each other. For one solution, the *electric* field is polarized along the Z crystal axis. This is called the Z-polarized wave and the other solution, with the electric field along the Y crystal axis is called the Y-polarized wave. The phase velocities of the two solutions are different,

$$(V_p)_{Z\text{-polarized}} = (\mu_0 \epsilon_{zz})^{-1/2}$$
$$(V_p)_{Y\text{-polarized}} = (\mu_0 \epsilon_{xx})^{-1/2};$$

that is, they are *nondegenerate*. In this case the medium is called *birefringent*, and superposition of the Z-polarized and Y-polarized waves does not give an unchanging linear, circular or elliptical polarization pattern (Fig. 3.4). The polarization pattern changes constantly, as in Fig. 3.8.

(b) Acoustic Waves. Since the fields vary only with the x coordinate, only the partial derivative $\partial/\partial x$ is non zero in (1.53) and (2.36). For a lossless medium with no body force sources, the equation of motion (4.20) and the strain-displacement relation (4.21) may be reduced to

$$\nabla \cdot \mathbf{T} = \rho\, \frac{\partial \mathbf{v}}{\partial t} \qquad\qquad \nabla_s \mathbf{v} = \mathbf{s} : \frac{\partial \mathbf{T}}{\partial t}$$

$$\frac{\partial}{\partial x} T_1 = \rho\, \frac{\partial v_x}{\partial t} \quad \text{(a)} \qquad \frac{\partial v_x}{\partial x} = s_{11}\frac{\partial T_1}{\partial t} + s_{12}\frac{\partial T_2}{\partial t} + s_{13}\frac{\partial T_3}{\partial t} \quad \text{(d)}$$

$$\frac{\partial}{\partial x} T_6 = \rho\, \frac{\partial v_y}{\partial t} \quad \text{(b)} \qquad 0 = s_{12}\frac{\partial T_1}{\partial t} + s_{11}\frac{\partial T_2}{\partial t} + s_{13}\frac{\partial T_3}{\partial t} \quad \text{(e)}$$

$$\frac{\partial}{\partial x} T_5 = \rho\, \frac{\partial v_z}{\partial t} \quad \text{(c)} \qquad 0 = s_{13}\frac{\partial T_1}{\partial t} + s_{13}\frac{\partial T_2}{\partial t} + s_{33}\frac{\partial T_3}{\partial t} \quad \text{(f)}$$

$$0 = s_{44}\frac{\partial}{\partial t} T_4 \quad \text{(g)}$$

$$\frac{\partial v_z}{\partial x} = s_{44}\frac{\partial}{\partial t} T_5 \quad \text{(h)}$$

$$\frac{\partial v_y}{\partial x} = s_{66}\frac{\partial}{\partial t} T_6 \quad \text{(i)}$$

$$(4.29)$$

by substituting the compliance matrix (4.23). It is directly evident from (4.29(g)) that $T_4 = 0$, except for a time-independent value that is not relevant to the problem of wave propagation.

Equations (4.29(b)), (4.29(c)), (4.29(h)), and (4.29(i)) are completely analogous to the electromagnetic equations, and may be treated in exactly the same way. Elimination of T_6 from (4.29(b)) and (4.29(i)) thus gives the wave equation

$$\frac{\partial^2}{\partial x^2} v_y = \rho\, s_{66} \frac{\partial^2}{\partial t^2} v_y; \qquad (4.30)$$

and elimination of T_5 from (4.29(c)) and (4.29(h)) gives

$$\frac{\partial^2}{\partial x^2} v_z = \rho\, s_{44} \frac{\partial^2}{\partial t^2} v_z ,$$

(4.31)

In (4.30), the solutions are Y-polarized shear waves with phase velocity

$$(V_p)_{Y\text{-polarized}} = (\rho\, s_{66})^{-1/2}$$

(4.32)

and

$$T_{6\pm} = \mp \left(\frac{\rho}{s_{66}}\right)^{1/2} v_{y\pm} ,$$

(4.33)

where the upper signs refer to a positive-traveling wave and the lower signs to a negative-traveling wave. In (4.31), the solutions are Z-polarized shear waves with

$$(V_p)_{Z\text{-polarized}} = (\rho\, s_{44})^{-1/2}$$

(4.34)

and

$$T_{5\pm} = \mp \left(\frac{\rho}{s_{44}}\right)^{1/2} v_{z\pm} .$$

(4.35)

Since the inversion of (4.23) gives

$$c_{44} = 1/s_{44}$$

and

$$c_{66} = \tfrac{1}{2}(c_{11} - c_{12}) = 1/s_{66} ,$$

the phase velocities may also be expressed in terms of the stiffnesses as

$$(V_p)_{Y\text{-polarized}} = \left(\frac{c_{11} - c_{12}}{2\rho}\right)^{1/2}$$

$$(V_p)_{Z\text{-polarized}} = \left(\frac{c_{44}}{\rho}\right)^{1/2}.$$

Time-harmonic field patterns are illustrated in Fig. 4.3a and b.

All of the acoustic equations have now been used except (4.29(a)), (4.29(d)), (4.29(e)), and (4.29(f)). The first of these involves only T_1 and v_x but the others contain extra variables T_2 and T_3. There are, therefore, *three* stress variables that must be eliminated in order to obtain a wave equation for v_x. As the equations stand, this is a rather tedious task and the analysis is considerably simplified by rewriting the strain-displacement relation in the following way,

$$\mathbf{c}:\nabla_s\mathbf{v} = \mathbf{c}:\mathbf{s}:\frac{\partial \mathbf{T}}{\partial t} = \frac{\partial \mathbf{T}}{\partial t} .$$

In matrix form, the first three rows of this symbolic equation are

$$\frac{\partial T_1}{\partial t} = c_{11} \frac{\partial v_x}{\partial x}$$

(4.36)

$$\frac{\partial T_2}{\partial t} = c_{12} \frac{\partial v_x}{\partial x}$$

(4.37)

$$\frac{\partial T_3}{\partial t} = c_{13} \frac{\partial v_x}{\partial x} .$$

(4.38)

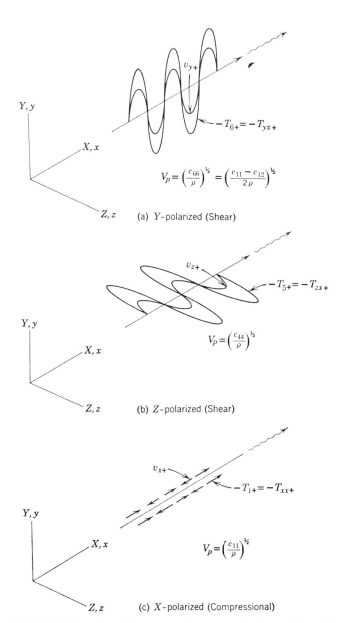

v_{y+}

$-T_{6+} = -T_{yx+}$

$$V_p = \left(\frac{c_{66}}{\rho}\right)^{\frac{1}{2}} = \left(\frac{c_{11} - c_{12}}{2\rho}\right)^{\frac{1}{2}}$$

(a) Y-polarized (Shear)

v_{z+}

$-T_{5+} = -T_{zx+}$

$$V_p = \left(\frac{c_{44}}{\rho}\right)^{\frac{1}{2}}$$

(b) Z-polarized (Shear)

v_{x+}

$-T_{1+} = -T_{xx+}$

$$V_p = \left(\frac{c_{11}}{\rho}\right)^{\frac{1}{2}}$$

(c) X-polarized (Compressional)

FIGURE 4.3. **Acoustic propagation along in the positive X direction of a hexagonal medium.**

It is now a simple matter to eliminate T_1 from (4.29(a)) and (4.36). This gives the wave equation

$$\frac{\partial^2 v_x}{\partial x^2} = (\rho/c_{11}) \frac{\partial^2 v_x}{\partial t^2}. \tag{4.39}$$

Solutions to (4.39) are compressional waves† with

$$T_{1\pm} = \mp (\rho c_{11})^{1/2} v_{x\pm} \tag{4.40}$$

and phase velocity

$$V_p = (c_{11}/\rho)^{1/2}. \tag{4.41}$$

The remaining stress components T_2 and T_3 can now be calculated from (4.37) and (4.38). That is,

$$T_{2\pm} = \mp \frac{c_{12}}{V_p} v_{x\pm}$$

$$T_{3\pm} = \mp \frac{c_{13}}{V_p} v_{x\pm}, \tag{4.42}$$

where the upper and lower signs are interpreted as in previous examples. As in Example 5 of Chapter 3, these "extra" stress components do not play an active part in the propagation.

Summarizing, there are *three* kinds of acoustic uniform plane waves that can propagate along the X crystal axis of a hexagonal crystal. Two of these waves (shear waves) have the particle displacement velocity polarized *transverse* to the propagation direction and the third wave (compressional wave) has *longitudinal* polarization; that is, the particle displacement velocity is along the propagation direction. For the two shear waves the transverse polarizations are along the Y and Z crystal axes, respectively. Polarization directions for the three waves are therefore mutually perpendicular and, in this case, all waves are nondegenerate. That is, there are three different values of phase velocity,

$$(V_p)_{\substack{Y\text{-polarized}\\ \text{shear}}} = (\rho\, s_{66})^{-1/2} = \left(\frac{c_{66}}{\rho}\right)^{1/2}$$

$$(V_p)_{\substack{Z\text{-polarized}\\ \text{shear}}} = (\rho\, s_{44})^{-1/2} = \left(\frac{c_{44}}{\rho}\right)^{1/2}$$

$$(V_p)_{\text{compressional}} = \left(\frac{c_{11}}{\rho}\right)^{1/2}.$$

In this case the shear waves are said to be *birefringent*, as in Example 9 of Chapter 3, but the complete set of acoustic waves is called *trirefringent*.

† The particle velocity displacement **v** is polarized along the propagation direction $\hat{x}$.

EXAMPLE 2.   Uniform Plane Wave Propagation in the XZ-plane of a Hexagonal Crystal.

(a) *Electromagnetic Waves.* In this example the propagation is at an arbitrary angle η in the XZ-plane† (Fig. 4.4). Since the permittivity matrix relates the electric displacement vector **D** to the electric field vector **E**, it is a second rank tensor and

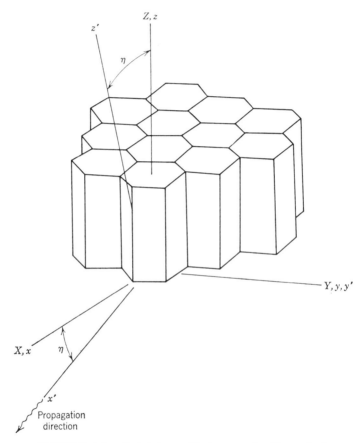

FIGURE 4.4.   **Rotated coordinate axes for Example 2.**

† It is evident from (4.22) that the permittivity matrix $[\epsilon]$ is unchanged by arbitrary co-ordinate rotations about the *Z-axis* in Fig. 4.1. Although it is not at all obvious from the form of the stiffness matrix $[c]$, or from the lattice symmetry, a simple application of the transformation (3.39) shows that $[c]$ is also unchanged by any Z-axis rotation. This means that all hexagonal crystals are *uniaxial* in their electrical and elastic properties and that the results of this example will apply to *any* propagation direction in the crystal.

must transform in the same way as strain and stress. That is,

$$[\epsilon'] = [a][\epsilon]\widetilde{[a]}.$$

In this case, the coordinate transformation matrix is

$$[a] = \begin{bmatrix} \cos \eta & 0 & -\sin \eta \\ 0 & 1 & 0 \\ \sin \eta & 0 & \cos \eta \end{bmatrix}$$

from Example 7 in Chapter 3 and the transformed permittivity matrix is therefore

$$[\epsilon'] = \begin{bmatrix} \epsilon'_{xx} & 0 & \epsilon'_{xz} \\ 0 & \epsilon'_{yy} & 0 \\ \epsilon'_{xz} & 0 & \epsilon'_{zz} \end{bmatrix} \tag{4.43}$$

with

$$\epsilon'_{xx} = \epsilon_{xx} \cos^2 \eta + \epsilon_{zz} \sin^2 \eta$$

$$\epsilon'_{xz} = (\epsilon_{xx} - \epsilon_{zz}) \cos \eta \sin \eta$$

$$\epsilon'_{yy} = \epsilon_{xx}$$

$$\epsilon'_{zz} = \epsilon_{xx} \sin^2 \eta + \epsilon_{zz} \cos^2 \eta.$$

Part (b) of Example 1 showed that analysis of the acoustic waves was greatly simplified by rewriting the stress-displacement relation (4.21) for the lossless case in the form

$$\mathbf{c} : \nabla_s \mathbf{v} = \frac{\partial}{\partial t} \mathbf{T}. \tag{4.44}$$

In the present example a corresponding simplification is realized by rewriting the curl of $\mathbf{H}$ equation as

$$\boldsymbol{\epsilon}^{-1} \cdot \nabla \times \mathbf{H} = \frac{\partial}{\partial t} \mathbf{E}, \tag{4.45}$$

where ϵ^{-1} is represented by the inverse of the permittivity matrix (4.43). That is,

$$[\epsilon']^{-1} = \begin{bmatrix} \dfrac{\epsilon'_{zz}}{\epsilon'_{xx}\epsilon'_{zz} - (\epsilon'_{xz})^2} & 0 & -\dfrac{\epsilon'_{xz}}{\epsilon'_{xx}\epsilon'_{zz} - (\epsilon'_{xz})^2} \\ \\ 0 & (\epsilon'_{yy})^{-1} & 0 \\ \\ -\dfrac{\epsilon'_{xz}}{\epsilon'_{xx}\epsilon'_{zz} - (\epsilon'_{xz})^2} & 0 & \dfrac{\epsilon'_{xx}}{\epsilon'_{xx}\epsilon'_{zz} - (\epsilon'_{xz})^2} \end{bmatrix}$$

It is also advantageous to introduce time-harmonic solutions at the beginning of the analysis and to use the exponential form of wave function $e^{i(\omega t - kx')}$. This means that

$$\frac{\partial}{\partial x'} \rightarrow -ik$$

$$\frac{\partial}{\partial y'} \rightarrow 0$$

$$\frac{\partial}{\partial z'} \rightarrow 0$$

in (4.11), and

$$\frac{\partial}{\partial t} \rightarrow i\omega.$$

Maxwell's equations therefore reduce to

$$-\nabla \times \mathbf{E} = \mu_0 \frac{\partial \mathbf{H}}{\partial t} \qquad\qquad \epsilon^{-1} \cdot \nabla \times \mathbf{H} = \frac{\partial \mathbf{E}}{\partial t}$$

$$0 = i\omega\mu_0 H_{x'} \quad \text{(a)} \qquad\qquad \frac{ik\,\epsilon'_{xz}}{\epsilon'_{xx}\epsilon'_{zz} - (\epsilon'_{xz})^2} H_{y'} = i\omega E_{z'} \quad \text{(d)}$$

$$-ikE_{z'} = i\omega\mu_0 H_{y'} \quad \text{(b)} \qquad\qquad ik(\epsilon'_{yy})^{-1} H_{z'} = i\omega E_{y'} \quad \text{(e)}$$

$$ikE_{y'} = i\omega\mu_0 H_{z'} \quad \text{(c)} \qquad\qquad -\frac{ik\,\epsilon'_{xx}}{\epsilon'_{xx}\epsilon'_{zz} - (\epsilon'_{xz})^2} H_{y'} = i\omega E_{z'} \quad \text{(f)}$$

$$(4.46)$$

It is again true, as in Example 1, that $H_{x'} = 0$; but, from (4.46(d)), $E_{x'}$ is no longer zero. Thus *only* the magnetic field is transverse to the propagation direction. Solutions of this kind are called *quasitransverse waves*.†

The analysis follows the procedure used in Example 1. Equations involving the same field components are grouped and the electric field is then eliminated. From (4.46(b)), (4.46(d)), and (4.46(f)) one obtains a *quasitransverse* wave with **E** polarized in the XZ plane

$$H_{y'\pm} \sim e^{i(\omega t \mp kx')}$$

$$E_{z'\pm} = \mp \frac{k}{\omega} \frac{\epsilon'_{xx}}{\epsilon'_{xx}\epsilon'_{zz} - (\epsilon'_{xz})^2} H_{y'\pm}$$

$$E_{x'\pm} = \pm \frac{k}{\omega} \frac{\epsilon'_{xz}}{\epsilon'_{xx}\epsilon'_{zz} - (\epsilon'_{xz})^2} H_{y'\pm}$$

$$V_p = \frac{\omega}{k} = \left(\frac{\epsilon'_{xx}}{\mu_0(\epsilon'_{xx}\epsilon'_{zz} - (\epsilon'_{xz})^2)} \right)^{1/2}; \tag{4.47}$$

and, from (4.46(c)) and (4.46(e)), a *pure transverse* wave with **E** polarized along the Y axis,

$$H_{z'\pm} \sim e^{i(\omega t \mp kx')}$$

$$E_{y'\pm} = \pm \frac{k}{\omega} (\epsilon'_{yy})^{-1} H_{z'\pm}$$

$$V_p = \frac{\omega}{k} (\mu_0\epsilon'_{yy})^{-1/2}. \tag{4.48}$$

In (4.47) and (4.48) the upper signs refer to positive-traveling waves and the lower signs to negative-traveling waves.

There are again two electromagnetic wave solutions, both having **E** and **H** perpendicular to each other but with **E** perpendicular to **k** for only one of the solutions (Fig. 4.5). The polarization angle χ between **E** and **k** in Fig. 4.5a is obtained from (4.47). Using the expressions (4.43) for the permittivity components, one finds that

$$\chi = \tan^{-1} \frac{E_{z'}}{E_{x'}} = \tan^{-1} \left(\frac{\epsilon_{xx} \cos^2 \eta + \epsilon_{zz} \sin^2 \eta}{(\epsilon_{xx} - \epsilon_{zz}) \cos \eta \sin \eta} \right). \tag{4.49}$$

When $\eta \to 0$, $\chi \to \pi/2$ and (4.47) becomes identical with the Z-polarized solution of Example 1. Solution (4.48) is identical with the Y-polarized solution of Example 1 for all values of η. For propagation along the Z (hexagonal) crystal axis ($\eta = \pi/2$), (4.47) again becomes pure transverse and has a phase velocity degenerate with solution (4.48). These characteristics are typical of electromagnetic wave propagation in anisotropic media. The two solutions are generally quasitransverse and nondegenerate but may become transverse and/or degenerate for special propagation directions.

† An electromagnetic wave is called quasitransverse when either **E** or **H** is nonperpendicular to the propagation direction.

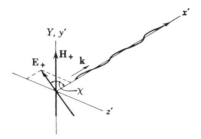

(a) ZX-polarized (quasitransverse)

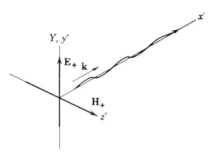

(b) Y-polarized (pure transverse)

FIGURE 4.5. Electromagnetic wave propagation in the positive direction along the rotated coordinate axis x' in Fig. 4.4.

(**b**) *Acoustic Waves.*   In the rotated coordinate system of Fig. 4.4 the elastic stiffness matrix assumes the form

$$
[c] =
\begin{bmatrix}
c'_{11} & c'_{12} & c'_{13} & 0 & c'_{15} & 0 \\
c'_{12} & c'_{22} & c'_{23} & 0 & c'_{25} & 0 \\
c'_{13} & c'_{23} & c'_{33} & 0 & c'_{35} & 0 \\
0 & 0 & 0 & c'_{44} & 0 & c'_{46} \\
c'_{15} & c'_{25} & c'_{35} & 0 & c'_{55} & 0 \\
0 & 0 & 0 & c'_{46} & 0 & c'_{66}
\end{bmatrix}
\tag{4.50}
$$

If a wave function $e^{i(\omega t - kx')}$ is assumed and (4.50) is substituted into (4.44), the acoustic field equations may be reduced to

$$\nabla \cdot \mathbf{T} = \rho \frac{\partial \mathbf{v}}{\partial t} \qquad\qquad \frac{\partial \mathbf{T}}{\partial t} = \mathbf{c} : \nabla_s \mathbf{v}$$

$$-ik T_{1'} = i\omega \rho v_{x'} \quad \text{(a)} \qquad i\omega T_{1'} = -ik(c'_{11}v_{x'} + c'_{15}v_{z'}) \quad \text{(d)}$$

$$-ik T_{6'} = i\omega \rho v_{y'} \quad \text{(b)} \qquad i\omega T_{2'} = -ik(c'_{12}v_{x'} + c'_{25}v_{z'}) \quad \text{(e)}$$

$$-ik T_{5'} = i\omega \rho v_{z'} \quad \text{(c)} \qquad i\omega T_{3'} = -ik(c'_{13}v_{x'} + c'_{35}v_{z'}) \quad \text{(f)}$$

$$i\omega T_{4'} = -ik c'_{46}v_{y'} \quad \text{(g)}$$

$$i\omega T_{5'} = -ik(c'_{15}v_{x'} + c'_{55}v_{z'}) \quad \text{(h)}$$

$$i\omega T_{6'} = -ik c'_{66}v_{y'}. \quad \text{(i)}$$

$$(4.51)$$

These are very much more complicated than the acoustic equations of Example 1. Nevertheless, the same procedure of grouping equations and eliminating variables can still be followed.

In (4.51) the only equations involving $v_{y'}$ are (b), (g), (i), and these are decoupled from all of the other equations. Elimination of $T_{6'}$ from (b) and (i) gives

$$\omega^2 \rho v_{y'} = k^2 c'_{66}v_{y'}.$$

From this, one concludes that

$$v_{y'\pm} \sim e^{i(\omega t \mp kx')}$$

$$T_{6'\pm} = \mp \frac{\omega}{k} \rho v_{y'\pm}$$

$$V_p = \frac{\omega}{k} = \left(\frac{c'_{66}}{\rho}\right)^{1/2}, \qquad (4.52)$$

where the upper and lower signs are to be interpreted as in the previous examples, and the "extra" stress component $T_{4'}$ is calculated from (g); that is,

$$T_{4'\pm} = \mp \frac{k}{\omega} c'_{46}v_{y'}.$$

This is a *pure transverse* (or *pure shear*) wave with the particle displacement velocity **v** polarized along $\hat{\mathbf{y}}'$.

Further examination of (4.51) shows that the remaining equations are all coupled together, but that they may be reduced to two equations in $v_{x'}$ and $v_{z'}$ by eliminating $T_{1'}$ and $T_{5'}$ from (a), (c), (d), and (h). That is,

$$-\frac{\omega^2 \rho}{k} v_{x'} = -k(c'_{11}v_{x'} + c'_{15}v_{z'})$$

$$-\frac{\omega^2 \rho}{k} v_{z'} = -k(c'_{15}v_{x'} + c'_{55}v_{z'}).$$

After dividing both equations by k and noting that

$$\frac{\omega^2}{k^2} = V_p^2,$$

one has

$$(c'_{11} - \rho V_p^2)v_{x'} + c'_{15}v_{z'} = 0$$
$$c'_{15}v_{x'} + (c'_{55} - \rho V_p^2)v_{z'} = 0. \tag{4.53}$$

The requirement for nontrivial solutions to exist is that the determinant of the system of equations must be zero; that is,

$$(c'_{11} - \rho V_p^2)(c'_{55} - \rho V_p^2) - c'_{15}c'_{15} = 0. \tag{4.54}$$

This gives two allowed values of $V_p^2 = \omega^2/k^2$,

$$(V_p^2)_{\mathrm{I}}, (V_p^2)_{\mathrm{II}} = \frac{c'_{11} + c'_{55} \pm \sqrt{(c'_{11} - c'_{55})^2 + 4(c'_{15})^2}}{2\rho}, \tag{4.55}$$

each corresponding to an acoustic wave propagating along x'. Both of these waves have x' and z' components of $\mathbf{v}$ and are, consequently, neither pure transverse nor pure longitudinal. When $\eta \to 0$ in (4.50), $c'_{15} = 0$ and it follows from (4.55) that

$$(V_p^2)_{\mathrm{I}} \to \frac{c_{11}}{\rho}.$$

This corresponds to the pure longitudinal (or pure compressional) wave of Example 1, and Solution I is therefore called a *quasilongitudinal* (or *quasicompressional*) wave. Solution II, on the other hand, reduces to a pure transverse (or pure shear) wave when $\eta \to 0$ and is called a *quasitransverse* (or *quasishear*)wave.[†] Parts (e) and (f) of (4.51) have not yet been used. These equations simply give "extra" stress components, corresponding to the stress $T_{4'}$ that was associated with the pure shear wave in (4.52).

Although solutions I and II are not pure longitudinal and pure transverse, their particle displacement velocities are, nevertheless, still perpendicular to each other (Fig. 4.6). This is easily proved by performing some simple manipulations on (4.53). The terms containing V_p^2 are first transferred to the right-hand sides of the equations

$$c'_{11}v_{x'} + c'_{15}v_{z'} = \rho V_p^2 v_{x'}$$
$$c'_{15}v_{x'} + c'_{55}v_{z'} = \rho V_p^2 v_{z'}.$$

These equations are satisfied by both solutions I and II. In the first case,

$$c'_{11}(v_{x'})_{\mathrm{I}} + c'_{15}(v_{z'})_{\mathrm{I}} = \rho(V_p^2)_{\mathrm{I}}(v_{x'})_{\mathrm{I}} \tag{4.56}$$
$$c'_{15}(v_{x'})_{\mathrm{I}} + c'_{55}(v_{z'})_{\mathrm{I}} = \rho(V_p^2)_{\mathrm{I}}(v_{z'})_{\mathrm{I}}. \tag{4.57}$$

If (4.56) is multiplied by $(v_{x'})_{\mathrm{II}}$ and added to $(v_{z'})_{\mathrm{II}}$ times (4.57), the result is

$$c'_{11}(v_{x'})_{\mathrm{I}}(v_{x'})_{\mathrm{II}} + c'_{15}((v_{z'})_{\mathrm{I}}(v_{x'})_{\mathrm{II}} + (v_{x'})_{\mathrm{I}}(v_{z'})_{\mathrm{II}}) + c'_{55}(v_{z'})_{\mathrm{I}}(v_{z'})_{\mathrm{II}} = \rho(V_p^2)_{\mathrm{I}}(\mathbf{v}_{\mathrm{I}} \cdot \mathbf{v}_{\mathrm{II}}). \tag{4.58}$$

Repetition with reversed subscripts I and II gives

$$c'_{11}(v_{x'})_{\mathrm{II}}(v_{x'})_{\mathrm{I}} + c'_{15}((v_{z'})_{\mathrm{II}}(v_{x'})_{\mathrm{I}} + (v_{x'})_{\mathrm{II}}(v_{z'})_{\mathrm{I}}) + c'_{55}((v_{z'})_{\mathrm{II}}(v_{z'})_{\mathrm{I}}) = \rho(V_p^2)_{\mathrm{II}}(\mathbf{v}_{\mathrm{I}} \cdot \mathbf{v}_{\mathrm{II}}); \tag{4.59}$$

[†] The polarization direction of an acoustic wave is determined by the particle velocity field.

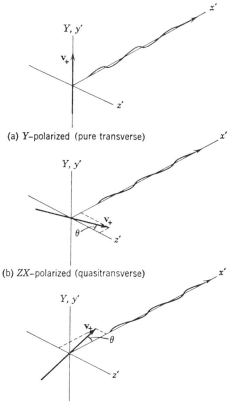

(a) Y-polarized (pure transverse)

(b) ZX-polarized (quasitransverse)

(c) ZX-polarized (quasilongitudinal)

FIGURE 4.6. **Acoustic wave propagation in the positive direction along the rotated coordinate axis x' in Fig. 4.4.**

and

$$\rho((V_p^2)_\text{I} - (V_p^2)_\text{II})(\mathbf{v}_\text{I} \cdot \mathbf{v}_\text{II}) = 0 \qquad (4.60)$$

is obtained by subtracting (4.59) from (4.58). In other words

$$\mathbf{v}_\text{I} \cdot \mathbf{v}_\text{II} = 0$$

when $(V_p^2)_\text{I} \neq (V_p^2)_\text{II}$. *This means that the particle displacement velocities are mutually perpendicular. If the phase velocities are degenerate ($V_{p\text{I}} = V_{p\text{II}}$), the two solutions I and II may be combined to give arbitrary polarizations, and one can therefore always find two linear combinations with $\mathbf{v}$'s that are perpendicular to each other.*

Figure 4.6 shows the $\mathbf{v}$ fields for the three kinds of acoustic uniform plane waves propagating along $\hat{\mathbf{x}}'$. There is one pure transverse wave, one quasilongitudinal wave, and one quasitransverse wave; and the polarization directions are mutually orthogonal among the three solutions. In Chapter 7 it will be shown that this is true even when none of the waves is pure transverse or pure longitudinal, as is generally the case. Special directions for which the wave solutions become pure transverse and pure longitudinal are called *pure mode directions* and can often be deduced from symmetry considerations (Section 7.I).

E. BOUNDARY CONDITIONS, REFLECTIONS, AND CHARACTERISTIC IMPEDANCES

There are very few practical problems in electromagnetism and acoustics that do not involve boundaries between regions with different material properties. To solve problems of this kind one must know how to join or match the fields across the discontinuity surfaces between different media. It is first necessary to find field solutions within each of the homogeneous regions and then to construct a complete solution for the problem by joining the individual solutions according to the appropriate *boundary conditions* at the discontinuities. In this section, boundary conditions will be derived for both the electromagnetic field and the acoustic field. Use of these boundary conditions will be demonstrated with some simple examples.

For the electromagnetic field, boundary conditions associated with discontinuities between media with different material properties are derived by integrating the field equations (4.4) or (4.5) over the area of a rectangular loop enclosing the boundary (Fig. 4.7). The area integrals of the curl operations are then converted to line integrals and the width of the loop is allowed to vanish. This leads to the result

$$\hat{n} \times \mathbf{E} = \hat{n} \times \mathbf{E}'$$
$$\hat{n} \times \mathbf{H} = \hat{n} \times \mathbf{H}', \qquad (4.61)$$

where $\hat{n}$ is a unit vector normal to the boundary. In (4.61), unprimed field quantities are evaluated at the boundary in one medium and primed quantities are evaluated in the other medium.

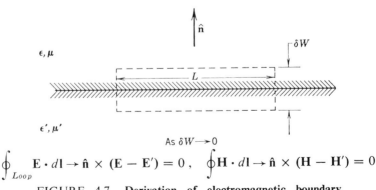

$$\oint_{Loop} \mathbf{E} \cdot d\mathbf{l} \rightarrow \hat{n} \times (\mathbf{E} - \mathbf{E}') = 0, \quad \oint \mathbf{H} \cdot d\mathbf{l} \rightarrow \hat{n} \times (\mathbf{H} - \mathbf{H}') = 0$$

FIGURE 4.7. Derivation of electromagnetic boundary conditions.

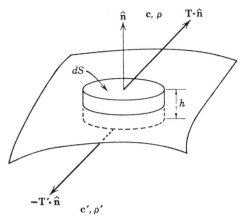

FIGURE 4.8. **Derivation of stress boundary conditions.**

In acoustic problems, the interfaces between different solid media are usually firmly bonded together so that there is no slippage of one with respect to the other. This means that the particle displacement velocity must be continuous across the discontinuity surface. That is,[†]

$$\mathbf{v} = \mathbf{v}'. \tag{4.62}$$

It should be noted that, in contrast to the electromagnetic case, the material interface actually moves. This should, strictly speaking, be taken into account in evaluating the boundary condition (4.62). However, the higher order terms that occur are of no importance in the linearized small-signal theory of interest here. Motion of the boundary is therefore ignored. Boundary conditions for the stress field are derived by assuming a small disklike volume, enclosing an area dS of the interface surface (Fig. 4.8). The external traction forces on the disk are

$$\mathbf{T} \cdot \hat{\mathbf{n}} \, dS$$

on the upper face and

$$-\mathbf{T}' \cdot \hat{\mathbf{n}} \, dS$$

on the lower face. Traction forces on the sides are of order h, the height of the disk, and the body force is $\mathbf{F}h \, dS$. From (2.15),

$$(\mathbf{T} - \mathbf{T}') \cdot \mathbf{n} \, dS + \mathbf{F}h \, dS = \left(\frac{\rho + \rho'}{2} \right) \frac{\partial \mathbf{v}}{\partial t} \, h \, dS.$$

As $h \to 0$ the body force and inertia terms drop out and the boundary condition

$$\mathbf{T} \cdot \hat{\mathbf{n}} = \mathbf{T}' \cdot \hat{\mathbf{n}} \tag{4.63}$$

is obtained. This states that the traction force must be continuous across the

[†] At a boundary between a solid and a nonviscous liquid the liquid moves freely parallel to the boundary, and condition (4.62) applies only to the normal component of particle velocity.

boundary. It should be noted that the acoustic boundary conditions (4.62) and (4.63) each involve *three* vector components, whereas the electromagnetic boundary conditions (4.61) each involve only *two* vector components.

One of the most important boundary value problems in electromagnetism and acoustics is also the simplest. This is the case of a uniform plane wave incident upon a plane discontinuity surface. In this situation the interface boundary conditions cannot be satisfied by the incident wave alone, and one or more reflected and transmitted waves are generated. If the incident wave impinges obliquely upon the discontinuity surface, the transmitted waves are bent (or *refracted*) into different directions. When the second medium at the interface is birefringent, like the anisotropic electromagnetic medium in Example 1(a), there are two transmitted waves traveling in different directions. This is called *birefringence* (or *double refraction*). In the case of an anisotropic acoustic medium, *trirefringence* (or *triple refraction*) may occur. There are then three different transmitted waves. Problems involving obliquely incident waves are beyond the scope of this chapter and will have to be deferred to Chapter 9 in Volume II. The simpler problem of normal incidence can, however, be used as an elementary illustration of boundary condition equations and will also serve as an introduction to the basic concepts of wave scattering theory.

EXAMPLE 3. Reflection of Normally Incident Uniform Plane Waves at a Plane Interface Between Two Hexagonal Crystal Media. The geometrical configuration is illustrated in Fig. 4.9. In both media the crystal axes have the same orientation, and the interface is normal to the X crystal axis. The coordinate axes x, y, z are oriented parallel to the crystal axes, with $+x$ pointing in the

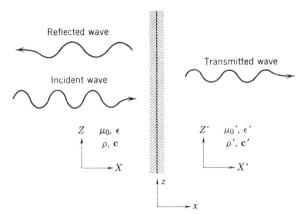

FIGURE 4.9. **Reflection of a normally incident wave at a plane boundary.**

propagation direction of the incident wave. Material properties and field components in the first medium are denoted by unprimed symbols and in the second medium by primed symbols, as in Fig. 4.7 and Fig. 4.8. The four electromagnetic boundary conditions at the discontinuity surface $x = 0$ are then

$$E_y(0) = E'_y(0) \qquad \text{(a)}$$
$$E_z(0) = E'_z(0) \qquad \text{(b)}$$
$$H_y(0) = H'_y(0) \qquad \text{(c)}$$
$$H_z(0) = H'_z(0) \qquad \text{(d)} \quad (4.64)$$

from (4.61). For the acoustic case the normal vector appearing in the stress boundary condition (4.63) is $\hat{\mathbf{n}} = \hat{\mathbf{x}}$, and therefore

$$\mathbf{T} \cdot \hat{\mathbf{n}} = \mathbf{T} \cdot \hat{\mathbf{x}} = \hat{\mathbf{x}} T_{xx} + \hat{\mathbf{y}} T_{yx} + \hat{\mathbf{z}} T_{zx}.$$

From (4.62) and (4.63), the six acoustic boundary conditions are thus

$$v_x(0) = v'_x(0) \qquad \text{(a)}$$
$$v_y(0) = v'_y(0) \qquad \text{(b)}$$
$$v_z(0) = v'_z(0) \qquad \text{(c)}$$
$$T_{xx}(0) = T'_{xx}(0) \qquad \text{(d)}$$
$$T_{yx}(0) = T'_{yx}(0) \qquad \text{(e)}$$
$$T_{zx}(0) = T'_{zx}(0). \qquad \text{(f)} \quad (4.65)$$

(a) **Electromagnetic Waves.** The electromagnetic wave solutions appropriate to this problem were found in Example 1, Part (a). Suppose that the incident wave is the Z-polarized solution given by (4.26). If the electric field for a time-harmonic incident wave is written in exponential notation as

$$\mathbf{E}_{\text{incident}} = \hat{\mathbf{z}} A e^{i(\omega t - kx)}, \qquad (4.66)$$

where A is the electric field amplitude, the corresponding magnetic field is

$$\mathbf{H}_{\text{incident}} = -\hat{\mathbf{y}} \frac{A}{(\mu_0/\epsilon_{zz})^{1/2}} e^{i(\omega t - kx)} .$$

The fields of a transmitted wave with the same polarization are

$$\mathbf{E}_{\text{transmitted}} = \hat{\mathbf{z}} B' e^{i(\omega t - kx')}$$
$$\mathbf{H}_{\text{transmitted}} = -\hat{\mathbf{y}} \frac{B'}{(\mu_0/\epsilon'_{zz})^{1/2}} e^{i(\omega t - k'x)}. \qquad (4.67)$$

Both of these waves have a z component of electric field and a y component of magnetic field, and the boundary conditions (b) and (c) in (4.64) require that these be equal at $x = 0$. Since the incident amplitude A in (4.66) is specified independently, only the transmitted amplitude B' can be adjusted to suit the boundary conditions and this does not allow simultaneous matching of (b) and (c). One must therefore add a reflected wave of the same polarization and amplitude B. This has electric and magnetic fields

$$\mathbf{E}_{\text{reflected}} = \hat{z} B e^{i(\omega t + kx)}$$

$$\mathbf{H}_{\text{reflected}} = \hat{y} \frac{B}{(\mu_0/\epsilon_{zz})^{1/2}} e^{i(\omega t + kx)}.$$

The total electric and magnetic fields in the first medium are then

$$\mathbf{E}_{\text{incident}} + \mathbf{E}_{\text{reflected}}$$

$$\mathbf{H}_{\text{incident}} + \mathbf{H}_{\text{reflected}} .$$

These fields and the transmitted fields (4.67) must be equal at $x = 0$. One then has

$$A + B = B' \tag{4.68}$$

$$-\frac{1}{(\mu_0/\epsilon_{zz})^{1/2}} (A - B) = -\frac{1}{(\mu_0/\epsilon'_{zz})^{1/2}} B'. \tag{4.69}$$

These equations may be combined to eliminate B', giving

$$\frac{B}{A} = \frac{(\mu_0/\epsilon'_{zz})^{1/2} - (\mu_0/\epsilon_{zz})^{1/2}}{(\mu_0/\epsilon'_{zz})^{1/2} + (\mu_0/\epsilon_{zz})^{1/2}}, \tag{4.70}$$

and B' can then be found by substituting this result into (4.68). That is

$$B'/A = \frac{2(\mu_0/\epsilon'_{zz})^{1/2}}{(\mu_0/\epsilon'_{zz})^{1/2} + (\mu_0/\epsilon_{zz})^{1/2}} \tag{4.71}$$

The quantity

$$R_E(0) = \frac{E_{\text{reflected}}(0)}{E_{\text{incident}}(0)} = \frac{B}{A} \tag{4.72}$$

is called the *electric field reflection coefficient* at $x = 0$, and

$$T_E(0) = \frac{E'_{\text{transmitted}}(0)}{E_{\text{incident}}(0)} = \frac{B'}{A} \tag{4.73}$$

is the *electric field transmission coefficient*. Corresponding reflection and transmission coefficients for the *magnetic field* are

$$R_H(0) = \frac{H_{\text{reflected}}(0)}{H_{\text{incident}}(0)} = -R_E(0) \tag{4.74}$$

and

$$T_H(0) = \frac{H'_{\text{transmitted}}(0)}{H_{\text{incident}}(0)} = \frac{(\mu_0/\epsilon_{zz})^{1/2}}{(\mu_0/\epsilon'_{zz})^{1/2}} T_E(0). \tag{4.75}$$

In exactly the same way, but using boundary conditions (a) and (d) in (4.64), the reflection and transmission coefficients for the Y-polarized incident wave are found to be

$$R_E(0) = \frac{(\mu_0/\epsilon'_{xx})^{1/2} - (\mu_0/\epsilon_{xx})^{1/2}}{(\mu_0/\epsilon'_{xx})^{1/2} + (\mu_0/\epsilon_{xx})^{1/2}} \tag{4.76}$$

and

$$T_E(0) = \frac{2(\mu_0/\epsilon'_{xx})^{1/2}}{(\mu_0/\epsilon'_{xx})^{1/2} + (\mu_0/\epsilon_{xx})^{1/2}} . \tag{4.77}$$

The boundary conditions (4.64) have thus been completely "used up" by the wave solutions available. *This equivalence between the number of boundary conditions and the number of plane wave solutions occurs in all scattering problems and will prove to be a useful guideline for solving the complicated acoustic and piezoelectric scattering problems encountered in Chapter 9.*

(b) Acoustic Waves. In this case the relevant uniform plane wave solutions are given in Example 1, Part (b) and the procedure is exactly the same. Consider the compressional wave solutions given by (4.40) to (4.42). For the incident wave

$$
(v_x)_I = A e^{i(\omega t - kx)}
$$
$$
(T_1)_I = -(\rho c_{11})^{1/2} A e^{i(\omega t - kx)}
$$
$$
(T_2)_I = -\frac{c_{12}}{V_p} A e^{i(\omega t - kx)}
$$
$$
(T_3)_I = -\frac{c_{13}}{V_p} A e^{i(\omega t - kx)}, \tag{4.78}
$$

for the reflected wave

$$
(v_x)_R = B e^{i(\omega t + kx)}
$$
$$
(T_1)_R = (\rho c_{11})^{1/2} B e^{i(\omega t + kx)}
$$
$$
(T_2)_R = (c_{12}/V_p) B e^{i(\omega t + kx)}
$$
$$
(T_3)_R = (c_{13}/V_p) B e^{i(\omega t + kx)}, \tag{4.79}
$$

and for the transmitted wave

$$
(v_x')_T = B' e^{i(\omega t - k'x)}
$$
$$
(T_1')_T = -(\rho' c_{11}')^{1/2} B' e^{i(\omega t - k'x)}
$$
$$
(T_2')_T = -\frac{c_{12}'}{V_p'} B' e^{i(\omega t - k'x)}
$$
$$
(T_3')_T = -\frac{c_{13}'}{V_p'} B' e^{i(\omega t - k'x)}. \tag{4.80}
$$

Only the stress component $T_1 = T_{xx}$ is involved in the boundary condition equations (4.65). Stress components $T_2 = T_{yy}$ and $T_3 = T_{zz}$ do not contribute to the traction force on the discontinuity surface at $x = 0$. From (a) and (d) in (4.65), one has therefore

$$
(v_x(0))_I + (v_x(0))_R = (v_x'(0))_T
$$
$$
(T_1(0))_I + (T_1(0))_R = (T_1'(0))_T.
$$

That is,

$$
A + B = B' \tag{4.81}
$$

$$
-(\rho c_{11})^{1/2}(A - B) = -(\rho' c_{11}')^{1/2} B'. \tag{4.82}
$$

Solution of these equations for B and B' gives the *particle velocity* reflection and transmission coefficients

$$R_v(0) = \frac{(v_x(0))_{\mathrm{R}}}{(v_x(0))_{\mathrm{I}}} = -\frac{(\rho'c_{11}')^{1/2} - (\rho c_{11})^{1/2}}{(\rho'c_{11}')^{1/2} + (\rho c_{11})^{1/2}} \tag{4.83}$$

$$T_v(0) = \frac{(v_x'(0))_{\mathrm{T}}}{(v_x(0))_{\mathrm{I}}} = \frac{2(\rho c_{11})^{1/2}}{(\rho'c_{11}')^{1/2} + (\rho c_{11})^{1/2}}. \tag{4.84}$$

It may be seen from (4.70) to (4.75) that these are analogous to the *magnetic field* scattering coefficients for the electromagnetic case. The *stress* reflection and transmission coefficients

$$R_{\mathrm{T}}(0) = \frac{(T_1(0))_{\mathrm{R}}}{(T_1(0))_{\mathrm{I}}} = -R_v(0) \tag{4.85}$$

$$T_{\mathrm{T}}(0) = \frac{(T_1'(0))_{\mathrm{T}}}{(T_1(0))_{\mathrm{I}}} = \frac{(\rho'c_{11}')^{1/2}}{(\rho c_{11})^{1/2}} T_v(0) \tag{4.86}$$

are equivalent to the *electric field* scattering coefficients (4.70) and (4.71).

For the Y-polarized shear wave, it is found in the same way that

$$R_v(0) = -\frac{(\rho'c_{66}')^{1/2} - (\rho c_{66})^{1/2}}{(\rho'c_{66}')^{1/2} + (\rho c_{66})^{1/2}} \tag{4.87}$$

$$T_v(0) = \frac{2(\rho c_{66})^{1/2}}{(\rho'c_{66}')^{1/2} + (\rho c_{66})^{1/2}}, \tag{4.88}$$

and for the Z-polarized shear wave, that

$$R_v(0) = -\frac{(\rho'c_{44}')^{1/2} - (\rho c_{44})^{1/2}}{(\rho'c_{44}')^{1/2} + (\rho c_{44})^{1/2}} \tag{4.89}$$

$$T_v(0) = \frac{2(\rho c_{44})^{1/2}}{(\rho'c_{44}')^{1/2} + (\rho c_{44})^{1/2}}. \tag{4.90}$$

It will be noted that the reflection and transmission coefficients in Example 3 are completely specified by parameters such as $(\mu_0/\epsilon_{ii})^{1/2}$ for the electromagnetic case and by parameters such as $(\rho c_{ii})^{1/2} = \rho V_p$ for the acoustic case. In electromagnetic theory the quantities $(\mu_0/\epsilon_{ii})^{1/2}$, which give the ratio of the electric to the magnetic field intensities for a uniform plane wave solution (see (4.26) and (4.28)), have dimension

$$\frac{\text{volts/meter}}{\text{amperes/meter}} = \text{ohms}$$

and are called *intrinsic* (or *characteristic*) *impedances*† of the electromagnetic

† The term *wave impedance* is also used.

plane wave solutions. Typical values are in the range of 100 ohms. The corresponding acoustic quantities $(\rho c_{ii})^{1/2} = \rho V_p$ describe the ratio ($-$traction force/particle displacement velocity) and are called *specific* (or *characteristic*) *impedances* for the acoustic plane wave solutions. In MKS units, characteristic acoustic impedances are measured in watts/m²/(m/s)² (or MKS rayls). Typical material parameters in the range $\rho \approx 4000$ kg/m³ and $c_{ii} \approx 10^{11}$ newtons/m² give an acoustic impedance of order $Z_a \approx 2 \times 10^7$ MKS rayls.

PROBLEMS

1. Consider a volume V totally enclosed by a surface S through which a current density

$$\mathbf{J} = \boldsymbol{\sigma} \cdot \mathbf{E} + \mathbf{J}_s$$

is flowing. Because electrical charge is conserved, the total *outward* current flow $(I = \oint_s \mathbf{J} \cdot \hat{\mathbf{n}} \, dS$, where $\hat{\mathbf{n}}$ is the unit vector in the outward normal direction to the surface S) must equal the rate of decrease of the total charge within S. Show that the *equation of charge conservation* (4.12) follows from this statement.

2. Show that (4.13) and (4.14) can be derived from (4.4), (4.5), and (4.12).

3. Consider an elementary volume element

$$dV = dx \, dy \, dz$$

in an unstrained solid medium. Using the concepts developed in Chapter 1, show that the volume of this element changes to

$$dx \, dy \, dz(1 + S_{xx})(1 + S_{yy})(1 + S_{zz})$$

when the medium is strained. S_{xx}, S_{yy}, S_{zz} are diagonal components of the strain matrix at the position of the volume element.

4. Using the fact that the total mass of the volume element in Problem 3 is conserved, show that the change in mass density due to strain is given by

$$\delta\rho = -\rho_0(S_{xx} + S_{yy} + S_{zz})$$

when second order terms in $\delta\rho$, S_{xx}, S_{yy}, S_{zz} are ignored. Here, ρ_0 is the density of the unstrained medium. From this result, derive the linearized *equation of mass conservation for a homogeneous medium*

$$\rho_0 \nabla \cdot \mathbf{v} = -\frac{\partial \rho}{\partial t}.$$

5. The acoustic field equations (4.15), (4.17) and the conservation equation given in Problem 4 can be used to derive relations corresponding to the divergence equations (4.13), (4.14) for the electromagnetic field. For example, if (4.17) is expanded in rectangular coordinates one has

$$\frac{\partial v_x}{\partial x} = \frac{\partial S_{xx}}{\partial t} \qquad \frac{1}{2}\left(\frac{\partial v_x}{\partial y} + \frac{\partial v_y}{\partial x}\right) = \frac{\partial S_{xy}}{\partial t}$$

$$\frac{\partial v_y}{\partial y} = \frac{\partial S_{yy}}{\partial t} \qquad \text{etc.}$$

$$\frac{\partial v_z}{\partial z} = \frac{\partial S_{zz}}{\partial t}$$

Take second derivatives of the first equation with respect to y, the second equation with respect to x, and the fourth equation with respect to x and y. From this, show that

$$\frac{\partial^2 S_{xx}}{\partial y^2} + \frac{\partial^2 S_{yy}}{\partial x^2} = 2\frac{\partial^2 S_{xy}}{\partial x\,\partial y}.$$

Derive the additional relations

$$\frac{\partial^2 S_{yy}}{\partial z^2} + \frac{\partial^2 S_{zz}}{\partial y^2} = 2\frac{\partial^2 S_{yz}}{\partial y\,\partial z}$$

$$\frac{\partial^2 S_{zz}}{\partial x^2} + \frac{\partial^2 S_{xx}}{\partial z^2} = 2\frac{\partial^2 S_{zx}}{\partial z\,\partial x}$$

$$\frac{\partial^2 S_{xx}}{\partial y\,\partial z} = \frac{\partial}{\partial x}\left(-\frac{\partial S_{yz}}{\partial x} + \frac{\partial S_{zx}}{\partial y} + \frac{\partial S_{xy}}{\partial z}\right)$$

$$\frac{\partial^2 S_{yy}}{\partial z\,\partial x} = \frac{\partial}{\partial y}\left(\frac{\partial S_{yz}}{\partial x} - \frac{\partial S_{zx}}{\partial y} + \frac{\partial S_{xy}}{\partial z}\right)$$

$$\frac{\partial^2 S_{zz}}{\partial x\,\partial y} = \frac{\partial}{\partial z}\left(\frac{\partial S_{yz}}{\partial x} + \frac{\partial S_{zx}}{\partial y} - \frac{\partial S_{xy}}{\partial z}\right).$$

These are the *Saint-Venant strain compatibility equations.*

6. Derive the boundary condition equations (4.61). Use the electromagnetic field equations to show that an equivalent set of boundary conditions is

$$\hat{\mathbf{n}} \cdot \mathbf{B} = \hat{\mathbf{n}} \cdot \mathbf{B}'$$

$$\hat{\mathbf{n}} \cdot \mathbf{D} = \hat{\mathbf{n}} \cdot \mathbf{D}',$$

provided there is no charge layer on the boundary.

7. Show that the acoustic boundary condition equations for a perfectly *frictionless* boundary between two media are

$$\hat{n} \cdot v = \hat{n} \cdot v'$$

$$\hat{n} \cdot (T \cdot \hat{n}) = \hat{n} \cdot (T' \cdot \hat{n})$$

$$\hat{n} \times (T \cdot \hat{n}) = \hat{n} \times (T' \cdot \hat{n}) = 0.$$

8. Suppose that the second medium in part (b) of Example 3 is vacuum. Show that the boundary condition equations are now

$$T \cdot \hat{x} = 0$$

and use these conditions to calculate the reflection coefficients for compressional and shear waves.

9. Assume that the second medium in part (b) of Example 3 is perfectly rigid. Find appropriate boundary condition equations and calculate the reflection coefficients for compressional and shear waves.

10. Take the boundary in part (b) of Example 3 to be completely frictionless, as in Problem 7. Find the reflection and transmission coefficients for compressional and shear waves.

11. Evaluate the scattering coefficients $R_v(0)$ and $T_v(0)$ in part (b) of Example 3 for all combinations of cadmium selenide, cadmium sulfide, and zinc oxide media. Use material constants from Appendix 2. (Neglect the piezoelectric effect; that is, use the stiffness constants given in part A-5 of Appendix 2.)

12. Consider the transmission of a compressional wave through the layer shown in the figure below. All three media are hexagonal, with the

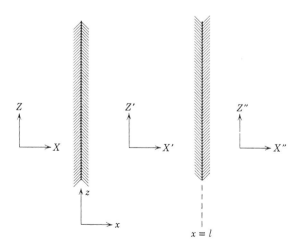

orientations indicated. Show that the transmission coefficient is

$$T_v(0) = \frac{B''}{A} = \frac{e^{ik''l}}{\dfrac{(\rho''c_{11}'')^{1/2} + (\rho c_{11})^{1/2}}{(\rho c_{11})^{1/2}} \dfrac{\cos k'l}{2} + i \dfrac{\rho'c_{11}' + (\rho''c_{11}'')^{1/2}(\rho c_{11})^{1/2}}{(\rho c_{11})^{1/2}(\rho'c_{11}')^{1/2}} \dfrac{\sin k'l}{2}}$$

where

$$(v_x)_I = Ae^{i(\omega t - kx)}$$

is the incident wave and

$$(v_x)_T = B''e^{i(\omega t - k''x)}$$

is the transmitted wave. Prove that $|T_v(0)|$ is unity when the first and third media are identical and l is equal to an integral number of half wavelengths in the second medium. Derive similar expressions for Y- and Z-polarized shear waves.

REFERENCES

Electromagnetic Fields

1. R. E. Collin, *Field Theory of Guided Waves*, Ch. 1, McGraw-Hill, New York, 1960.

2. C. C. Johnson, *Field and Wave Electrodynamics*, Ch. 1, McGraw-Hill, New York, 1965.

3. R. F. Harrington, *Time-harmonic Electromagnetic Fields*, Ch. 1, McGraw-Hill, New York, 1961.

4. S. Ramo, J. R. Whinnery and T. van Duzer, *Fields and Waves in Communication Electronics*, Wiley, New York, 1965.

Acoustic Fields

5. L. M. Brekhovskikh, *Waves in Layered Media*, pp. 168–171, Academic Press, New York, 1960.

6. K. G. Budden, *The Wave-guide Mode Theory of Wave Propagation*, pp. 18–19, Logos Press, London, 1961.

7. P. C. Chou and N. J. Pagano, *Elasticity—Tensor, Dyadic and Engineering Approaches*, Ch. 4, van Nostrand, New York, 1967.

8. I. S. Sokolnikoff, *Mathematical Theory of Elasticity*, pp. 72–83, McGraw-Hill, New York, 1946.

CHAPTER 5

POWER FLOW AND ENERGY BALANCE

A. ENERGY CONSERVATION RELATIONS

No theory of wave propagation can be considered complete until it has formulated an energy conservation relation. This should interrelate the power supplied by sources of the waves, the power lost through dissipation mechanisms, the rate of change of energy stored in the wave fields, and the power flow in the waves. In electromagnetism a relation of this kind, called *Poynting's Theorem*, is derived by performing some simple manipulations of the Maxwell equations (4.4) and (4.5). Using the electromagnetic-acoustic analogy established in Chapter 4, a corresponding acoustic energy-power relation may be obtained by following a procedure that parallels the electromagnetic derivation in a rather remarkable manner. This will be demonstrated in the present chapter. To emphasize the strong similarity of the two types of wave fields, the acoustic relation obtained in this way will be called the *acoustic Poynting's Theorem*.

B. POYNTING'S THEOREM FOR THE ELECTROMAGNETIC FIELD†

Poynting's Theorem for the electromagnetic field is obtained from Maxwell's equations by taking the scalar product of (4.4) with $-\mathbf{H}$,

$$\mathbf{H} \cdot (\nabla \times \mathbf{E}) = -\mathbf{H} \cdot \frac{\partial \mathbf{B}}{\partial t},$$

and adding to it the scalar product of (4.5) with $-\mathbf{E}$,

$$-\mathbf{E} \cdot (\nabla \times \mathbf{H}) = -\mathbf{E} \cdot \frac{\partial \mathbf{D}}{\partial t} - \mathbf{E} \cdot \mathbf{J}_c - \mathbf{E} \cdot \mathbf{J}_s.$$

The result,

$$\mathbf{H} \cdot (\nabla \times \mathbf{E}) - \mathbf{E} \cdot (\nabla \times \mathbf{H}) = -\mathbf{H} \cdot \frac{\partial \mathbf{B}}{\partial t} - \mathbf{E} \cdot \frac{\partial \mathbf{D}}{\partial t} - \mathbf{E} \cdot \mathbf{J}_c - \mathbf{E} \cdot \mathbf{J}_s$$

is converted to

$$\nabla \cdot (\mathbf{E} \times \mathbf{H}) = -\mathbf{H} \cdot \frac{\partial \mathbf{B}}{\partial t} - \mathbf{E} \cdot \frac{\partial \mathbf{D}}{\partial t} - \mathbf{E} \cdot \mathbf{J}_c - \mathbf{E} \cdot \mathbf{J}_s \qquad (5.1)$$

by using the vector identity

$$\nabla \cdot (\mathbf{E} \times \mathbf{H}) = \mathbf{H} \cdot \nabla \times \mathbf{E} - \mathbf{E} \cdot \nabla \times \mathbf{H}. \qquad (5.2)$$

Integration of (5.1) over a volume V and use of the divergence theorem gives

$$\oint_S (\mathbf{E} \times \mathbf{H}) \cdot \hat{\mathbf{n}} \, dS = -\int_V \left(\mathbf{H} \cdot \frac{\partial \mathbf{B}}{\partial t} + \mathbf{E} \cdot \frac{\partial \mathbf{D}}{\partial t} \right) dV$$
$$-\int_V (\mathbf{E} \cdot \mathbf{J}_c) \, dV + \int_V (-\mathbf{E} \cdot \mathbf{J}_s) \, dV, \quad (5.3)$$

where S is the surface enclosing V and $\hat{\mathbf{n}}$ is a unit vector in the *outward* normal direction to the surface. This is Poynting's Theorem for electromagnetic fields.

Terms in (5.3) may be identified with either power flow or energy storage in the field. Consider first a volume totally enclosed by a perfectly conducting boundary along S. Also assume that the conductivity is everywhere zero within V. The surface integral in (5.3) vanishes because

$$(\mathbf{E} \times \mathbf{H}) \cdot \hat{\mathbf{n}} = (\hat{\mathbf{n}} \times \mathbf{E}) \cdot \mathbf{H}$$

---

† This chapter is restricted to media with negligible piezoelectricity and magnetostriction. Power flow and energy balance in piezoelectric media is treated in Section 8.G.

and the tangential electric field $\hat{n} \times \mathbf{E}$ is zero at a perfectly conducting surface; the second volume integral in (5.3) vanishes because $\mathbf{J}_c = 0$. Poynting's Theorem then takes the form

$$\int_V \left(\mathbf{H} \cdot \frac{\partial \mathbf{B}}{\partial t} + \mathbf{E} \cdot \frac{\partial \mathbf{D}}{\partial t} \right) dV = \int_V (-\mathbf{E} \cdot \mathbf{J}_s) \, dV. \tag{5.4}$$

On the right-hand side of (5.4),

$$\int_V - \mathbf{E} \cdot \mathbf{J}_s \, dV = P_s$$

is the power supplied by sources moving charges against the electric field. Since the system is completely isolated and lossless, the power supplied must be stored in the fields. By energy conservation, the left-hand side of (5.4) is therefore the rate of change of stored energy in the system

$$\frac{\partial}{\partial t} U = \int_V \left(\mathbf{H} \cdot \frac{\partial \mathbf{B}}{\partial t} + \mathbf{E} \cdot \frac{\partial \mathbf{D}}{\partial t} \right) dV. \tag{5.5}$$

The terms under the integral (5.5) are identified as the rate of change of magnetic stored energy density u_H and of electric stored energy density u_E. Substitution of (4.7) for $\mathbf{B}$ gives

$$\frac{\partial u_H}{\partial t} = \mathbf{H} \cdot \boldsymbol{\mu} \cdot \frac{\partial \mathbf{H}}{\partial t} = H_i \mu_{ij} \frac{\partial H_j}{\partial t} ,$$

and therefore

$$du_H = \mu_{ij} H_i \, dH_j , \tag{5.6}$$

where the differentials are taken with the spatial variables held constant. Similarly,

$$du_E = \epsilon_{ij} E_i \, dE_j . \tag{5.7}$$

According to (5.6) and the chain rule of partial differentiation

$$\frac{\partial u_H}{\partial H_j} = \mu_{ij} H_i . \tag{5.8}$$

By interchanging the subscript labels in (5.6), one finds also that

$$\frac{\partial u_H}{\partial H_i} = \mu_{ji} H_j . \tag{5.9}$$

If the magnetic energy density is determined entirely by the magnetic field intensity,[†] u_H is a function only of the field components H_x, H_y, H_z. In this

---

† This is not always true for certain kinds of magnetic materials—for example, ferrites.

case the second partial derivatives of u_H are independent of the order of differentiation. From (5.8) and (5.9), then,

$$\frac{\partial^2 u_H}{\partial H_i \, \partial H_j} = \mu_{ij} = \frac{\partial^2 u_H}{\partial H_j \, \partial H_i} = \mu_{ji},$$

showing that $\boldsymbol{\mu}$ is a symmetric tensor. Pairs of terms having unequal subscripts in (5.6) can therefore be combined,

$$du_H = \tfrac{1}{2}(\mu_{ij}H_i \, dH_j + \mu_{ji}H_j \, dH_i)$$
$$= \tfrac{1}{2}\mu_{ij}(H_i \, dH_j + H_j \, dH_i) = \tfrac{1}{2}\mu_{ij} \, d(H_i H_j). \quad (5.10)$$

Since the differentials in (5.10) are taken with spatial variables held constant, integration with respect to time gives the magnetic stored energy density

$$u_H = \tfrac{1}{2}\mu_{ij}H_i H_j = \tfrac{1}{2}\mathbf{H} \cdot \boldsymbol{\mu} \cdot \mathbf{H} \quad (5.11)$$

as a function of spatial position. A similar argument shows that $\boldsymbol{\epsilon}$ is a symmetric tensor and that the electrical stored energy density is

$$u_E = \tfrac{1}{2}\mathbf{E} \cdot \boldsymbol{\epsilon} \cdot \mathbf{E}. \quad (5.12)$$

If the conduction current $\mathbf{J}_c$ is retained in (5.3), a term

$$P_d = \int_V \mathbf{E} \cdot \boldsymbol{\sigma} \cdot \mathbf{E} \, dV \quad (5.13)$$

is added to the left-hand side of (5.4). The integrand in (5.13), a scalar product of the electric field strength and the conduction current density, is the density of conductive power loss; and P_d is therefore the total conductive power loss in the system. Symmetry of the conductivity matrix $\boldsymbol{\sigma}$ will be demonstrated in Section 5.F.

With these identifications, Poynting's Theorem for the general case can be written as

$$\oint_S (\mathbf{E} \times \mathbf{H}) \cdot \hat{\mathbf{n}} \, dS + \frac{\partial U}{\partial t} + P_d = P_s. \quad (5.14)$$

From energy conservation, the surface integral on the left must therefore be interpreted as the total power flow *outward* through the closed surface S. The vector quantity

$$\mathbf{P} = (\mathbf{E} \times \mathbf{H}) \quad (5.15)$$

in the surface integral is identified as power flow density and is called the *Poynting vector*. In MKS units $\mathbf{E}$ and $\mathbf{H}$ have dimensions of volts/m and amps/m, respectively, and the Poynting vector is therefore given in watts/m². The stored energy densities u_H and u_E are measured in joules/m³ and the dissipated power P_d in (5.13) is in watts.

It should be noted that Poynting's Theorem does not apply to static fields, since the assumption of a time variation is essential to the derivation. Another qualification is that $\mathbf{P} \cdot \hat{\mathbf{n}}$ can be rigorously identified with power flow density through a surface only if it is integrated over a completely closed surface. A technique for dealing with this restriction will be shown in the following example.

EXAMPLE 1.   One elementary application of Poynting's Theorem is the calculation of power flow in a uniform plane wave. In Example 1(a) of Chapter 4 the two electromagnetic waves propagating along the X axis of a hexagonal crystal were found to have fields given by (4.26), (4.27), and (4.28). If the solutions are time-harmonic,

$$H_{y\pm} = A_Z \cos \omega(t \mp x/V_p)$$

$$E_{z\pm} = \mp \left(\frac{\mu_0}{\epsilon_{zz}}\right)^{1/2} H_{y\pm} \tag{5.16}$$

for the Z-polarized wave, and

$$H_{z\pm} = A_Y \cos \omega(t \mp x/V_p)$$

$$E_{y\pm} = \pm \left(\frac{\mu_0}{\epsilon_{xx}}\right)^{1/2} H_{z\pm} \tag{5.17}$$

for the Y-polarized wave. The upper and lower signs refer to propagation in the $+x$ and $-x$ directions, respectively; A_Z and A_Y are arbitrary amplitudes.

From (5.16) the Poynting vector for the Z-polarized wave is then

$$\mathbf{P}_{Z\text{-polarized}} = -\hat{\mathbf{x}} E_{z\pm} H_{y\pm} = \pm \hat{\mathbf{x}} \left(\frac{\mu_0}{\epsilon_{zz}}\right)^{1/2} A_Z^2 \cos^2 \omega(t \mp x/V_p); \tag{5.18}$$

and for the Y-polarized wave it is

$$\mathbf{P}_{Y\text{-polarized}} = \hat{\mathbf{x}} E_{y\pm} H_{z\pm} = \pm \hat{\mathbf{x}} \left(\frac{\mu_0}{\epsilon_{xx}}\right)^{1/2} A_Y^2 \cos^2 \omega(t \mp x/V_p). \tag{5.19}$$

In both cases the Poynting vector is in the $+x$ direction for a wave propagating in that direction and in the $-x$ direction for the negative-traveling waves. The interpretation of $\mathbf{P}$ as a power flow density is thus consistent with the physical concept of energy transport in a travelling wave. It was noted above, however, that this interpretation is strictly correct only if the power flow is calculated for a completely closed surface. In practical problems one can always overcome this difficulty by using a suitably chosen closed surface. For the present example the surface illustrated in Fig. 5.1 may be chosen. Since the Poynting vector $\mathbf{P}$ is parallel to the x axis, $\mathbf{P} \cdot \hat{\mathbf{n}}$ is zero on the sides of the surface and the only contribution to the surface integral in (5.14) comes from the end faces S_1 and S_2. If x_2 is allowed to recede to infinity and an infinitesimal amount of attenuation is assumed in the medium, the fields at S_2 go to zero and the total power fed *into* the volume enclosed by the surface is

$$-\mathbf{P} \cdot \hat{\mathbf{n}} S_1 = \mathbf{P} \cdot \hat{\mathbf{x}} S_1 .$$

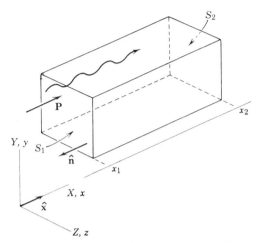

FIGURE 5.1.   Calculation of the power flow density in an electromagnetic wave traveling along the x axis. X, Y, Z are crystal axes and x, y, z are coordinate axes.

In other words, the power flow density flowing through the area S_1 is rigorously given by

$$\mathbf{P} \cdot \hat{\mathbf{x}} = P_x. \tag{5.20}$$

It is seen from (5.18) and (5.19) that the instantaneous power flow density P_x is not a constant and actually goes to zero at spatial positions and instants of time where

$$\cos \omega(t \mp x/V_p) = 0.$$

The time-average power flow density $(P_x)_{\text{AV}}$, on the other hand, is proportional to

$$(\cos^2 \omega(t \mp x/V_p))_{\text{AV}} = \tfrac{1}{2}$$

and is independent of spatial position. This is to be expected in a lossless medium, because there is no loss of energy and the power flow density must, consequently, be the same at all positions along the propagation path.

To understand the origin of the spatial variations in the instantaneous power flow, one must consider the stored energy densities (5.11) and (5.12). For the Z-polarized wave propagating in the $+x$ direction

$$u_H = \frac{\mu_0}{2} A_Z^2 \cos^2 \omega(t - x/V_p)$$

$$u_E = \frac{\epsilon_{zz}}{2} A_Z^2 \left(\frac{\mu_0}{\epsilon_{zz}}\right) \cos^2 \omega(t - x/V_p) = u_H .\dagger \tag{5.21}$$

† The stored energy in the wave is thus equally divided between the two storage mechanisms. It will be seen in the subsequent examples of this chapter and in Section O of Chapter 10 in Volume II, that this is a property of all types of traveling waves.

If Poynting's Theorem (5.14) is now applied to the surface and volume in Fig. 5.1, the result is

$$(\mathbf{P} \cdot \hat{\mathbf{x}})_{x_2} = (\mathbf{P} \cdot \hat{\mathbf{x}})_{x_1} - \frac{\partial}{\partial t} \int_{x_1}^{x_2} (u_H + u_E) \, dx. \tag{5.22}$$

When u_H and u_E are substituted from (5.21) and the integration is performed, it is found that the difference in the instantaneous power flows at x_1 and x_2 is exactly accounted for by the instantaneous time rate of change of stored energy in the enclosed volume.

EXAMPLE 2.   The fields for uniform electromagnetic plane waves traveling at an arbitrary angle η in the XZ plane of a hexagonal crystal were given by (4.47) and (4.48) of Example 2(a) in Chapter 4. For the *pure transverse* Y-polarized solution (4.48),

$$H_{z' \pm} = A_Y \cos (\omega t \mp kx')$$

$$E_{y' \pm} = \pm \left(\frac{\mu_0}{\epsilon_{xx}} \right)^{1/2} H_{z' \pm} \tag{5.23}$$

when the real part of the complex exponential is used. These fields have the same form as (5.18) in the previous example and the Poynting vector calculation is exactly the same. That is

$$\mathbf{P}_{Y\text{-polarized}} = \pm \hat{\mathbf{x}}' \left(\frac{\mu_0}{\epsilon'_{yy}} \right)^{1/2} A_Y^2 \cos^2 (\omega t \mp kx'), \tag{5.24}$$

and the Poynting vector is again directed normal to the planes of constant phase ($x' = $ constant).

For the *quasitransverse* XZ-polarized wave (4.47) the situation is different. Here $\mathbf{H}$ lies along the y' direction but, from (4.49), $\mathbf{E}$ does not lie along the z' direction. Owing to this canted orientation of $\mathbf{E}$, the Poynting vector $\mathbf{P} = \mathbf{E} \times \mathbf{H}$ is rotated through an angle $\pi/2 - \chi$ away from the wave vector $\mathbf{k}$ (Fig. 5.2) and its magnitude is given by

$$P_\pm = E_\pm H_\pm$$

where

$$E_\pm = (E_{x\pm}^2 + E_{z\pm}^2)^{1/2}.$$

Since the direction of $\mathbf{P}$ is normal to the plane defined by $\mathbf{E}$ and $\mathbf{H}$, this deflection of $\mathbf{P}$ away from $\mathbf{k}$ always occurs when an electromagnetic wave is quasitransverse.

As in the previous example, $\mathbf{P} \cdot \hat{\mathbf{n}}$ is interpreted as the power flow density across a surface in the direction of its normal $\hat{\mathbf{n}}$, and the direction of maximum power flow therefore lies along $\mathbf{P}$. In Fig. 5.2 this means that the power flow in the wave is not in the same direction as the motion of the planes of constant phase, which travel parallel to $\mathbf{k}$.

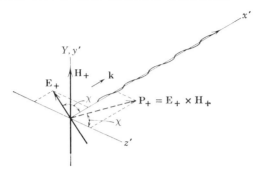

FIGURE 5.2. Poynting vector calculation for the quasitransverse electromagnetic wave propagating in the positive direction along the rotated coordinate axis x' in Fig. 4.5a.

C. POYNTING'S THEOREM FOR THE ACOUSTIC FIELD

In the previous chapters several different kinds of tensor products were introduced,

$$\mathbf{T} \cdot \hat{\mathbf{n}} = T_{ij}n_j$$
$$\mathbf{c}:\mathbf{S} = c_{ijkl}S_{kl} = c_{IJ}S_J \,,$$

and so on. Summation over repeated subscripts is assumed. The examples given show that a single dot product designates summation over a single full subscript (i), while a double dot product designates summation over pairs of full subscripts (kl) or summation over a single abbreviated subscript (J). To derive the acoustic Poynting's Theorem, the double dot product of two second rank tensors is required. This is defined as

$$\mathbf{A}:\mathbf{B} = A_{ij}B_{ij} \,,$$

with summation over the repeated subscripts. For symmetric tensors (such as the strain tensor in general and the stress tensor in nonpolar media) terms with unequal subscripts can be grouped in pairs and abbreviated subscripts used. Thus, if $\mathbf{A} = \mathbf{T}$ and $\mathbf{B} = \mathbf{S}$ the double dot product will contain terms such as

$$T_{yz}S_{yz} + T_{zy}S_{zy} = 2T_{yz}S_{yz} = T_4S_4 \,,$$

since $S_4 = 2S_{yz}$. For symmetric $\mathbf{T}$, then,

$$\mathbf{T}:\mathbf{S} = T_{ij}S_{ij} = T_IS_I. \tag{5.25}$$

The acoustic Poynting Theorem is obtained from the acoustic field equations by following a procedure similar to that used in the electromagnetic case. The single dot product of (4.15) is taken with $-\mathbf{v}$,

$$-\mathbf{v} \cdot (\nabla \cdot \mathbf{T}) = -\mathbf{v} \cdot \frac{\partial \mathbf{p}}{\partial t} + \mathbf{v} \cdot \mathbf{F},$$

and added to the double dot product of (4.17) with $-\mathbf{T}$,

$$-\mathbf{T}:\nabla_s\mathbf{v} = -\mathbf{T}:\frac{\partial \mathbf{S}}{\partial t}.$$

The result,

$$-\mathbf{v} \cdot (\nabla \cdot \mathbf{T}) - \mathbf{T}:(\nabla_s\mathbf{v}) = -\mathbf{v} \cdot \frac{\partial \mathbf{p}}{\partial t} - \mathbf{T}:\frac{\partial \mathbf{S}}{\partial t} + \mathbf{v} \cdot \mathbf{F},$$

is converted to

$$-\nabla \cdot (\mathbf{v} \cdot \mathbf{T}) = -\mathbf{v} \cdot \frac{\partial \mathbf{p}}{\partial t} - \mathbf{T}:\frac{\partial \mathbf{S}}{\partial t} + \mathbf{v} \cdot \mathbf{F} \tag{5.26}$$

by using the identity†

$$\nabla \cdot (\mathbf{v} \cdot \mathbf{T}) = \mathbf{v} \cdot (\nabla \cdot \mathbf{T}) + \mathbf{T}:\nabla_s\mathbf{v} \tag{5.27}$$

for *symmetric* second rank tensors $\mathbf{T}$. As in the electromagnetic derivation, integration is carried out over a volume V and the divergence theorem is used to introduce an integral over the bounding surface S,

$$\oint_S (-\mathbf{v} \cdot \mathbf{T}) \cdot \hat{\mathbf{n}}\, dS = -\int_V \mathbf{v} \cdot \frac{\partial \mathbf{p}}{\partial t}\, dV - \int_V \mathbf{T}:\frac{\partial \mathbf{S}}{\partial t}\, dV + \int_V \mathbf{v} \cdot \mathbf{F}\, dV. \tag{5.28}$$

Terms in (5.28) may be identified by using the same procedure followed in the electromagnetic case. The boundary S is first assumed to be stress-free, $\mathbf{T} \cdot \hat{\mathbf{n}} = 0$, and the elastic damping is taken to be zero. Under the first of these conditions, the surface integral in (5.28) goes to zero and

$$\int_V \left(\mathbf{v} \cdot \frac{\partial \mathbf{p}}{\partial t} + \mathbf{T}:\frac{\partial \mathbf{S}}{\partial t}\right) dV = \int_V \mathbf{v} \cdot \mathbf{F}\, dV. \tag{5.29}$$

On the right-hand side of (5.29) $\int_V \mathbf{v} \cdot \mathbf{F}\, dV = P_s$ is the power supplied to the volume V by the sources. Since there is no energy loss in the system, the left-hand side is the rate of change of stored energy in the system, $\partial U/\partial t$.

The first term under the left-hand integral in (5.29) is immediately identified as the rate of change of kinetic energy density,

$$\frac{\partial}{\partial t} u_v = \rho\mathbf{v} \cdot \frac{\partial \mathbf{v}}{\partial t} = \frac{\partial}{\partial t}(\tfrac{1}{2}\rho v^2), \tag{5.30}$$

---

† This identity is the acoustic counterpart of (5.2) for the electromagnetic field. See Problem 1 at the end of the chapter.

and the other term must therefore be identified as the rate of change of the elastic stored energy density, u_S. For a lossless medium,

$$T_I = c_{IJ}S_J$$

and the rate of change of elastic stored energy (*strain energy*) is

$$\frac{\partial u_S}{\partial t} = \mathbf{T}:\frac{\partial \mathbf{S}}{\partial t} = T_I \frac{\partial S_I}{\partial t} = c_{IJ}S_J \frac{\partial S_I}{\partial t}. \tag{5.31}$$

Accordingly,

$$du_S = c_{IJ}S_J dS_I. \tag{5.32}$$

As in the case of (5.6) and (5.7), the differentials in (5.32) are taken with the spatial variables held constant. Since u_S is a function only of the strain field, the argument used in Section 5.B to demonstrate symmetry of μ_{ij} can be repeated here to prove that

$$c_{IJ} = c_{JI} ; \tag{5.33}$$

and the compliance matrix, which is the inverse of the stiffness matrix, will also be symmetric,

$$s_{IJ} = s_{JI} . \tag{5.34}$$

For a lossless medium, then, the strain energy density can be written as

$$u_S = \tfrac{1}{2}S_I c_{IJ}S_J \tag{5.35}$$

by following the same steps used in proceeding from (5.6) to (5.11).

For a viscously damped medium the second term on the left-hand side of (5.29) contains both stored energy and power loss terms. In this case the stress field can be eliminated by using (4.3), giving†

$$\int_V \mathbf{T}:\frac{\partial \mathbf{S}}{\partial t} \, dV = \int_V \mathbf{S}:\mathbf{c}:\frac{\partial \mathbf{S}}{\partial t} \, dV + \int_V \frac{\partial \mathbf{S}}{\partial t}:\boldsymbol{\eta}:\frac{\partial \mathbf{S}}{\partial t} \, dV. \tag{5.36}$$

The first term on the right is simply the rate of change of $\int_V u_S \, dV$, from (5.31), and the second term is the total viscous power loss in the system,

$$P_d = \int_V \frac{\partial \mathbf{S}}{\partial t}:\boldsymbol{\eta}:\frac{\partial \mathbf{S}}{\partial t} \, dV. \tag{5.37}$$

Symmetry of the viscosity tensor $\boldsymbol{\eta}$ will be demonstrated in Section 5.F.

With these identifications, the acoustic Poynting's Theorem can be written for the general case as

$$\oint_S (-\mathbf{v} \cdot \mathbf{T}) \cdot \hat{\mathbf{n}} \, dS + \frac{\partial U}{\partial t} + P_d = P_s , \tag{5.38}$$

† Problem 3 at the end of the chapter.

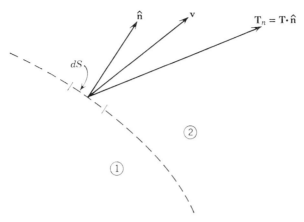

Power flow from ① to ② is $-\mathbf{v} \cdot \mathbf{T} \cdot \hat{\mathbf{n}}\, dS = \mathbf{P} \cdot \hat{\mathbf{n}}\, dS$

FIGURE 5.3. **Power flow and the acoustic Poynting vector.**

where

$$U = \int_V (u_v + u_s)\, dV$$

is the total kinetic and elastic stored energy in V, and the surface integral must therefore be identified as the total power flow *outward* through the closed surface S. The integrand

$$\mathbf{P} \cdot \hat{\mathbf{n}} = -\mathbf{v} \cdot \mathbf{T} \cdot \hat{\mathbf{n}} \tag{5.39}$$

is interpreted as the power flow density in the direction of $\hat{\mathbf{n}}$, and

$$\mathbf{P} = -\mathbf{v} \cdot \mathbf{T} \tag{5.40}$$

is called the *acoustic Poynting vector*. With $\mathbf{v}$ in m/sec and $\mathbf{T}$ in newtons/m², the acoustic Poynting vector has the appropriate dimensionality watts/m². From (5.30) and (5.35) the stored energy densities u_V and u_S have dimension joules/m³ = newtons/m² and P_d in (5.37) is in watts.

The minus sign appearing in (5.39) can be explained by referring to Fig. 5.3. From elementary mechanical considerations the power delivered to an object is the force applied times the velocity of the point of application. In Fig. 5.3, therefore, the power delivered through dS from medium (1) to medium (2) is the dot product of (force exerted by (1) on (2)) times (particle velocity at dS). According to the sign conventions established in Section 2.B,

(Force exerted *by (1) on (2)*) $= -\mathbf{T} \cdot \hat{\mathbf{n}}\, dS,$

and the power delivered through dS *from (1) to (2)* is therefore

$$-\mathbf{v} \cdot \mathbf{T} \cdot \hat{\mathbf{n}}\, dS.$$

This shows that interpretation of the acoustic Poynting vector as a power density has a concrete physical basis. There is therefore no need of restricting its application to completely closed surfaces, as in the electromagnetic case.

EXAMPLE 3. The three acoustic wave solutions for propagation along the X axis of a hexagonal crystal were given in Example 1(b) of Chapter 4. For the Y-polarized shear wave, the fields are

$$v_{y\pm} = A_Y \cos \omega(t \mp x/V_p)$$

$$T_{6\pm} = \mp \left(\frac{\rho}{s_{66}}\right)^{1/2} v_{y\pm} \tag{5.41}$$

from (4.30) and (4.33). According to (5.40) the Poynting vector for acoustic fields is

$$\begin{aligned}\mathbf{P} &= -\mathbf{v} \cdot \mathbf{T} \\ &= \hat{\mathbf{x}}(-v_x T_{xx} - v_y T_{yx} - v_z T_{zx}) \\ &+ \hat{\mathbf{y}}(-v_x T_{xy} - v_y T_{yy} - v_z T_{zy}) \\ &+ \hat{\mathbf{z}}(-v_x T_{xz} - v_y T_{yz} - v_z T_{zz}).\end{aligned} \tag{5.42}$$

In (5.41) the only nonzero field components are $v_{y\pm}$ and $T_{6\pm} = T_{xy\pm} = T_{yx\pm}$. Therefore

$$\mathbf{P}_{Y\text{-polarized}} = -\hat{\mathbf{x}}v_{y\pm}T_{yx\pm} = \pm\hat{\mathbf{x}}\left(\frac{\rho}{s_{66}}\right)^{1/2} A_Y^2 \cos^2 \omega(t \mp x/V_p). \tag{5.43}$$

Similar calculations for the Z-polarized shear wave ((4.31) and (4.35)) and the compressional wave ((4.39) to (4.42)) give Poynting vectors

$$(P_x)_{Z\text{-polarized}} = -v_{z\pm}T_{zx\pm} = \pm\left(\frac{\rho}{s_{44}}\right)^{1/2} A_Z^2 \cos^2 \omega(t \mp x/V_p) \tag{5.44}$$

and

$$(P_x)_{\text{compressional}} = -v_{x\pm}T_{xx\pm} = \pm(\rho c_{11})^{1/2} A_X^2 \cos^2 \omega(t \mp x/V_p). \tag{5.45}$$

It should be noted that the "extra" stress components (4.42) in the compressional case do not contribute to the power flow.

As in Example 1, the instantaneous power density again varies along x, and this variation is related to the stored energy in the wave. For the Y-polarized shear wave the kinetic stored energy density is

$$u_v = A_Y^2 \frac{\rho}{2} \cos^2 \omega(t \mp x/V_p) \tag{5.46}$$

from (5.30), and the strain stored energy density

$$u_s = \tfrac{1}{2}c_{66}S_{6\mp}^2 = A_Y^2 \frac{\rho}{2} \cos^2 \omega(t \mp x/V_p) \tag{5.47}$$

is obtained by using

$$u_{y\pm} = \frac{A_Y \sin \omega(t \mp x/V_p)}{\omega}$$

$$S_6 = \frac{\partial u_y}{\partial x}$$

and

$$V_p = (\rho s_{66})^{-1/2} = (c_{66}/\rho)^{1/2}.$$

As in Example 1, application of Poynting's Theorem shows that the difference in instantaneous power flow at two points x_1 and x_2 on the propagation path is equal to the rate of change of stored energy

$$\frac{\partial}{\partial t} \int_{x_1}^{x_2} (u_v + u_s)\, dV.$$

EXAMPLE 4. Solutions for uniform acoustic plane waves traveling at an arbitrary angle η in the XZ plane of a hexagonal crystal were given in Example 2(b) of Chapter 4. For the Y-polarized *pure shear* solution (4.52), the fields may be written as

$$v_{y'\pm} = A_Y \cos (\omega t \mp kx')$$
$$T_{6'\pm} = \mp (\rho c'_{66})^{1/2} v_{y'\pm}$$
$$T_{4'\pm} = \mp \left(\frac{\rho}{c'_{66}}\right)^{1/2} c'_{46} v_{y'\pm}. \tag{5.48}$$

The Poynting vector is therefore

$$\mathbf{P}_{Y\text{-polarized}} = \hat{\mathbf{x}}'(-v_{y'\pm} T_{y'x'\pm}) + \hat{\mathbf{z}}'(-v_{y'\pm} T_{y'z'\pm})$$
$$= \pm \left(\hat{\mathbf{x}}'(\rho c'_{66})^{1/2} + \hat{\mathbf{z}}'\left(\frac{\rho}{c'_{66}}\right)^{1/2} c'_{46}\right) A_Y^2 \cos^2 (\omega t \mp kx') \tag{5.49}$$

In the *quasishear* and *quasilongitudinal* solutions to the simultaneous equations (4.53) both the $v_{x'}$ and $v_{z'}$ components of particle velocity are present, and examination of (4.51) shows that the stress components are $T_{1'} = T_{x'x'}$, $T_{2'} = T_{y'y'}$, $T_{3'} = T_{z'z'}$, $T_{5'} = T_{x'z'} = T_{z'x'}$. For both solutions the Poynting vector will, therefore, be of the form

$$\mathbf{P} = \hat{\mathbf{x}}'(-v_{x'} T_{x'x'} - v_{z'} T_{z'x'}) + \hat{\mathbf{z}}'(-v_{x'} T_{x'z'} - v_{z'} T_{z'z'}).$$

Using (4.51(a), (c), and (f)) this can be expressed entirely in terms of particle velocity components. That is

$$\mathbf{P} = \hat{\mathbf{x}}' \frac{\omega\rho}{k}\left((v_{x'})^2 + (v_{z'})^2\right) + \hat{\mathbf{z}}'\left\{\left(\frac{\omega\rho}{k} + \frac{k}{\omega} c'_{13}\right) v_{x'} v_{z'} + \frac{k}{\omega} c'_{35} (v_{z'})^2\right\} \tag{5.50}$$

D. PHYSICAL REALIZABILITY CONDITIONS FOR c AND s

If the zero strain reference state is a point of static equilibrium it must correspond to a state of minimum strain energy. The strain energy must therefore always increase as the medium is deformed. Consequently, the strain energy density $u_S = \frac{1}{2}S_I c_{IJ} S_J$ can never become negative and goes to zero only when the strain components are all zero. In mathematics such a quadratic expression is defined as a *positive definite quadratic function* and it is shown that the following constraints on $[c]$ must be satisfied to ensure positive definiteness,†

$$c_{II} > 0$$

$$\begin{vmatrix} c_{II} & c_{IJ} \\ c_{IJ} & c_{JJ} \end{vmatrix} > 0$$

$$\cdot$$
$$\cdot$$
$$\cdot$$

$$\det [c_{IJ}] > 0 \qquad (5.51)$$

Analogous restrictions apply to the compliance matrix elements s_{IJ}.

EXAMPLE 5. As an illustration, physical realizability conditions will be applied to the hexagonal compliance matrix (4.23). The first condition $s_{II} > 0$ gives

$$s_{11} > 0$$
$$s_{33} > 0$$
$$s_{44} > 0$$
$$s_{66} = 2(s_{11} - s_{12}) > 0.$$

From the second order determinants

$$\begin{vmatrix} s_{II} & s_{IJ} \\ s_{IJ} & s_{JJ} \end{vmatrix} > 0,$$

the additional conditions

$$\begin{vmatrix} s_{11} & s_{12} \\ s_{12} & s_{11} \end{vmatrix} = s_{11}^2 - s_{12}^2 > 0$$

$$\begin{vmatrix} s_{11} & s_{13} \\ s_{13} & s_{33} \end{vmatrix} = s_{11}s_{33} - s_{13}^2 > 0$$

$$s_{11}s_{44} > 0$$
$$s_{11}s_{66} > 0$$

---

† See, for example, Section 13.5–6 in G. A. Korn and T. M. Korn, *Mathematical Handbook for Scientists and Engineers*, McGraw-Hill, New York, 1961.

are obtained. Only the first two of these are independent of the conditions $s_{II} > 0$. From the third order determinants, one has first of all that

$$
\begin{vmatrix} s_{11} & s_{12} & s_{13} \\ s_{12} & s_{11} & s_{13} \\ s_{13} & s_{13} & s_{33} \end{vmatrix} = s_{11}(s_{11}s_{33} - s_{13}^2) - s_{12}(s_{12}s_{33} - s_{13}^2) + s_{13}(s_{12}s_{13} - s_{13}s_{11})
$$

$$
= (s_{11}^2 - s_{12}^2)s_{33} - 2s_{13}^2(s_{11} - s_{12}) > 0. \tag{5.52}
$$

No other independent conditions are obtained from the remaining third order determinants or from the fourth, fifth, and sixth order determinants.

The physical realizability conditions obtained above may be simplified by removing redundancies. First of all,

$$
s_{66} = (s_{11} - s_{12}) > 0
$$

and

$$
s_{11}^2 - s_{12}^2 > 0
$$

may be combined into

$$
s_{11} > |s_{12}|. \tag{5.53}
$$

This also implicitly includes the condition $s_{11} > 0$. After removing the factor $s_{11} - s_{12}$ from the third order condition (5.52), one obtains

$$
(s_{11} + s_{12})s_{33} - 2s_{13}^2 > 0,
$$

or

$$
(s_{11} + s_{12})s_{33} > 2s_{13}^2. \tag{5.54}
$$

Conditions (5.53) and (5.54), taken together, imply both of the conditions

$$
s_{33} > 0
$$

and

$$
s_{11}s_{33} - s_{13}^2 > 0.
$$

A *sufficient* set of physical realizability conditions is therefore given by (5.53), (5.54), and

$$
s_{44} > 0 \tag{5.55}
$$

For zinc oxide

$$
s_{11} = 7.82 \times 10^{-12} \text{ m}^2/\text{newton}
$$
$$
s_{33} = 6.93 \times 10^{-12} \text{ m}^2/\text{newton}
$$
$$
s_{44} = 23.6 \times 10^{-12} \text{ m}^2/\text{newton}
$$
$$
s_{12} = -3.40 \times 10^{-12} \text{ m}^2/\text{newton}
$$
$$
s_{13} = -2.20 \times 10^{-12} \text{ m}^2/\text{newton}
$$

from Part A.4 of Appendix 2. Substitution of these numbers into the independent physical realizability conditions gives

$$
7.82 > 3.40
$$

from (5.53)

$$
(7.82 - 3.40)(6.93) > 2(2.20)^2
$$

from (5.54), and

$$
23.6 > 0
$$

from (5.55).

FIGURE 5.4.   Sign conventions for voltage and current at circuit terminals.

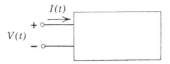

E.   COMPLEX POWER FLOW

In circuit theory the *instantaneous* power flow into a pair of circuit terminals, with positive voltage and current defined as in Fig. 5.4, is

$$P(t) = V(t)I(t). \tag{5.56}$$

This applies to voltages and currents with arbitrary waveforms. For time-harmonic voltages and currents, circuit calculations are greatly simplified by using the complex exponential notation described in Example 11 of Chapter 3, and this is now standard practice. That is,

$$V(t) = |V_0| \cos(\omega t + \phi_V) = \frac{V_0 e^{i\omega t} + V_0^* e^{-i\omega t}}{2} = \mathscr{R}e\, V_0 e^{i\omega t}$$

$$I(t) = |I_0| \cos(\omega t + \phi_I) = \frac{I_0 e^{i\omega t} + I_0^* e^{-i\omega t}}{2} = \mathscr{R}e\, I_0 e^{i\omega t} \tag{5.57}$$

where the complex numbers

$$V_0 = |V_0| e^{i\phi_V}$$

$$I_0 = |I_0| e^{i\phi_I}$$

describe both the amplitudes and phases of the voltage and current.

In order to gain full advantage of the simplification provided by complex notation, it is important to have a means for calculating power flow without reconverting to real variables. Derivation of an expression for power flow is much easier if there is only one phase angle in (5.57). The reference time $t = 0$ is therefore chosen such that $\phi_V = 0$. In this case, V_0 is real ($V_0^* = V_0$). Substitution into (5.56) then gives for the instantaneous power,

$$P(t) = \frac{\mathscr{R}e\, V_0(I_0)^*}{2} + \frac{V_0 I_0 e^{i2\omega t} + V_0(I_0)^* e^{-i2\omega t}}{4}$$

$$= \frac{\mathscr{R}e\, V_0(I_0)^*}{2}(1 + \cos 2\omega t) + \frac{\mathscr{I}m\, V_0(I_0)^*}{2} \sin 2\omega t. \tag{5.58}$$

This contains both a pulsating unidirectional power flow

$$\mathscr{R}e\, \frac{V_0(I_0)^*}{2}(1 + \cos 2\omega t),$$

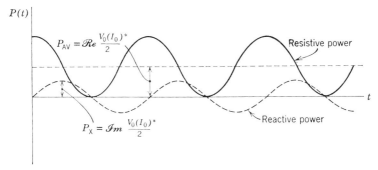

FIGURE 5.5.   **Definition of complex power flow.**

which supplies resistive losses and has a time-average value

$$P_{AV} = \mathscr{R}e\, \frac{V_0(I_0)^*}{2}\,; \qquad (5.59)$$

and a periodic power flow

$$\mathscr{I}m\, \frac{V_0(I_0)^*}{2}\, \sin 2\omega t,$$

which feeds in and extracts the energy stored in the coils and capacitors of the circuit and has a peak value defined by

$$P_X = \mathscr{I}m\, \frac{V_0(I_0)^*}{2}\,. \qquad (5.60)$$

The instantaneous power waveform (Fig. 5.5) is completely specified by the average resistive power (5.59) and the peak reactive power (5.60). These quantities are seen to be just the real and imaginary parts of the quantity

$$P = \frac{V_0(I_0)^*}{2} = P_{AV} + iP_X, \qquad (5.61)$$

which is defined as the *complex* power flow.† Although this result has been derived for the special case where V_0 is real, the interpretation of (5.61) is clearly independent of the choice of time zero.

F.   COMPLEX POYNTING'S THEOREMS
FOR TIME-HARMONIC FIELDS

The Poynting's Theorems derived in Sections 5.B and 5.C relate to instantaneous power flow and energy storage. Since the complex notation of the

---

† The same symbol is normally used for both complex and instantaneous power. If a distinction must be made, the latter is written with an explicit time dependence, as in (5.56).

preceding section is also useful in field theory (Example 11 of Chapter 3) it is important to derive Poynting's Theorems for complex power flow as well. This derivation will be given only for the electromagnetic case. The acoustic theorem is obtained in an entirely analogous way and will be stated without proof.

F.1   Electromagnetic Fields

In complex notation the electromagnetic field variables in (4.4) to (4.8) are represented by the real parts of

$$\mathbf{E}e^{i\omega t}, \mathbf{H}e^{i\omega t}, \mathbf{B}e^{i\omega t}, \mathbf{D}e^{i\omega t}, \text{etc.} \tag{5.62}$$

The complex field amplitudes in (5.62) may be distinguished from the instantaneous variables $\mathbf{E}(t)$ etc., by use of a subscript 0, as in (5.57). It is, however, not usually necessary to make this distinction and the subscript will therefore be omitted in what follows.

To derive the complex Poynting's Theorem for electromagnetic fields the scalar product of (4.4) is first taken with $-\mathbf{H}^*$,

$$\mathbf{H}^* \cdot \nabla \times \mathbf{E} = -i\omega \mathbf{H}^* \cdot \mathbf{B}.$$

This is added to the scalar product of the complex conjugate of (4.5) with $-\mathbf{E}$

$$-\mathbf{E} \cdot \nabla \times \mathbf{H}^* = i\omega \mathbf{E} \cdot \mathbf{D}^* - \mathbf{E} \cdot \mathbf{J}_c^* - \mathbf{E} \cdot \mathbf{J}_s^*.$$

The sum is then reduced by using the identity (5.2) and multiplied by $\frac{1}{2}$. After integrating over a volume V and using the divergence theorem, one then has

$$\oint_S \frac{\mathbf{E} \times \mathbf{H}^* \cdot \hat{\mathbf{n}}}{2} dS = i\omega \int_V \left(\frac{\mathbf{E} \cdot \mathbf{D}^*}{2} - \frac{\mathbf{H}^* \cdot \mathbf{B}}{2} \right) dV$$

$$- \int_V \frac{\mathbf{E} \cdot \mathbf{J}_c^*}{2} dV - \int_V \frac{\mathbf{E} \cdot \mathbf{J}_s^*}{2} dV. \tag{5.63}$$

This is the *complex Poynting's Theorem* for electromagnetic fields.[†]

Comparison of the last integral on the right-hand side of (5.63) with the integral on the right-hand side of (5.4) shows that

$$- \int_V \frac{\mathbf{E} \cdot \mathbf{J}_s^*}{2} dV = P_s \tag{5.64}$$

is the complex power supplied by the sources. From (5.11) the instantaneous

---

† Note that (5.63) is valid when the time variation is $e^{i\omega t}$, while (5.3) is valid for arbitrary time dependence.

magnetic stored energy density is

$$u_H(t) = \tfrac{1}{2}\mathbf{H}(t) \cdot \boldsymbol{\mu} \cdot \mathbf{H}(t). \tag{5.65}$$

The quantity

$$\frac{\mathbf{H}^* \cdot \mathbf{B}}{2} = \frac{\mathbf{H}^* \cdot \boldsymbol{\mu} \cdot \mathbf{H}}{2}$$

in (5.63) is pure real and can therefore be identified as *twice* the time average of the magnetic stored energy density (5.65). Accordingly, the *peak* magnetic stored energy density is

$$(u_H)_{\text{peak}} = \frac{\mathbf{H}^* \cdot \mathbf{B}}{2}.$$

In the same way, it is seen that

$$(u_E)_{\text{peak}} = \frac{\mathbf{E} \cdot \mathbf{D}^*}{2}.$$

The first integral on the right-hand side of (5.63) therefore gives

$$i\omega \int_V \left(\frac{\mathbf{E} \cdot \mathbf{D}^*}{2} - \frac{\mathbf{H}^* \cdot \mathbf{B}}{2} \right) dV = i\omega (U_{E_{\text{peak}}} - U_{H_{\text{peak}}}). \tag{5.66}$$

where U_E and U_H are the total stored energies within V.

In (5.13), the integrand

$$\mathbf{E}(t) \cdot \boldsymbol{\sigma} \cdot \mathbf{E}(t) = \mathbf{E}(t) \cdot \mathbf{J}_c(t)$$

is the instantaneous conductive power loss per unit volume. Because $\boldsymbol{\sigma}$ is pure real, $\mathbf{J}_c(t)$ is in phase with $E(t)$. The integrand

$$\frac{\mathbf{E} \cdot \mathbf{J}_c^*}{2}$$

in (5.63) is therefore a pure real quantity and represents the time-average conductive power loss per unit volume. Therefore

$$\int_V \frac{\mathbf{E} \cdot \mathbf{J}_c^*}{2} \, dV = \int_V \frac{\mathbf{E} \cdot \boldsymbol{\sigma} \cdot \mathbf{E}^*}{2} \, dV = (P_d)_{\text{AV}}, \tag{5.67}$$

is the average conductive power loss in the system. From the definition of complex power, this must always be a positive real quantity. Consequently, the integral

$$\frac{\mathbf{E} \cdot \boldsymbol{\sigma} \cdot \mathbf{E}^*}{2} = \frac{E_i E_j^* \sigma_{ij}}{2}$$

is always positive real, and it follows that the conductivity matrix is symmetric,

$$\sigma_{ij} = \sigma_{ji}. \tag{5.68}$$

After combining these results, one can write the complex Poynting's Theorem as

$$\oint_S \frac{(\mathbf{E} \times \mathbf{H}^*)}{2} \cdot \hat{\mathbf{n}} \, dS - i\omega \{(U_E)_{\text{peak}} - (U_H)_{\text{peak}}\} + (P_d)_{\text{AV}} = P_s. \quad (5.69)$$

From conservation of complex power, the surface integral is then the total complex power flow through S—the real part being average power and the imaginary part being peak reactive power—and the *complex Poynting vector* is defined as

$$\mathbf{P} = \frac{\mathbf{E} \times \mathbf{H}^*}{2}. \quad (5.70)$$

F.2   Acoustic Fields

In this case the complex Poynting's Theorem is found to be

$$\oint_S \frac{-\mathbf{v}^* \cdot \mathbf{T}}{2} \cdot \hat{\mathbf{n}} \, dS = i\omega \int_V \left(\frac{\mathbf{T}:\mathbf{S}^*}{2} - \frac{\mathbf{v}^* \cdot \mathbf{p}}{2} \right) dV + \int_V \frac{\mathbf{v}^* \cdot \mathbf{F}}{2} \, dV, \quad (5.71)$$

corresponding to (5.63). Here

$$\int_V \frac{\mathbf{v}^* \cdot \mathbf{F}}{2} \, dV = P_s \quad (5.72)$$

is the complex power supplied by the sources, and

$$\int_V \frac{\mathbf{v}^* \cdot \mathbf{p}}{2} \, dV = \int_V \frac{\rho \, |v|^2}{2} \, dV = (U_v)_{\text{peak}} \quad (5.73)$$

is the peak kinetic stored energy in the system. Substitution of (4.3) for $\mathbf{T}$ in the first term on the right-hand side of (5.71) gives

$$i\omega \int_V \frac{\mathbf{T}:\mathbf{S}^*}{2} \, dV = i\omega(U_S)_{\text{peak}} - (P_d)_{\text{AV}},$$

where

$$(U_S)_{\text{peak}} = \int_V \frac{\mathbf{S}:\mathbf{c}:\mathbf{S}^*}{2} \, dV \quad (5.74)$$

is the peak strain energy in the system and

$$(P_d)_{\text{AV}} = \omega^2 \int_V \frac{\mathbf{S}:\boldsymbol{\eta}:\mathbf{S}^*}{2} \, dV \quad (5.75)$$

is the time-average power loss due to viscous damping.

With these identifications, the complex acoustic Poynting's Theorem can be written as

$$\oint_S \frac{(-\mathbf{v}^* \cdot \mathbf{T}) \cdot \hat{\mathbf{n}}}{2} \, dS - i\omega \{(U_S)_{\text{peak}} - (U_v)_{\text{peak}}\} + (P_d)_{\text{AV}} = P_s. \quad (5.76)$$

The complex acoustic Poynting vector is then defined as

$$\mathbf{P} = \frac{-\mathbf{v}^* \cdot \mathbf{T}}{2}. \tag{5.77}$$

As in the electromagnetic case, proof that the viscosity matrix is symmetric,

$$\eta_{IJ} = \eta_{JI}, \tag{5.78}$$

follows from the positive real nature of $(P_d)_{AV}$ in (5.75).

EXAMPLE 6. In complex exponential notation the fields of the Y-polarized shear wave (5.41) in Example 3 are

$$v_{y\pm} = A_Y e^{i\omega(t \mp x/V_p)}$$

$$T_{6\pm} = \mp \left(\frac{\rho}{s_{66}}\right)^{1/2} v_{y\pm}. \tag{5.79}$$

The complex acoustic Poynting vector (5.77) is therefore

$$\mathbf{P}_{Y\text{-polarized}} = -\hat{\mathbf{x}} \frac{v_{y\pm}^* T_{yx\pm}}{2} = \pm \hat{\mathbf{x}} \frac{1}{2} \left(\frac{\rho}{s_{66}}\right)^{1/2} |A_Y|^2. \tag{5.80}$$

This is pure real and therefore represents average power flow density. Because the medium is lossless the average power flow is independent of position along the path of propagation. From (5.73) and (5.74) the peak kinetic and elastic stored energy densities are

$$(u_v)_{\text{peak}} = \frac{\rho |v|^2}{2} = \frac{\rho}{2} |A_Y|^2$$

$$(u_s)_{\text{peak}} = \frac{\mathbf{S} : \mathbf{c} : \mathbf{S}^*}{2} = \frac{\mathbf{T} : \mathbf{s} : \mathbf{T}^*}{2} = \frac{s_{66}}{2} |T_6|^2 = \frac{\rho}{2} |A_Y|^2 \tag{5.81}$$

in agreement with (5.46) and (5.47).

An interesting relationship exists between the average stored energy density

$$u_{AV} = \frac{(u_v)_{\text{peak}} + (u_s)_{\text{peak}}}{2} = \frac{\rho}{2} |A_Y|^2, \tag{5.82}$$

from (5.81), the average power flow density

$$P_{AV} = \frac{1}{2} \left(\frac{\rho}{s_{66}}\right)^{1/2} |A_Y|^2,$$

from (5.80), and the propagation velocity

$$V_p = (\rho s_{66})^{-1/2},$$

from (4.32). It is easily checked by direct substitution that these quantities are related according to the equation

$$P_{AV} = V_p u_{AV}. \tag{5.83}$$

A physical interpretation of this result is easily obtained by referring to Fig. 5.1 and assuming that the wave is traveling in the $+x$ direction. The total average stored energy in the enclosed volume is

$$U_{AV} = u_{AV}(x_2 - x_1)S_1 \, .$$

This energy is supplied to the volume by power flow through S_1 during the period of time τ required for the propagating wave to fill the volume, namely

$$\tau = \frac{(x_2 - x_1)}{V_p} \, .$$

It therefore follows that

$$P_{AV}\tau S_1 = \frac{P_{AV}}{V_p}(x_2 - x_1)S_1 = u_{AV}(x_2 - x_1)S_1 \, ,$$

which is exactly equivalent to (5.83). In Section F of Chapter 7 a relationship of this kind will be shown to apply for acoustic waves of the most general kind.

EXAMPLE 7. Example 11 of Chapter 3 dealt with a [100]-polarized shear wave propagating along the [010] direction of a lossy cubic crystal. From (3.65) and (3.66) the particle velocity and stress fields for this case may be taken as

$$v_x = Ae^{-\alpha y}e^{i(\omega t - ky)}$$

$$T_6 = -\frac{(\alpha + ik)}{i\omega}(c_{44} + i\omega \eta_{44})v_x \tag{5.84}$$

for the positive-traveling wave. According to (3.72) and (3.73),

$$\alpha = \frac{\omega^2}{2}\left(\frac{\rho}{c_{44}}\right)^{1/2}\left(\frac{\eta_{44}}{c_{44}}\right) \tag{5.85}$$

and

$$k = \omega\left(\frac{\rho}{c_{44}}\right)^{1/2}\left(1 + \frac{3}{4}\left(\frac{\omega \eta_{44}}{c_{44}}\right)^2\right)^{-1/2} \tag{5.86}$$

for a low-loss material. The complex Poynting vector,

$$\mathbf{P} = \hat{\mathbf{y}}\frac{1}{2}\frac{(\alpha + ik)}{i\omega}(c_{44} + i\omega \eta_{44})\,|A|^2\,e^{-2\alpha y}, \tag{5.87}$$

is now complex and decays exponentially along the direction of propagation. This is easily understood from the fact that the wave is constantly losing energy to the viscous damping in the medium and must therefore carry less and less power the further it travels.

From (5.87) the average power flow density is

$$P_{AV} = \mathscr{R}e\, P_y \approx \frac{1}{2}\frac{k}{\omega}c_{44}\,|A|^2\,e^{-2\alpha y} = \tfrac{1}{2}(\rho c_{44})^{1/2}\,|A|^2\,e^{-2\alpha y} \tag{5.88}$$

to a first order approximation in the viscosity coefficient η_{44}. In the same order of approximation the average stored energy density is

$$u_{\mathrm{AV}} = \frac{\rho |A|^2}{2} e^{-2\alpha y}, \tag{5.89}$$

and the average viscous power dissipation *per unit volume* is

$$(P_d)_{\mathrm{AV}} = \omega^2 \frac{\eta_{44}}{c_{44}} u_{\mathrm{AV}}, \tag{5.90}$$

from (5.75). One can now easily check by direct substitution that the attenuation factor (5.85) is

$$\alpha = \frac{(P_d)_{\mathrm{AV}}}{2 P_{\mathrm{AV}}}. \tag{5.91}$$

Like the velocity relationship (5.83) this equation has a simple physical interpretation. The rate of decrease of average power density with distance must equal the average power dissipation *per unit volume*. Thus,

$$P_{\mathrm{AV}} = P_0 e^{-2\alpha y},$$

from (5.88), and therefore

$$\frac{\partial P_{\mathrm{AV}}}{\partial y} = -(P_d)_{\mathrm{AV}} = -2\alpha P_{\mathrm{AV}},$$

which is equivalent to (5.91). The relationship (5.91) should therefore apply to all kinds of wave propagation. In Chapter 12 attenuation expressions of this kind are derived for Rayleigh wave propagation by applying formal perturbation theory to the problem.

EXAMPLE 8.   In Example 3 of Chapter 4 an illustration was given of wave scattering at a material discontinuity lying normal to the propagation direction. For compressional waves propagating along the X-direction in a hexagonal crystal medium the particle velocity reflection and transmission coefficients are given by (4.83) and (4.84). If the second medium is free space ($\rho' = 0$) there is no transmitted wave and the reflection coefficient is

$$R_v(0) = 1,$$

from (4.83). The total acoustic field in the first medium is then obtained from (4.78) and (4.79). That is,

$$v_x = 2A e^{i\omega t} \cos kx$$
$$T_1 = 2i(\rho c_{11})^{1/2} A e^{i\omega t} \sin kx$$
$$T_2 = 2i(c_{12}/V_p) A e^{i\omega t} \sin kx$$
$$T_3 = 2i(c_{13}/V_p) A e^{i\omega t} \sin kx. \tag{5.92}$$

To calculate the instantaneous power density at any plane $x = $ constant, the real parts of the fields (5.92) are taken and then substituted into (5.40), giving

$$\mathbf{P}(t) = -\mathbf{v} \cdot \mathbf{T} = -\hat{x} v_x T_{xx} = -\hat{x} v_x T_1$$
$$= \hat{x} 4 (\rho c_{11})^{1/2} A^2 \cos \omega t \cos kx \sin \omega t \sin kx$$
$$= \hat{x} (\rho c_{11})^{1/2} A^2 \sin 2\omega t \sin 2kx. \qquad (5.93)$$

Since there is no dissipation and the acoustic wave is totally reflected, the *average* power flow is zero everywhere. Another characteristic of (5.93) is that the *instantaneous* power flow is identically zero at certain values of x. This is easily understood by considering the stored energy in the system. From (5.30) the instantaneous kinetic energy density is

$$u_v(t) = \tfrac{1}{2} \rho v_x^2 = 2\rho A^2 \cos^2 \omega t \cos^2 kx;$$

and from (5.35) the instantaneous strain energy density is

$$u_s(t) = \tfrac{1}{2} c_{11} S_1^2 = 2\rho A^2 \sin^2 \omega t \sin^2 kx,$$

which is obtained by using

$$u_x = \frac{2}{\omega} A \sin \omega t \cos kx$$

$$S_1 = \frac{\partial u_x}{\partial x}$$

and

$$\frac{k}{\omega} = \frac{1}{V_p} = \left(\frac{\rho}{c_{11}} \right)^{1/2}.$$

The total instantaneous stored energy in the volume shown in Fig. 5.6 is therefore

$$U(t) = S_1 \int_{-x_1}^{0} (u_v(t) + u_s(t)) \, dx = \rho S_1 A^2 \left(x_1 + \frac{\cos 2\omega t \sin 2kx_1}{2k} \right) \qquad (5.94)$$

When $\sin 2kx_1 = 0$, U is independent of time. In this case there is simply an interchange of kinetic and strain energy within the volume, and this is exactly the condition for the power density P_x to be zero at the end surface S_1. When the rate of change of the total stored energy is zero, the total power flow through the surface enclosing the volume must be zero. Since the power density is always zero at the stress-free end surface S_2 and also at the side surfaces, this requires that there be zero power density at the end surface S_1.

The complex Poynting vector for this situation is

$$\mathbf{P} = -\hat{x} \frac{v_x^* T_1}{2} = -\hat{x} 2i (\rho c_{11})^{1/2} |A|^2 \cos kx \sin kx$$
$$= -\hat{x} i (\rho c_{11})^{1/2} |A|^2 \sin 2kx. \qquad (5.95)$$

Since the average power flow is zero, the Poynting vector is purely imaginary and has a magnitude equal to the amplitude of the periodic instantaneous power flow given by (5.93).

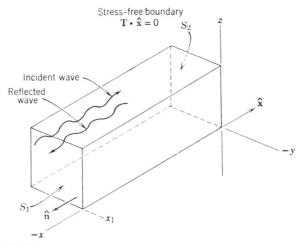

FIGURE 5.6.   Calculation of the power-energy relation in an acoustic wave that is totally reflected by a stress-free boundary at the plane $x = 0$.

PROBLEMS

1. Verify the identities (5.2) and (5.27) by expanding in rectangular Cartesian coordinates.

2. Prove that the stiffness and compliance matrices are symmetric ($c_{IJ} = c_{JI}, s_{IJ} = s_{JI}$).

3. By expanding in terms of abbreviated subscript components, show that

$$\left(\mathbf{c}:\mathbf{S}\right):\frac{\partial \mathbf{S}}{\partial t} = \mathbf{S}:\mathbf{c}:\frac{\partial \mathbf{S}}{\partial t}$$

and

$$\left(\boldsymbol{\eta}:\frac{\partial \mathbf{S}}{\partial t}\right):\frac{\partial \mathbf{S}}{\partial t} = \frac{\partial \mathbf{S}}{\partial t}:\boldsymbol{\eta}:\frac{\partial \mathbf{S}}{\partial t} \; .$$

4. Using (5.51), derive physical realizability conditions for the cubic stiffness matrix of Example 3 and the trigonal stiffness matrix of Problem 1 in Chapter 3.

5. It was noted in Section A of Chapter 3 that the elastic limit for relatively rigid materials occurs at strains of order 10^{-3}. Electrical field breakdown in solids requires typical field strengths of order 10^7 volts/m. Using these estimates and the material constants given in Appendix 2, compare the breakdown values of strain and electrical stored energies in diamond, fused silica, pyrex glass, and strontium titanate.

6. Using a *peak* power density of 10^3 W/m² and material constants for cadmium selenide, cadmium sulfide, and zinc oxide, calculate [from (5.30), (5.35), and (5.40)] the peak values of particle velocity and stress, and the *average* stored energy density for the three plane wave solutions of part (b) Example 1 in Chapter 4. (Neglect the piezoelectric effect; that is, use the stiffness constants given in part A.5 of Appendix 2.)

7. Calculate the phase velocities in cadmium selenide, cadmium sulfide, and zinc oxide for the three plane wave solutions in the previous problem. Use (5.83) to calculate the *average* stored energy density corresponding to a *peak* power density of 10^3 W/m³.

8. Following the analysis of Example 4, calculate and graph the Poynting vector direction in cadmium sulfide as a function of the wave vector direction η. Neglect the piezoelectric effect.

9. In Example 3(b) of Chapter 4 show that the particle velocity field, the stress field, and the complex Poynting vector for the compressional wave solution are

$$v_x = Ae^{i\omega t}(e^{-ikx} + R_v(0)e^{ikx})$$

$$T_1 = -(\rho c_{11})^{1/2}Ae^{i\omega t}(e^{-ikx} - R_v(0)e^{ikx})$$

and

$$\mathbf{P} = \hat{\mathbf{x}}\,\frac{(\rho c_{11})^{1/2}\,|A|^2}{2}\,\{1 - |R_v(0)|^2 - i2R_v(0)\sin 2kx\}.$$

10. Using the result obtained in Problem 9 and the requirement that power be conserved at the scattering interface, show that

$$(\rho c_{11})^{1/2}\{1 - |R_v(0)|^2\} = (\rho'c_{11}')^{1/2}\,|T_v(0)|^2.$$

Verify that the scattering coefficients obtained in Example 3(b) of Chapter 4 do satisfy this condition.

REFERENCES

Stored Energy

1. P. C. Chou and N. J. Pagano, *Elasticity—Tensor, Dyadic and Engineering Approaches*, pp. 145–151, van Nostrand, New York, 1967.

2. J. F. Nye, *Physical Properties of Crystals*, pp. 57–60, 74, 136–137, 142, Oxford, England, 1964.

3. J. A. Stratton, *Electromagnetic Theory*, pp. 104–114, 118–130, McGraw-Hill, New York, 1941.

Poynting's Theorem

4. R. F. Harrington, *Time-harmonic Electromagnetic Fields*, pp. 9–12, McGraw-Hill, New York, 1961.

5. G. Nadeau, *Introduction to Elasticity*, pp. 61–62, Holt, Rinehart and Winston, New York, 1964.

6. S. Ramo, J. R. Whinnery, T. van Duzer, *Fields and Waves in Communication Electronics*, pp. 242–245, Wiley, New York, 1965.

7. J. A. Stratton, *Electromagnetic Theory*, pp. 131–135, McGraw-Hill, New York, 1941.

8. R. N. Thurston, "Wave Propagation in Fluids and Solids," pp. 12–14, in *Physical Acoustics*, Vol. 1A, W. P. Mason Ed., Academic Press, New York, 1964.

Complex Poynting's Theorem

9. R. F. Harrington, *Time-harmonic Electromagnetic Fields*, pp. 19–26, McGraw-Hill, New York, 1961.

10. J. A. Stratton, *Electromagnetic Theory*, pp. 135–137, McGraw-Hill, New York, 1941.

Chapter 6

ACOUSTIC PLANE WAVES IN ISOTROPIC SOLIDS

A. THE ACOUSTIC WAVE EQUATION AND THE CHRISTOFFEL EQUATION

The previous chapters have illustrated a number of acoustic plane wave solutions. In the present chapter a unified treatment of this topic will be given for the special case of propagation in isotropic media. The following chapter will present a similar treatment of the anisotropic case.

In all of the examples considered up to this point, plane wave solutions have been obtained by writing out the acoustic field equations (4.15) and (4.17) in component form and grouping the resulting equations into independent sets. This procedure is very helpful for gaining physical insight into the properties of the solutions and for comparing acoustic solutions with electromagnetic solutions, but it is not the most efficient approach. It is much simpler to eliminate unnecessary variables by manipulating the field equations in symbolic form. This is a very useful method in electromagnetism, where wave equations for the $\mathbf{E}$ or $\mathbf{H}$ fields are derived by eliminating either $\mathbf{H}$ or $\mathbf{E}$ from Maxwell's equations and the constitutive relations. In the same way, acoustic wave equations may be derived by eliminating either $\mathbf{T}$ or $\mathbf{v}$ from the acoustic field equations and constitutive relations. Usually the stress field $\mathbf{T}$ is eliminated, since it involves six field components rather than three.

From (4.20) and (4.21) the lossless acoustic field equations are

$$\nabla \cdot \mathbf{T} = \rho \frac{\partial \mathbf{v}}{\partial t} - \mathbf{F} \tag{6.1}$$

$$\nabla_s \mathbf{v} = \mathbf{s} : \frac{\partial \mathbf{T}}{\partial t} . \tag{6.2}$$

The stress field is eliminated by first differentiating (6.1) with respect to t,

$$\nabla \cdot \frac{\partial \mathbf{T}}{\partial t} = \rho \frac{\partial^2 \mathbf{v}}{\partial t^2} - \frac{\partial \mathbf{F}}{\partial t} . \tag{6.3}$$

Multiplication of (6.2) by the stiffness gives

$$\mathbf{c} : \nabla_s \mathbf{v} = \frac{\partial \mathbf{T}}{\partial t} , \tag{6.4}$$

and the wave equation for $\mathbf{v}$ is found by substituting this into (6.3),

$$\nabla \cdot \mathbf{c} : \nabla_s \mathbf{v} = \rho \frac{\partial^2 \mathbf{v}}{\partial t^2} - \frac{\partial \mathbf{F}}{\partial t} . \tag{6.5}$$

In matrix form with abbreviated subscripts, this is

$$\nabla_{iK} c_{KL} \nabla_{Lj} v_j = \rho \frac{\partial^2 v_i}{\partial t^2} - \frac{\partial}{\partial t} F_i , \tag{6.6}$$

where the matrix-differential operators ∇_{iK} and ∇_{Lj} have been defined in (2.36) and (1.53), respectively.

In a source-free region ($\mathbf{F} = 0$), a uniform plane wave propagating along the direction

$$\hat{\mathbf{l}} = \hat{\mathbf{x}} l_x + \hat{\mathbf{y}} l_y + \hat{\mathbf{z}} l_z \tag{6.7}$$

has fields proportional to $e^{i(\omega t - k \hat{\mathbf{l}} \cdot \mathbf{r})}$. In this case operators ∇_{iK} and ∇_{Lj} in (6.6) may be replaced by matrices $-ik_{iK}$ and $-ik_{Lj}$ respectively, where

$$-ik_{iK} = -ikl_{iK} \rightarrow -ik \begin{bmatrix} l_x & 0 & 0 & 0 & l_z & l_y \\ 0 & l_y & 0 & l_z & 0 & l_x \\ 0 & 0 & l_z & l_y & l_x & 0 \end{bmatrix} \tag{68}$$

and

$$-ik_{Lj} = -ikl_{Lj} = -ik \begin{bmatrix} l_x & 0 & 0 \\ 0 & l_y & 0 \\ 0 & 0 & l_z \\ 0 & l_z & l_y \\ l_z & 0 & l_x \\ l_y & l_x & 0 \end{bmatrix} \qquad (6.9)$$

The wave equation (6.6), with $\mathbf{F} = 0$, then reduces to

$$k^2(l_{iK}c_{KL}l_{Lj})v_j = k^2\Gamma_{ij}v_j = \rho\omega^2v_i . \qquad (6.10)$$

This is called the *Christoffel equation;* and the matrix on the left-hand side,

$$\Gamma_{ij} = l_{iK}c_{KL}l_{Lj} \qquad (6.11)$$

is called the *Christoffel matrix.* Its elements are functions only of the plane wave propagation direction and the stiffness constants of the medium.

The formulation of the Christoffel equation in (6.10) applies to uniform plane waves† in both isotropic and anisotropic media. After derivation of the compliance and stiffness isotropy conditions in the following section, the Christoffel matrix (6.11) will be calculated for an isotropic medium and plane wave solutions will then be obtained from (6.10). The anisotropic case is considered in the next chapter.

B.   ELASTIC ISOTROPY CONDITIONS

First, it is necessary to determine the restrictions imposed by elastic isotropy on the compliance and stiffness matrices. This derivation follows the same type of symmetry arguments used in the more difficult anisotropic case and will be carried out in some detail so as to provide an elementary preview of Chapter 7.

In an isotropic medium the three coordinate axes, x, y, z and the three coordinate planes yz, xz, xy are equivalent. Consequently, the response of the medium must be the same for a compressive stress applied along any axis. Therefore

$$s_{11} = s_{22} = s_{33}$$
$$s_{12} = s_{13} = s_{23} .$$

† Often called *bulk waves.*

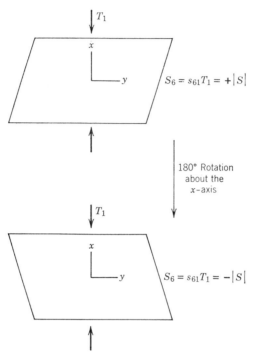

FIGURE 6.1. **Symmetry argument showing that $s_{61} = -s_{61} = 0$ in an isotropic medium.**

Also, the shear strain produced in a coordinate plane by a shear stress applied in that plane must be the same for all coordinate planes,

$$s_{44} = s_{55} = s_{66}.$$

Next, suppose that a compressive stress T_1 applied along the x axis produces a shear strain S_6 in the xy plane. Since the medium is isotropic, an equally valid solution to the applied stress problem is one in which the medium is rotated $180°$ about the x axis (Fig. 6.1). This reverses the sign of S_6, which can therefore only be zero. That is,

$$s_{16} = 0.$$

Similarly

$$s_{14} = s_{15} = s_{24} = s_{25} = s_{26} = s_{34} = s_{35} = s_{36} = 0;$$

and arguments of the same kind, using an applied shear stress in a coordinate plane, show that

$$s_{45} = s_{46} = s_{56} = 0.$$

By exactly the same kind of arguments it is easily shown that the same conditions apply to the stiffness matrix. When these conditions are imposed upon a general stiffness matrix, the result

$$[c] = \begin{bmatrix} c_{11} & c_{12} & c_{12} & 0 & 0 & 0 \\ c_{12} & c_{11} & c_{12} & 0 & 0 & 0 \\ c_{12} & c_{12} & c_{11} & 0 & 0 & 0 \\ 0 & 0 & 0 & c_{44} & 0 & 0 \\ 0 & 0 & 0 & 0 & c_{44} & 0 \\ 0 & 0 & 0 & 0 & 0 & c_{44} \end{bmatrix} \tag{6.12}$$

is obtained.

It has not yet been determined whether the above conditions are sufficient to ensure isotropy of the medium. This could be tested by checking to see if (6.12) is unchanged by an arbitrary rotation of coordinate axes. One would suspect that this is not so, from the fact that (6.12) is identical with the cubic crystal stiffness matrix in Example 3 of Chapter 3. Fortunately, this stiffness matrix has already been transformed by an arbitrary rotation about the z axis in Example 6 of Chapter 3, and (3.43) shows that the condition

$$c_{12} = c_{11} - 2c_{44} \tag{6.13}$$

is required for invariance of $[c]$ with respect to this rotation. This same condition is also obtained for rotation about the y axis† and, since an arbitrary rotation may be produced by successive rotations about z, y, and again about z, this guarantees that (6.13) is the general elastic isotropy condition.

From (6.12) and (6.13) it follows that an isotropic medium has only two independent elastic constants. These are often taken to be the *Lamé constants* λ and μ, defined by

$$\lambda = c_{12}$$

$$\mu = c_{44}.$$

For wave problems a more suitable choice is c_{11} and c_{44}, which will be seen in the next section to relate directly to the velocities of compressional and shear waves in the medium.

† Example 7 in Chapter 3.

C. THE CHRISTOFFEL EQUATION FOR ISOTROPIC SOLIDS

When the stiffness matrix given by (6.12) and (6.13) is substituted into (6.11), the result is

$$[\Gamma_{ij}] = \begin{bmatrix} c_{11}l_x^2 + c_{44}(1 - l_x^2) & (c_{12} + c_{44})l_x l_y & (c_{12} + c_{44})l_x l_z \\ \\ (c_{12} + c_{44})l_y l_x & c_{11}l_y^2 + c_{44}(1 - l_y^2) & (c_{12} + c_{44})l_y l_z \\ \\ (c_{12} + c_{44})l_z l_x & (c_{12} + c_{44})l_z l_y & c_{11}l_z^2 + c_{44}(1 - l_z^2) \end{bmatrix}$$

$$c_{12} = c_{11} - 2c_{44} \tag{6.14}$$

It is not at all obvious from direct inspection that this matrix will lead to isotropic propagation characteristics, but this is a necessary consequence of the isotropy of the stiffness matrix.

EXAMPLE 1. Compressional and Shear Waves in Lossless Isotropic Solids. In an isotropic medium, wave solutions are the same for all directions of propagation and one can therefore pick a direction that reduces the analysis to its simplest terms. Assume that

$$\hat{\mathbf{l}} = \hat{\mathbf{z}}.$$

Then (6.14) becomes a purely diagonal matrix, and the Christoffel equation (6.10) reduces to

$$k^2 \begin{bmatrix} c_{44} & 0 & 0 \\ 0 & c_{44} & 0 \\ 0 & 0 & c_{11} \end{bmatrix} \begin{bmatrix} v_x \\ v_y \\ v_z \end{bmatrix} = \rho\omega^2 \begin{bmatrix} v_x \\ v_y \\ v_z \end{bmatrix} \tag{6.15}$$

This separates into the three independent equations

$$k^2 c_{44} v_x = \rho\omega^2 v_x$$
$$k^2 c_{44} v_y = \rho\omega^2 v_y$$
$$k^2 c_{11} v_z = \rho\omega^2 v_z . \tag{6.16}$$

According to (6.16), the x-polarized z-propagating shear wave solution

$$\mathbf{v}_s = \hat{\mathbf{x}} v e^{i(\omega t - kz)} \tag{6.17}$$

and the y-polarized z-propagating solution

$$\mathbf{v}_{s'} = \hat{\mathbf{y}} v e^{i(\omega t - kz)} \tag{6.18}$$

must both satisfy a dispersion relation

$$k^2 c_{44} = \rho\omega^2. \tag{6.19}$$

This is the same as the shear wave dispersion relation for propagation along a cubic crystal axis, given by (3.24) in Example 4 of Chapter 3. As in that example, the shear waves are pure modes and are degenerate. This means that they can be combined to produce circular and elliptical polarizations. In the isotropic case these solutions are valid for *all* propagation directions. If propagation is along the arbitrary direction (6.7), the shear wave solutions

$$\mathbf{v}_s = \hat{\mathbf{a}} v e^{i(\omega t - k\hat{\mathbf{l}}\cdot\mathbf{r})}$$
$$(\hat{\mathbf{a}} \cdot \hat{\mathbf{l}} = 0) \tag{6.20}$$

and

$$\mathbf{v}_{s'} = \hat{\mathbf{a}} \times \hat{\mathbf{l}} v e^{i(\omega t - k\hat{\mathbf{l}}\cdot\mathbf{r})} \tag{6.21}$$

still satisfy the dispersion relation (6.19). The strain and stress fields associated with these waves are obtained by applying the field equation (4.17) and the constitutive relation (4.3), with $\eta = 0$.

From (6.16), the compressional wave solution

$$\mathbf{v}_l = \hat{\mathbf{z}} v e^{i(\omega t - kz)} \tag{6.22}$$

satisfies a dispersion relation

$$k^2 c_{11} = \rho\omega^2. \tag{6.23}$$

This is the same as (3.29), which was obtained in Example 5 of Chapter 3 for compressional wave propagation along a cubic crystal axis. As in the shear wave case, (6.22) applies for *all* propagation directions in an isotropic medium. The compressional wave solution for propagation in the direction (6.7) is thus

$$\mathbf{v}_l = \hat{\mathbf{l}} v e^{i(\omega t - k\hat{\mathbf{l}}\cdot\mathbf{r})} \tag{6.24}$$

and associated strain and stress fields may be calculated from (4.17) and (4.3).

EXAMPLE 2.   Lossy Isotropic Solids.   For fields varying as $e^{i\omega t}$ the lossy Hooke's Law relation (4.3) may be expressed in terms of a complex stiffness matrix

$$c_{IJ} \rightarrow c_{IJ} + i\omega\eta_{IJ}. \tag{6.25}$$

Using the same arguments employed in Section B one finds that the η_{IJ}'s must satisfy the same isotropy conditions as the c_{IJ}'s. There are therefore three different damping coefficients for an isotropic medium (η_{11}, η_{12}, η_{44}). Two of these are independent and the third is determined by the isotropy relation,

$$\eta_{12} = \eta_{11} - 2\eta_{44}. \tag{6.26}$$

When (6.25) is substituted into the Christoffel matrix (6.14), the dispersion relations (6.19) and (6.23) become

$$k^2(c_{44} + i\omega\eta_{44}) = \rho\omega^2 \tag{6.27}$$

and

$$k^2(c_{11} + i\omega\eta_{11}) = \rho\omega^2. \tag{6.28}$$

The first of these is identical with (3.68) in Example 11 of Chapter 3, and exact solutions are given by (3.69) and (3.71). For low loss materials these may be approximated by (3.72) and (3.73). The same solutions, with

$$\begin{aligned} c_{44} &\rightarrow c_{11} \\ \eta_{44} &\rightarrow \eta_{11}, \end{aligned} \tag{6.29}$$

apply to the compressional wave case (6.28).

It is rather common practice to express acoustic losses in terms of the acoustic Q, which was defined for some special cases in Table 3.1. From (6.16) and (3.72) the attenuation of a shear wave in an isotropic solid can thus be expressed as

$$\alpha_s = \frac{\omega}{2}\left(\frac{\rho}{c_{44}}\right)^{1/2}\left(\frac{\omega\eta_{44}}{c_{44}}\right) = \frac{1}{2}\frac{k}{Q_s} \tag{6.30}$$

where

$$Q_s = Q_{44} = \frac{c_{44}}{\omega\eta_{44}}.$$

Similarly, the longitudinal wave attenuation is

$$\alpha_l = \frac{1}{2}\frac{k}{Q_l}, \tag{6.31}$$

with

$$Q_l = Q_{11} = \frac{c_{11}}{\omega\eta_{11}}.$$

According to Part A.5 of Appendix 2, yttrium aluminum garnet crystals very nearly satisfy the elastic isotropy condition (6.13). From Section E of Chapter 3

$$c_{44} = 11.5 \times 10^{10} \text{ newtons/m}^2$$

and

$$\eta_{44} = 37.2 \times 10^{-5} \text{ newton sec/m}^2$$

for this material, and the shear wave acoustic Q in (6.30) is therefore

$$Q_s = \frac{c_{44}}{\omega \eta_{44}} = \frac{0.309 \times 10^{15}}{\omega} .$$ (6.32)

At 100 MHz ($\omega = 6.28 \times 10^8$)

$$Q_s = 0.492 \times 10^6$$

and at 1000 MHz

$$Q_s = 0.492 \times 10^5.$$

These are values that are achieved by only a very few single crystal materials. A more typical value for isotropic materials is

$$Q_s = 3.90 \times 10^3$$

at 1000 MHz, for fused silica (Fig. 3.11).

D. TRANSMISSION LINE MODEL FOR ISOTROPIC SOLIDS

Up to this point only freely propagating plane waves in source-free media have been considered. The important problem of wave excitation by distributed sources has been ignored. As in electromagnetism, this problem is treated very conveniently by means of an electrical transmission line model.

In an isotropic medium the propagation direction can be taken as the z axis without any loss of generality. Acoustic damping will be ignored initially and introduced at a later point. With these assumptions, the first acoustic field equation (4.20) has a matrix representation

$$
\begin{bmatrix}
0 & 0 & 0 & 0 & \dfrac{\partial}{\partial z} & 0 \\
0 & 0 & 0 & \dfrac{\partial}{\partial z} & 0 & 0 \\
0 & 0 & \dfrac{\partial}{\partial z} & 0 & 0 & 0
\end{bmatrix}
\begin{bmatrix}
T_1 \\ T_2 \\ T_3 \\ T_4 \\ T_5 \\ T_6
\end{bmatrix}
= \rho \frac{\partial}{\partial t}
\begin{bmatrix}
v_x \\ v_y \\ v_z
\end{bmatrix}
-
\begin{bmatrix}
F_x \\ F_y \\ F_z
\end{bmatrix}
$$ (6.33)

The second field equation (4.21), with $\tau = 0$, is modified by multiplying with c and then has the matrix representation

$$
\begin{bmatrix}
c_{11} & c_{12} & c_{12} & 0 & 0 & 0 \\
c_{12} & c_{11} & c_{12} & 0 & 0 & 0 \\
c_{12} & c_{12} & c_{11} & 0 & 0 & 0 \\
0 & 0 & 0 & c_{44} & 0 & 0 \\
0 & 0 & 0 & 0 & c_{44} & 0 \\
0 & 0 & 0 & 0 & 0 & c_{44}
\end{bmatrix}
\begin{bmatrix}
0 \\
0 \\
\dfrac{\partial v_z}{\partial z} \\
\dfrac{\partial v_y}{\partial z} \\
\dfrac{\partial v_x}{\partial z} \\
0
\end{bmatrix}
= \frac{\partial}{\partial t}
\begin{bmatrix}
T_1 \\
T_2 \\
T_3 \\
T_4 \\
T_5 \\
T_6
\end{bmatrix}
\tag{6.34}
$$

The derivatives $\partial/\partial x$ and $\partial/\partial y$ are zero in (6.33) and (6.34) because uniform plane wave fields propagating along z are functions of the z coordinate only. For the same reason the body force components F_x, F_y, F_z in (6.33) are assumed to be functions of z alone.

Equations (6.33) and (6.34) break down into three independent sets involving stress components T_5, T_4, T_3; namely,

$$
\frac{\partial}{\partial z} T_5 = \rho \frac{\partial v_x}{\partial t} - F_x
$$

$$
c_{44} \frac{\partial v_x}{\partial z} = \frac{\partial}{\partial t} T_5
$$

$$
\tag{6.35}
$$

$$
\frac{\partial}{\partial z} T_4 = \rho \frac{\partial v_y}{\partial t} - F_y
$$

$$
c_{44} \frac{\partial v_y}{\partial z} = \frac{\partial}{\partial t} T_4
$$

$$
\tag{6.36}
$$

and

$$
\frac{\partial}{\partial z} T_3 = \rho \frac{\partial v_z}{\partial t} - F_z
$$

$$
c_{11} \frac{\partial v_z}{\partial z} = \frac{\partial}{\partial t} T_3 .
$$

$$
\tag{6.37}
$$

In addition, there are three individual equations

$$\frac{\partial}{\partial t} T_1 = c_{12} \frac{\partial v_z}{\partial z}$$

$$\frac{\partial}{\partial t} T_2 = c_{12} \frac{\partial v_z}{\partial z}$$

$$\frac{\partial}{\partial t} T_6 = 0$$

which specify the remaining three stress components. When $F_x = F_y = F_z = 0$ the equations (6.35), (6.36), and (6.37) are satisfied by the traveling wave solutions (6.17), (6.18), and (6.22), respectively. The stress components T_1 and T_2 in (6.37) are the "extra" stress components associated with the compressional wave (Example 1(b) in Chapter 4) and do not take an active part in the wave motion.

Each of the sets of equations (6.35)–(6.37) is equivalent to the electrical transmission line equations

$$\frac{\partial V}{\partial z} = -L \frac{\partial I}{\partial t} + v_s$$

$$\frac{\partial I}{\partial z} = -C \frac{\partial V}{\partial t} ,$$

(6.38)

with the following identifications

Negative Stress	$-T_J$	$\longrightarrow$ Voltage	V
Particle Velocity	v_i	$\longrightarrow$ Current	I
Body Force per	F_i	$\longrightarrow$ Source Voltage	v_s
Unit Volume		per Unit Length	
Mass per	ρ	$\longrightarrow$ Inductance	L
Unit Volume		per Unit Length	
Inverse	c_{JJ}^{-1}	$\longrightarrow$ Capacitance	C
Stiffness		per Unit Length	
Coefficient			

This leads to the representation of an acoustic plane wave by the equivalent circuit of Fig. (6.2a), corresponding to acoustic equations written in the form

$$\frac{\partial}{\partial z} (-T_J) = -\rho \frac{\partial}{\partial t} v_i + F_i$$

$$\frac{\partial}{\partial z} v_i = -c_{JJ}^{-1} \frac{\partial}{\partial t} (-T_J).$$

(6.39)

These are exactly parallel in form to the electrical equations (6.38).

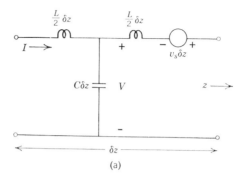

(a)

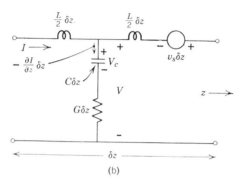

(b)

FIGURE 6.2. (a) Lossless transmission line section of infinitesimal length.
(b) Introduction of loss into the transmission line model for acoustic propagation in a viscously damped medium.

Dissipation is introduced into (6.39) by means of the substitution

$$c_{JJ} \rightarrow c_{JJ} + \eta_{JJ} \frac{\partial}{\partial t},$$

from (4.3). This gives

$$\frac{\partial}{\partial z}(-T_J) = -\rho \frac{\partial}{\partial t} v_i + F_i$$

$$\left(c_{JJ} + \eta_{JJ} \frac{\partial}{\partial t}\right) \frac{\partial v_i}{\partial z} = -\frac{\partial}{\partial t}(-T_J). \tag{6.40}$$

These equations can be modeled electrically by introducing a conductance
$G\,\delta z$ in series with the shunt arm of the transmission line (Fig. 6.2b). For a
line section of length δz the current in the shunt element is $-(\partial I/\partial z)\,\delta z$ and

the voltage across the capacitor $C\,\delta z$ is

$$V_c = V - (G\,\delta z)^{-1}\left(-\frac{\partial I}{\partial z}\,\delta z\right).$$

From this,

$$-\frac{\partial I}{\partial z}\,\delta z = C\,\delta z\,\frac{\partial V_c}{\partial t} = C\,\frac{\partial V}{\partial t}\,\delta z + \frac{C}{G}\,\frac{\partial^2 I}{\partial t\,\partial z}\,\delta z,$$

and

$$\left(C^{-1} + G^{-1}\frac{\partial}{\partial t}\right)\frac{\partial I}{\partial z} = -\frac{\partial V}{\partial t}.$$

The electrical equations corresponding to Fig. 6.2(b) are therefore

$$\frac{\partial V}{\partial z} = -L\frac{\partial I}{\partial t} + v_s$$

$$\left(C^{-1} + G^{-1}\frac{\partial}{\partial t}\right)\frac{\partial I}{\partial z} = -\frac{\partial V}{\partial t}. \tag{6.41}$$

Comparison of (6.40) and (6.41) shows that the following additional identification is required for transmission line modeling of dissipative acoustic propagation,

Inverse		Conductance	
Damping	$\eta_{JJ}^{-1} \rightarrow$	per Unit	G
Coefficient		Length	

Traveling wave solutions to the source-free form ($v_s = 0$) of the electrical transmission line equations (6.41) have the general form

$$V = Ae^{i(\omega t \mp kz)}$$

$$I = \pm\frac{A}{Z_0}\,e^{i(\omega t \mp kz)}, \tag{6.42}$$

where

$$k = \omega\left(\frac{LC}{1 + i\omega C/G}\right)^{1/2},$$

and

$$Z_0 = \left(\frac{L}{C}\right)^{1/2}(1 + i\omega C/G)^{1/2}$$

is the characteristic impedance of the line. The upper and lower signs refer to waves propagating in the $+z$ and $-z$ directions, respectively. The general expression for complex power flow (Section 5.E) in the $+z$ direction is

$$P = P_{AV} + iP_X = \tfrac{1}{2}VI^* = \frac{1}{2}\frac{|V|^2}{Z_0^*} = \tfrac{1}{2}Z_0\,|I|^2. \tag{6.43}$$

For pure traveling waves on a lossless line, the reactive power flow is zero and

$$P = P_{AV} = \pm \frac{1}{2} \frac{|V|^2}{Z_0^*} = \pm \tfrac{1}{2} Z_0 |I|^2. \tag{6.44}$$

By analogy, the traveling acoustic plane wave solutions to the source-free form ($F_i = 0$) of the acoustic equations (6.40) are therefore

$$-T_J = A e^{i(\omega t \mp kz)}$$
$$v_i = \pm \frac{A}{Z_a} e^{i(\omega t \mp kz)}, \tag{6.45}$$

where

$$k = \omega \left(\frac{\rho/c_{JJ}}{1 + \dfrac{i\omega\eta_{JJ}}{c_{JJ}}} \right)^{1/2}$$

and

$$Z_a = (\rho c_{JJ})^{1/2} \left(1 + \frac{i\omega\eta_{JJ}}{c_{JJ}} \right)^{1/2}$$

corresponds to the acoustic characteristic impedance defined for the lossless case in Example 3(b) of Chapter 4. Pursuing the analogy further, the complex power flow density in the *positive-traveling* wave is found to be

$$P = -\tfrac{1}{2} T_J v_i^* = \frac{1}{2} \frac{|T_J|^2}{Z_a^*} = \tfrac{1}{2} |v_i|^2 Z_a . \tag{6.46}$$

This is easily shown to be consistent with the complex Poynting's Theorem of Section 5.F.

Use of the transmission line model greatly simplifies wave scattering calculations of the kind illustrated in Example 3(b) of Chapter 4. (See Problems (8) to (15) at the end of this chapter.) The model may also be applied directly to wave excitation problems, but problems of this kind are more easily solved by transforming to *normal mode* variables.

E. EXCITATION OF PLANE WAVES BY DISTRIBUTED BODY FORCES

The lossless transmission line equations (6.38) may be written in matrix form as

$$\frac{\partial}{\partial z} \begin{bmatrix} 0 & 1 \\ 1 & 0 \end{bmatrix} \begin{bmatrix} V \\ I \end{bmatrix} = -\frac{\partial}{\partial t} \begin{bmatrix} C & 0 \\ 0 & L \end{bmatrix} \begin{bmatrix} V \\ I \end{bmatrix} + \begin{bmatrix} 0 \\ v_s \end{bmatrix} \tag{6.47}$$

Substitution of the traveling wave solution

$$V_m e^{i(\omega t - kz)}$$
$$I_m e^{i(\omega t - kz)}$$

into (6.47), with $v_s = 0$, leads to the matrix equation

$$\begin{bmatrix} 0 & 1 \\ 1 & 0 \end{bmatrix} \begin{bmatrix} V_m \\ I_m \end{bmatrix} = \frac{\omega}{k} \begin{bmatrix} C & 0 \\ 0 & L \end{bmatrix} \begin{bmatrix} V_m \\ I_m \end{bmatrix}. \tag{6.48}$$

Solutions to (6.48) are

$$\begin{bmatrix} V_+ \\ I_+ \end{bmatrix} = \begin{bmatrix} Z_0 \\ 1 \end{bmatrix}; \quad \left(\frac{\omega}{k}\right)_+ = \frac{1}{(LC)^{1/2}} \tag{6.49}$$

$$\begin{bmatrix} V_- \\ I_- \end{bmatrix} = \begin{bmatrix} -Z_0 \\ 1 \end{bmatrix}; \quad \left(\frac{\omega}{k}\right)_- = -\frac{1}{(LC)^{1/2}}, \tag{6.50}$$

where $Z_0 = (L/C)^{1/2}$ is the characteristic impedance. These solutions correspond to waves propagating toward $+z$ and $-z$ respectively.

If the matrix scalar product of (6.47) is taken with (6.49) and (6.50) in turn, two scalar equations

$$\frac{\partial}{\partial z}\left([V_m \ I_m]\begin{bmatrix} 0 & 1 \\ 1 & 0 \end{bmatrix}\begin{bmatrix} V \\ I \end{bmatrix}\right) = -\frac{\partial}{\partial t}\left([V_m \ I_m]\begin{bmatrix} C & 0 \\ 0 & L \end{bmatrix}\begin{bmatrix} V \\ I \end{bmatrix}\right) + [V_m \ I_m]\begin{bmatrix} 0 \\ v_s \end{bmatrix}$$

$$m = +, -$$

are obtained. After performing the matrix multiplications, one finds that

$$\frac{\partial}{\partial z}(V + IZ_0) = -\frac{1}{V_p}\frac{\partial}{\partial t}(V + IZ_0) + v_s \tag{6.51}$$

and

$$\frac{\partial}{\partial z}(V - IZ_0) = \frac{1}{V_p}\frac{\partial}{\partial t}(V - IZ_0) + v_s, \tag{6.52}$$

where

$$V_p = 1/(LC)^{1/2}$$

is the phase velocity. This is the normal mode form of the transmission line equations. The *normal mode amplitudes*, defined as

$$a_+(z, t) = V(z, t) + I(z, t)Z_0$$
$$a_-(z, t) = V(z, t) - I(z, t)Z_0, \tag{6.53}$$

correspond to waves propagating along $+z$ and $-z$ respectively. From (6.42),

$$I = +\frac{V}{Z_0} \qquad (a_- = 0) \tag{6.54}$$

for a positive traveling wave and

$$I = - \frac{V}{Z_0} \qquad (a_+ = 0) \qquad (6.55)$$

for a negative traveling wave. Because the positive and negative propagating waves are separated in this manner, normal mode transmission line equations are especially useful for calculating the excitation of waves by distributed sources.

For voltages and currents varying as $e^{i\omega t}$, the normal mode equations (6.51) and (6.52) become

$$\left(\frac{\partial}{\partial z} + \frac{i\omega}{V_p}\right) a_+ = v_s(z)$$
$$\left(\frac{\partial}{\partial z} - \frac{i\omega}{V_p}\right) a_- = v_s(z). \qquad (6.56)$$

These are very easily integrated to find the waves excited by a continuous distribution of voltage sources. Introduction of the auxiliary variables d_+ and d_-, defined by

$$a_+(z) = d_+(z) e^{-i(\omega/V_p)z}$$
$$a_-(z) = d_-(z) e^{i(\omega/V_p)z}, \qquad (6.57)$$

converts (6.56) to

$$\frac{\partial}{\partial z} d_+ = v_s(z) e^{i(\omega/V_p)z} \qquad (6.58)$$

$$\frac{\partial}{\partial z} d_- = v_s(z) e^{-i(\omega/V_p)z}. \qquad (6.59)$$

Suppose that the distribution of voltage sources extends only over a region bounded by $z = z_1$ on the left and $z = z_2$ on the right (Fig. 6.3). Integration

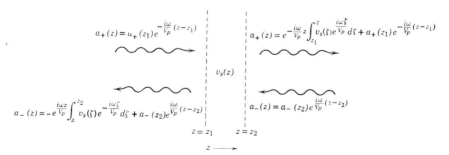

FIGURE 6.3.   Plane wave excitation by a source distribution $v_s(z)$.

of (6.58) from z_1 to any point z on the right-hand side of z_1 gives

$$d_+(z) - d_+(z_1) = \int_{z_1}^{z} v_s(\zeta) e^{i(\omega/V_p)\zeta}\, d\zeta, \qquad z > z_1,$$

where a dummy integration variable ζ has been introduced. Similarly, the integral of (6.59) from $z < z_2$ to z_2 is

$$d_-(z_2) - d_-(z) = \int_{z}^{z_2} v_s(\zeta) e^{-i(\omega/V_p)\zeta}\, d\zeta, \qquad z < z_2.$$

The normal mode amplitudes $a_+(z)$ and $a_-(z)$ are now obtained by means of the relations (6.57). That is,

$$a_+(z) = e^{-i(\omega/V_p)z} \int_{z_1}^{z} v_s(\zeta) e^{i(\omega/V_p)\zeta}\, d\zeta + a_+(z_1) e^{-i(\omega/V_p)(z-z_1)}$$
$$z > z_1. \quad (6.60)$$

$$a_-(z) = -e^{i(\omega/V_p)z} \int_{z}^{z_2} v_s(\zeta) e^{-i(\omega/V_p)\zeta}\, d\zeta + a_-(z_2) e^{i(\omega/V_p)(z-z_2)}$$
$$z < z_2. \quad (6.61)$$

These expressions have a very simple physical interpretation. The integrals represent the wave amplitudes excited by the sources and the other terms are waves propagating into the source region from the outside (Fig. 6.3). If the medium is *unbounded* and there are no sources outside of the region $z_1 < z < z_2$, then

$$a_+(z_1) = 0$$
$$a_-(z_2) = 0 \qquad\qquad (6.62)$$

and the solutions are

$$a_+(z) = e^{-i(\omega/V_p)z} \int_{z_1}^{z} v_s(\zeta) e^{i(\omega/V_p)\zeta}\, d\zeta, \qquad z > z_1, \quad (6.63)$$

$$a_-(z) = -e^{i(\omega/V_p)z} \int_{z}^{z_2} v_s(\zeta) e^{-i(\omega/V_p)\zeta}\, d\zeta, \qquad z < z_2. \quad (6.64)$$

When the medium is *bounded*, waves radiated outward from the sources are reflected back again, and the boundary conditions (6.62) no longer apply. A problem of this kind is considered in Example 13 of Chapter 8.

The time-harmonic normal mode equations (6.56) can be applied directly to acoustic problems by using the analogy derived in Section D. For longitudinal waves

$$\left(\frac{\partial}{\partial z} + \frac{i\omega}{V_l}\right) a_+ = F_z(z)$$
$$\left(\frac{\partial}{\partial z} - \frac{i\omega}{V_l}\right) a_- = F_z(z) \qquad (6.65)$$

where

$$a_+ = -T_3 + v_z Z_l \tag{6.66}$$
$$a_- = -T_3 - v_z Z_l$$

are the acoustic normal mode amplitudes,

$$V_l = \left(\frac{c_{11}}{\rho}\right)^{1/2} \tag{6.67}$$

is the longitudinal wave phase velocity, and

$$Z_l = (\rho c_{11})^{1/2} \tag{6.68}$$

is the longitudinal wave impedance. Similar equations are obtained for the two shear waves.

By analogy with (6.60) and (6.61), solutions to (6.65) are

$$a_+(z) = e^{-i(\omega/V_l)z} \int_{z_1}^{z} F_z(\zeta) e^{i(\omega/V_l)\zeta} \, d\zeta + a_+(z_1) e^{-i(\omega/V_l)(z-z_1)}, \quad z > z_1 \tag{6.69}$$

$$a_-(z) = -e^{i(\omega/V_l)z} \int_{z}^{z_2} F_z(\zeta) e^{-i(\omega/V_l)\zeta} \, d\zeta + a_-(z_2) e^{i(\omega/V_l)(z-z_2)}, \quad z < z_2. \tag{6.70}$$

Once $a_+(z)$ and $a_-(z)$ have been calculated, the stress and particle velocity fields are found by inverting relations (6.66). That is,

$$-T_3(z) = \frac{a_+(z) + a_-(z)}{2} \tag{6.71}$$

$$v_z(z) = \frac{a_+(z) - a_-(z)}{2Z_l} \tag{6.72}$$

EXAMPLE 3.   An unbounded, lossless, isotropic solid is excited by a z-directed time-harmonic body force distribution that is uniform along the x and y directions and has a constant value F in the region $|z| < |l|$, being zero at all other points (Fig. 6.4). That is,

$$F_z(z) = \begin{cases} F, & |z| < |l| \\ 0, & |z| > |l|. \end{cases} \tag{6.73}$$

Since the medium is unbounded, boundary conditions

$$a_+(-l) = a_-(+l) = 0$$

are applicable. The solutions obtained by substituting (6.73) into (6.69) and (6.70),

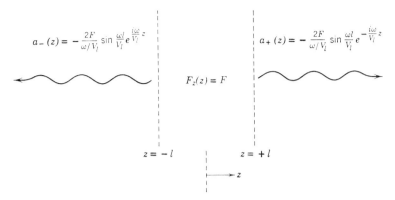

$$a_-(z) = -\frac{2F}{\omega/V_l}\sin\frac{\omega l}{V_l}e^{\frac{i\omega}{V_l}z}$$

$$a_+(z) = -\frac{2F}{\omega/V_l}\sin\frac{\omega l}{V_l}e^{-\frac{i\omega}{V_l}z}$$

$$F_z(z) = F$$

$$z = -l \qquad\qquad z = +l$$

$$\longrightarrow z$$

FIGURE 6.4.   Excitation of compressional waves by a uniform body force distribution between $-l$ and $+l$.

with $z_1 = -l$ and $z_2 = +l$, are then

$$a_+(z) = -\frac{iF}{\omega/V_l}e^{-i(\omega/V_l)z}\left(e^{i(\omega/V_l)z} - e^{-i(\omega/V_l)l}\right)$$

$$a_-(z) = -\frac{iF}{\omega/V_l}e^{i(\omega/V_l)z}\left(e^{-i(\omega/V_l)l} - e^{-i(\omega/V_l)z}\right) \qquad (6.74)$$

inside the excitation region ($|z| < l$). Outside the excitation region

$$a_+(z) = \frac{2F}{\omega/V_l}\sin\frac{\omega l}{V_l}e^{-i(\omega/V_l)z} \qquad z > +l \qquad (6.75)$$

$$a_-(z) = -\frac{2F}{\omega/V_l}\sin\frac{\omega l}{V_l}e^{i(\omega/V_l)z} \qquad z < -l,$$

as illustrated in Fig. 6.4.

According to (6.75) the acoustic radiation is zero when

$$\frac{\omega l}{V_l} = n\pi \qquad (6.76)$$

$$n = 1, 2, 3, \cdots$$

This is an interference effect. The integral expressions in (6.69) and (6.70) may be interpreted as superpositions of the waves radiated by a series of localized sources of strength $F_z(\zeta)\,d\zeta$. When $F_z(\zeta) = F$, all these localized sources have the same amplitude and phase, and two sources separated by a distance $\zeta_2 - \zeta_1$ radiate positive-traveling waves that are separated by a time-phase difference $\omega/V_l(\zeta_2 - \zeta_1)$ (Fig. 6.5). When the condition (6.76) is satisfied, the contributions of these localized sources all cancel out. This is easily demonstrated. If $n = 1$ the waves radiated by sources at $\zeta_1 = -l$ and $\zeta_2 = 0$ are 180° out of phase and therefore cancel. The same holds true for sources at $\zeta_1 = -l + \delta$ and $\zeta_2 = 0 + \delta$. In this way all of the sources

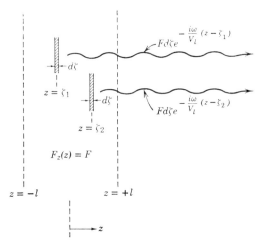

FIGURE 6.5. Positive-traveling waves radiated by two localized sources of strength $F\,d\zeta$.

can be grouped into canceling pairs. Arguments of the same kind apply to the cases $n = 2, 3$, etc. in (6.76).

EXAMPLE 4. Since the solutions (6.69) and (6.70) are formulated for a source distribution $F(z)$ that is arbitrary within the region $|z| < |l|$ and zero elsewhere, they may be applied directly to other problems. If

$$F(z) = \begin{cases} F \sin Kz, & |z| < |l| \\ 0, & |z| > |l| \end{cases} \tag{6.77}$$

and the medium is unbounded, the positive-traveling wave to the right of the excitation region is

$$a_+(z) = e^{-i(\omega/V_l)z} \int_{-l}^{l} F \sin K\zeta\, e^{i(\omega/V_l)\zeta}\, d\zeta \tag{6.78}$$

from (6.69). This is easily integrated by converting the sine function to a sum of exponential terms, giving

$$a_+(z) = F\frac{e^{-i(\omega/V_l)z}}{i}\left(\frac{\sin\left(\dfrac{\omega}{V_l} + K\right)l}{\left(\dfrac{\omega}{V_l} + K\right)} - \frac{\sin\left(\dfrac{\omega}{V_l} - K\right)l}{\left(\dfrac{\omega}{V_l} - K\right)}\right), \qquad z > +l \tag{6.79}$$

The corresponding negative-traveling wave solution is

$$a_-(z) = F\frac{e^{i(\omega/V_l)z}}{i}\left(\frac{\sin\left(\dfrac{\omega}{V_l} + K\right)l}{\left(\dfrac{\omega}{V_l} + K\right)} - \frac{\sin\left(\dfrac{\omega}{V_l} - K\right)l}{\left(\dfrac{\omega}{V_l} - K\right)}\right), \qquad z < -l. \tag{6.80}$$

$$z < -l$$

Solutions (6.79) and (6.80) are of particular interest when

$$K = \pm \frac{\omega}{V_l}.$$

Under these conditions, one or the other of the exponential terms in the source distribution

$$\sin Kz = \frac{e^{iKz} - e^{-iKz}}{2i}$$

is phase-synchronized with the positive- and negative-traveling waves. In terms of the physical picture given by Fig. 6.5 this means that all of the localized sources radiate in phase. When $\omega/V_l = -K$, for example, the first terms within the brackets in (6.79) and (6.80) become

$$\left(\frac{\sin \left(\frac{\omega}{V_l} + K \right) l}{\left(\frac{\omega}{V_l} + K \right)} \right)_{\frac{\omega}{V_l} + K = 0} = l.$$

The amplitudes of the radiated waves can thus be increased indefinitely by extending the length of the excitation region. In (6.75), by contrast, there is a maximum radiated amplitude $2F/(\omega/V_l)$ because of the interference effects.

Since the body force distribution $F_z(z)$ is zero outside the region $z_1 < z < z_2$, the limits of integration in (6.69) and (6.70) may be extended to $-\infty$ and $+\infty$; that is

$$a_+(z) = e^{-i(\omega/V_l)z} \int_{-\infty}^{\infty} F_z(\zeta) e^{i(\omega/V_l)\zeta} d\zeta + a_+(z_1) e^{-i(\omega/V_l)(z-z_1)} \qquad (6.81)$$

for $z > z_1$, and

$$a_-(z) = -e^{i(\omega/V_l)z} \int_{-\infty}^{\infty} F_z(\zeta) e^{-i(\omega/V_l)\zeta} d\zeta + a_-(z_2) e^{i(\omega/V_l)(z-z_2)} \qquad (6.82)$$

for $z < z_2$. This shows that the strengths of the radiated waves are proportional to the Fourier transform of the source distribution function $F_z(z)$. They can therefore be easily evaluated with the aid of Fourier transform tables.

PROBLEMS

1. Show that the isotropic compliance and stiffness constants satisfy relations

$$s_{11} = \frac{c_{11} + c_{12}}{(c_{11} - c_{12})(c_{11} + 2c_{12})}$$

$$s_{44} = \frac{1}{c_{44}}$$

$$s_{12} = s_{11} - \frac{s_{44}}{2}$$

and

$$c_{11} = \frac{s_{11} + s_{12}}{(s_{11} - s_{12})(s_{11} + 2s_{12})}$$

$$c_{44} = \frac{1}{s_{44}}$$

$$c_{12} = c_{11} - 2c_{44}.$$

Express the isotropic compliances in terms of the Lamé constants defined in Section B.

2. Using (5.51), derive the physical realizability conditions

$$c_{44} > 0, \qquad c_{11} > |c_{12}|, \qquad c_{11} + 2c_{12} > 0$$

for an isotropic solid. Show that the realizability conditions for the compliance constants are

$$s_{44} > 0, \quad s_{11} > |s_{12}|, \quad s_{11} + 2s_{12} > 0.$$

3. An isotropic parallelepiped is subject to a static pressure P on its ends and is unconstrained on its sides.

Show that the strain field is

$$S_1 = s_{12} \frac{P}{A}$$

$$S_2 = s_{12} \frac{P}{A}$$

$$S_3 = s_{11} \frac{P}{A}.$$

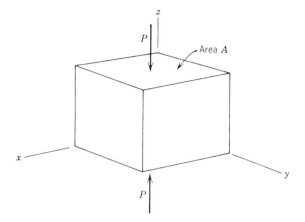

4. *Poisson's ratio* for an isotropic solid is defined in terms of the strains in Problem 3 as

$$\sigma = -\frac{S_1}{S_3} = -\frac{S_2}{S_3},$$

and *Young's modulus* is defined by

$$E = \frac{T_3}{S_3}.$$

Express σ and E in terms of c_{11}, c_{44} and, alternatively, in terms of the Lamé constants λ, μ. Show that σ must lie between -1 and $+\frac{1}{2}$.

5. In a solid subjected to the small signal strain field

$$S_1 = S_2 = S_3 = -|S|,$$

each volume element experiences a fractional change in volume equal to $-3|S|$ (Problem 3, Chapter 4), but no change in shape. This type of strain is called *pure dilatation*. Show that the corresponding stress field in an isotropic solid is defined by the hydrostatic pressure

$$T_1 = T_2 = T_3 = -(c_{11} + 2c_{12})|S| = -3K|S|,$$

where K is the *bulk modulus of elasticity*.

6. It has been shown that the elastic properties of an isotropic solid can be completely specified by two constants. Any two of the constants

$$c_{11}, c_{12} = \lambda, c_{44} = \mu, \sigma, E, K$$

can be used for this purpose. Show that

$$c_{11} = \frac{E(1 - \sigma)}{(1 + \sigma)(1 - 2\sigma)}$$

$$c_{12} = \frac{E\sigma}{(1 + \sigma)(1 - 2\sigma)}$$

$$c_{44} = \frac{E}{2(1 + \sigma)}$$

$$K = \frac{E}{3(1 - 2\sigma)}.$$

7. Using (6.14), express the Christoffel equation for an isotropic solid in the symbolic form

$$c_{11}\mathbf{k}(\mathbf{k} \cdot \mathbf{v}) - c_{44}\mathbf{k} \times (\mathbf{k} \times \mathbf{v}) = \rho\omega^2\mathbf{v}.$$

Show that this reduces to

$$c_{11}k^2\mathbf{v}_l = \rho\omega^2\mathbf{v}_l$$

for the compressional wave solution

$$\mathbf{v}_l = A_l\mathbf{k}e^{i(\omega t - \mathbf{k}\cdot\mathbf{r})},$$

and reduces to

$$c_{44}k^2\mathbf{v}_s = \rho\omega^2\mathbf{v}_s$$

for the shear wave solution

$$\mathbf{v}_s = A_s(\mathbf{k} \times \hat{\mathbf{a}})e^{i(\omega t - \mathbf{k}\cdot\mathbf{r})}$$

where $\mathbf{k}$ and $\hat{\mathbf{a}}$ have *arbitrary* directions.

8. A superposition of positive- and negative-traveling compressional waves

$$T_{3+} = Ae^{i(\omega t - kz)}$$

$$T_{3-} = Be^{i(\omega t + kz)},$$

is present in an isotropic solid. Show that the acoustic impedance at any point z is

$$Z(z) = -\frac{T_3}{v_z} = Z_a\left(\frac{1 + R_T(z)}{1 - R_T(z)}\right),$$

where

$$Z_a = (\rho c_{11})^{1/2}$$

is the characteristic acoustic impedance and

$$R_T(z) = \frac{T_{3-}(z)}{T_{3+}(z)} = \frac{B}{A}e^{i2kz}$$

is the stress reflection coefficient. Show also that

$$R_T(z) = \frac{Z(z) - Z_a}{Z(z) + Z_a},$$

and that these expressions apply to shear waves when the appropriate characteristic impedance Z_a is used.

9. A compressional wave is incident from the left on a rigidly bonded plane interface between two isotropic solids.

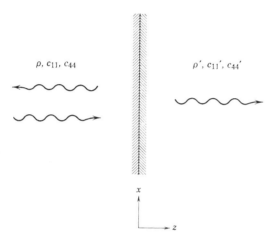

Use the acoustic boundary conditions at the interface to evaluate the acoustic impedance looking to the right at $z = 0$. From this, find the reflection coefficient $R_T(0)$. Compare this analysis with the method used in Example 3(b) of Chapter 4.

10. Starting from the impedance formula given in Problem 8, show that the acoustic impedances at any two points z_1 and z_2 are related by the expression

$$Z(z_1) = Z_a \left[\frac{Z(z_2) \cos k(z_2 - z_1) + iZ_a \sin k(z_2 - z_1)}{Z_a \cos k(z_2 - z_1) + iZ(z_2) \sin k(z_2 - z_1)} \right].$$

11. Two isotropic solids with material parameters ρ, c_{11}, c_{44} and ρ'', c_{11}'', c_{44}'' are bonded together by an isotropic slab with parameters ρ', c_{11}', c_{44}' and thickness l.

Use the formula derived in Problem 10 and the boundary conditions at $z = 0, l$ to show that the compressional wave impedance looking to the right at $z = 0$ is

$$Z(0) = (\rho' c_{11}')^{1/2} \left[\frac{(\rho'' c_{11}'')^{1/2} \cos k'l + i(\rho' c_{11}')^{1/2} \sin k'l}{(\rho' c_{11}')^{1/2} \cos k'l + i(\rho'' c_{11}'')^{1/2} \sin k'l} \right]$$

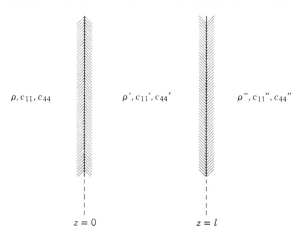

$$\rho, c_{11}, c_{44} \qquad \rho', c_{11}', c_{44}' \qquad \rho'', c_{11}'', c_{44}''$$

$$z = 0 \qquad z = l$$

Obtain an analogous formula for shear waves. If the medium ρ', c_{11}', c_{44}' is lossless, the *average* power flows at $z = 0$ and $z = l$ are equal. Use this principle to calculate the amplitude of the transmitted wave in the right-hand medium. Compare with the calculation in Problem 12 of Chapter 4.

12. Show that the input impedance at $z = 0$ in Problem 11 can be matched to the characteristic impedance of the left-hand medium by choosing

$$k'l = (2n + 1)\frac{\pi}{2}$$

$$(\rho'c_{11}')^{1/2} = \sqrt{(\rho c_{11})^{1/2}(\rho''c_{11}'')^{1/2}}$$

where n is any integer. Demonstrate that this also presents a matched imped-ance at $z = l$ to a wave incident from the right. Sections of this kind are called *anti-reflection layers*, or $(2n + 1)\lambda/4$ *impedance-matching transformers*.

13. Choosing $k'l$ and $(\rho'c_{11}')^{1/2}$ as defined in Problem 12, obtain an expres-sion for the normalized input impedance

$$Z_{IN} = \frac{Z(0)}{(\rho c_{11})^{1/2}} = \frac{Z(0)}{Z_a}$$

as a function of the impedance ratio

$$\frac{Z_a''}{Z_a} = \frac{(\rho''c_{11}'')^{1/2}}{(\rho c_{11})^{1/2}}$$

and

$$k'l = \omega\left(\frac{\rho'}{c_{11}'}\right)^{1/2}l.$$

Calculate and graph the normalized input impedance as a function of $k'l$ for impedance ratios 0.1, 0.5, 2, and 10.

14. In Problem 11, take $k'l = \pi/2$. The left-hand medium is fused silica, the central medium is heavy silicate flint glass, and the right-hand medium is light borate crown glass. Using material constants from Appendix 2, calculate the reflection coefficient $R_T(0)$ for a compressional wave incident from the left. Repeat the calculation for lucite on the left, polystyrene in the middle, and polyethylene on the right.

15. Consider the configuration

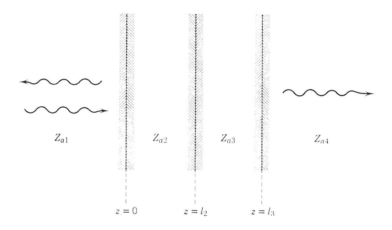

Z_{a1} Z_{a2} Z_{a3} Z_{a4}

$z = 0$ $z = l_2$ $z = l_3$

where l_2 is one-quarter wavelength in medium 2 and $l_3 - l_2$ is one-quarter wavelength in medium 3. Show that the impedance looking to the right at $z = 0$ is

$$Z(0) = \left(\frac{Z_{a2}}{Z_{a3}}\right)^2 Z_{a4}.$$

Calculate the reflection coefficient $R_T(0)$ for a wave incident from the left if medium 1 is light borate crown glass, medium 2 is lucite, medium 3 is polyethylene, and medium 4 is lucite. Compare with the reflection at a direct boundary between medium 1 and medium 4. Use material constants from Appendix 2.

16. Using either direct integration or a table of Fourier transforms, solve (6.65) with a driving force distribution

$$F_z(z) = \begin{cases} \dfrac{z}{l}\, F, & |z| < l \\ 0, & |z| > l \end{cases}$$

Graph the amplitudes of the positive- and negative-traveling radiated waves as a function of $\omega l / V_l$. Compare with the solution given by (6.75).

REFERENCES

Acoustic Wave Equation and Christoffel Equation

1. L. D. Landau and E. M. Lifshitz, *Theory of Elasticity*, pp. 98–100, Pergamon, New York, 1970.
2. W. P. Mason, *Physical Acoustics and the Properties of Solids*, pp. 368–369, van Nostrand, New York, 1958.
3. G. Nadeau, *Introduction to Elasticity*, pp. 209–212, Holt, Rinehart and Winston, New York, 1964.

Compressional and Shear Plane Waves

4. Reference 2, pp. 13–16.
5. Reference 3, pp. 212–215.

Transmissiom Line Theory and the Acoustic Analogy

6. Reference 2, pp. 30, 181–187.
7. A. A. Oliner, "Microwave Methods for Guided Elastic Waves", *IEEE Trans. on Microwave Theory and Techniques*, **MTT-17**, pp. 813–818 (1969).
8. S. Ramo, J. R. Whinnery, and T. van Duzer, *Fields and Waves in Communication Electronics*, pp. 23–58, Wiley, New York, 1965.

Normal Mode Equations

9. C. C. Johnson, *Field and Wave Electrodynamics*, pp. 124–126, McGraw-Hill, New York, 1965.
10. W. H. Louisell, *Coupled Mode and Parametric Electronics*, pp. 18–25, Wiley, New York, 1960.

Chapter 7

ACOUSTIC PLANE WAVES IN ANISOTROPIC SOLIDS

A. CRYSTAL SYMMETRIES

It was noted in Section 3.A that the general form of the compliance and stiffness matrices may be deduced by considering microscopic symmetry properties of the medium. Two general kinds of argument may be used. If two particular directions in a medium are symmetrically equivalent, then it can be said that identical stresses along these two directions must give rise to the same amount of strain. This shows that certain compliance constants must be equal. A more sophisticated argument is the following. If the medium itself is symmetric with respect to a particular transformation of coordinates, then the compliance and stiffness matrices must themselves be unchanged by the same transformation. This second technique was used in Section 6.B to prove that (6.12) is isotropic in its properties, and it is the usual approach to symmetry arguments for the anisotropic case. Symmetries for anisotropic media are, however, much more complicated than for the isotropic case and, before proceeding, it will be necessary to briefly classify these symmetries.

There are two basic kinds of symmetry operations that transform a crystal lattice into itself:

1. Translations, which displace the lattice as a whole,
2. Point transformations, which leave at least one point in the lattice unchanged.

Since the compliance and stiffness matrices relate stress and strain fields at the *same point* in a crystal, their symmetry properties can be obtained from the point symmetry transformations alone. These transformations include *rotations, reflections, inversions, rotation-inversions,* and *rotation-reflections.* The last category is equivalent to rotation-inversions and need not be considered further.

Rotation symmetries are defined in terms of the smallest rotation under which the lattice is symmetric. An *n*-fold rotation symmetry is defined as an operation with a minimum rotation angle $2\pi/n$; only 2-fold, 3-fold, 4-fold, and 6-fold rotations may occur among the symmetry operations of a crystal lattice.[†] In crystallography, a 3-fold axis is marked with a small triangle, and a rotation through $2\pi/3 = 120°$ is denoted by the symbol *3*. Two repetitions of this operation (that is, a rotation through $4\pi/3 = 240°$) are symbolized by 3^2. Three repetitions are denoted by $3^3 = 1$, where *1* is the identity operation (a rotation that returns the lattice to its original position). This is illustrated as follows:

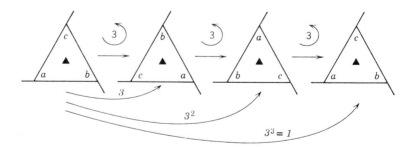

Corners of the triangular object have been lettered in order to mark the rotations, which would not otherwise be distinguishable. Further

† See, for example, References 3 and 4 at the end of the chapter.

applications of the rotation operation *3* will merely repeat the sequence shown.

The above example shows that successive applications of a symmetry operation produce other symmetry operations. The same is true for products of different symmetry operations. Mathematically, this relationship among symmetry operations is described by saying that they form a *group*. For the example given, the symmetry group consists of operations *1*, *3*, and *3²*. All of these operations can be generated by taking successive applications (or powers) of *3*, which is called the *generator of the group*. Since any group can be constructed from its generators, the symmetry of any crystal lattice can therefore be described by a set of group generators rather than listing all the symmetry operators of the group. To specify a particular symmetry group it is not necessary to show a symmetrical object and all of its transformations, as was done above. A simpler method is to construct a set of points which transform into each other under the group operations. This is called an *equivalent point diagram* for the group. The procedure is first to assume a point at some arbitrary position (not located on any symmetry axis or in a symmetry plane) and then to apply *all* symmetry operations of the group. Because of the conditions imposed on the choice of initial point, each symmetry operation will generate a new point. The number of points generated is therefore equal to the number of symmetry operations in the group. For the example given above, the initial point may be placed anywhere except on the 3-fold axis. The point diagram is then:

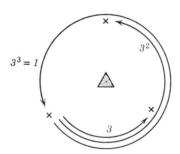

The convention used in these diagrams requires that a point above the plane of the page be indicated by an × and a point below the page by an ○.

Crystal point symmetry groups consist of those collections of rotations, reflections, and rotation-inversions under which the various crystal lattice structures occurring in nature remain invariant. Each crystal symmetry group is called a *class*, and the various classes are grouped into *systems* that will be seen to have certain physical properties in common. Equivalent point diagrams for all groups of this kind are listed in Table 7.1, which gives the generator operations for each group and defines a set of axes X, Y, Z (called *crystal axes*) relative to these operations. *Rotation* operations are indicated in these diagrams by the symbols:

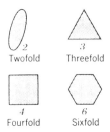

Mirror or reflection operations are denoted by the symbol m or $\bar{m}$. A mirror plane normal to the page is shown as

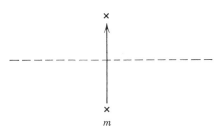

in the equivalent point diagrams, while a mirror plane lying in the page is usually marked $\bar{m}$ and is shown by superposed $\times$'s and $\bigcirc$'s. For example,

Inversion, or reflection through a point ($\mathbf{r} \to -\mathbf{r}$), is denoted by $\bar{I}$. This operation is represented diagramatically as

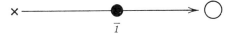

Rotation-inversion operations are denoted by $\bar{2}, \bar{3}, \bar{4}, \bar{6}$ and are indicated in the point diagrams by the symbols

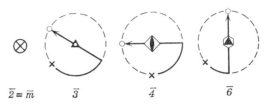

$$\bar{2} = \bar{m} \qquad \bar{3} \qquad \bar{4} \qquad \bar{6}$$

For the cubic groups, equivalent point diagrams are most easily visualized on the surface of a cube and they are shown in this way in the table.

Table 7.1 also lists the international symbols for the crystallographic point groups. Except for $\bar{6}m2$ and the cubic classes, generator operations for any class are given by the first few figures or letters in its international group symbol. *No groups of crystal point symmetries have more than three generators.* Only the generator axes and mirror planes have been shown on the point diagrams. There are, in general, a number of other axes, etc., some of which are indicated by additional entries in the group symbol. These are obvious from inspection of the point diagram. Table 7.2 gives the coordinate transformation matrices corresponding to the generator operations of all the symmetry groups, referred to the standard *crystal axes* X, Y, Z in Table 7.1. These are the basic tools used for deducing the symmetry characteristics of the compliance and stiffness matrices, or any other constitutive matrices for crystal media.

B. SYMMETRY CHARACTERISTICS OF s AND c

According to (3.38) and (3.37), the transformation laws for compliance and stiffness are

$$[s'] = [N][s][M]^{-1} \tag{7.1}$$

and

$$[c'] = [M][c][N]^{-1} \tag{7.2}$$

where the transformation matrices $[M]$ and $[N]$ are calculated from the coordinate transformation matrix $[a]$ by using (3.32) and (3.34). If the coordinate transformation is a *symmetry operation* of the crystal lattice, it follows from the preceding section that

$$[s'] = [s]$$
$$[c'] = [c] \tag{7.3}$$

TABLE 7.1. Point Symmetry Groups for Crystal Lattices

System	Class Symbol [†]	Point Diagram	Generator Operations
Triclinic			
	1		1
	$\overline{1}$		$\overline{1}$
Monoclinic			
	2		2
	m		m
	$\dfrac{2}{m}$		$2, m$
Orthorhombic			
	$2mm$		$2, m$
	222		$2, 2$

TABLE 7.1 (*Cont.*)

Class Symbol †	Point Diagram	Generator operations
Tetragonal		
$\widetilde{mmm}$		m, m, m
4		4
$\bar{4}$		$\bar{4}$
$\widetilde{\frac{4}{m}}$		$4, m$
422		$4, 2$
4mm		$4, m$
$\bar{4}2m$		$\bar{4}, 2$

TABLE 7.1 *(Cont.)*

System	Class Symbol †	Point Diagram	Generator operations
Tetragonal	(4/mmm)		4, m, m
Trigonal	3		3
	($\bar{3}$)		$\bar{3}$
	32		3, 2
	3m		3, m
	($\bar{3}m$)		$\bar{3}$, m
Hexagonal	6		6

TABLE 7.1 *(Cont.)*

System	Class Symbol †	Point Diagram	Generator Operations
Hexagonal	$\overline{6}$		$\overline{6}$
	$\dfrac{6}{m}$		$6, m$
	622		$6, 2$
	$6mm$		$6, m$
	$\overline{6}m2$		$\overline{6}, 2$
	$6/mmm$		$6, m, m$

TABLE 7.1 (*Cont.*)

System	Class Symbol †	Point Diagram	Generator Operations
Cubic			
	23		2, 3
	(m3)		2, 3, m
	432		4, 2
	(m3m)		4, 2, m
	$\overline{4}3m$		$\overline{4}, m$

† The circled crystal classes have inversion symmetry and cannot be piezoelectric (Section 8.E).

System and Class	Generator Matrices [a_{ij}]

Triclinic

1

$$1 \rightarrow \begin{bmatrix} 1 & 0 & 0 \\ 0 & 1 & 0 \\ 0 & 0 & 1 \end{bmatrix}$$

$\bar{1}$

$$\bar{1} \rightarrow \begin{bmatrix} -1 & 0 & 0 \\ 0 & -1 & 0 \\ 0 & 0 & -1 \end{bmatrix}$$

Monoclinic

2

$$2 \rightarrow \begin{bmatrix} -1 & 0 & 0 \\ 0 & 1 & 0 \\ 0 & 0 & -1 \end{bmatrix}$$

m

$$m \rightarrow \begin{bmatrix} 1 & 0 & 0 \\ 0 & -1 & 0 \\ 0 & 0 & 1 \end{bmatrix}$$

$\dfrac{2}{m}$

$$2 \rightarrow \begin{bmatrix} -1 & 0 & 0 \\ 0 & 1 & 0 \\ 0 & 0 & -1 \end{bmatrix}, \quad m \rightarrow \begin{bmatrix} 1 & 0 & 0 \\ 0 & -1 & 0 \\ 0 & 0 & 1 \end{bmatrix}$$

Orthorhombic

$2mm$

$$2 \rightarrow \begin{bmatrix} -1 & 0 & 0 \\ 0 & -1 & 0 \\ 0 & 0 & 1 \end{bmatrix}, \quad m \rightarrow \begin{bmatrix} 1 & 0 & 0 \\ 0 & -1 & 0 \\ 0 & 0 & 1 \end{bmatrix}$$

TABLE 7.2 (*Cont.*)

System and Class	Generator Matrices

Orthorhombic (*Cont.*)

222
$$2 \rightarrow \begin{bmatrix} -1 & 0 & 0 \\ 0 & -1 & 0 \\ 0 & 0 & 1 \end{bmatrix}, \quad 2 \rightarrow \begin{bmatrix} 1 & 0 & 0 \\ 0 & -1 & 0 \\ 0 & 0 & -1 \end{bmatrix}$$

mmm
$$m \rightarrow \begin{bmatrix} -1 & 0 & 0 \\ 0 & 1 & 0 \\ 0 & 0 & 1 \end{bmatrix}, \quad m \rightarrow \begin{bmatrix} 1 & 0 & 0 \\ 0 & -1 & 0 \\ 0 & 0 & 1 \end{bmatrix}, \quad m \rightarrow \begin{bmatrix} 1 & 0 & 0 \\ 0 & 1 & 0 \\ 0 & 0 & -1 \end{bmatrix}$$

Tetragonal

4
$$4 \rightarrow \begin{bmatrix} 0 & 1 & 0 \\ -1 & 0 & 0 \\ 0 & 0 & 1 \end{bmatrix}$$

$\bar{4}$
$$\bar{4} \rightarrow \begin{bmatrix} 0 & -1 & 0 \\ 1 & 0 & 0 \\ 0 & 0 & -1 \end{bmatrix}$$

$\dfrac{4}{m}$
$$4 \rightarrow \begin{bmatrix} 0 & 1 & 0 \\ -1 & 0 & 0 \\ 0 & 0 & 1 \end{bmatrix}, \quad m \rightarrow \begin{bmatrix} 1 & 0 & 0 \\ 0 & 1 & 0 \\ 0 & 0 & -1 \end{bmatrix}$$

422
$$4 \rightarrow \begin{bmatrix} 0 & 1 & 0 \\ -1 & 0 & 0 \\ 0 & 0 & 1 \end{bmatrix}, \quad 2 \rightarrow \begin{bmatrix} 1 & 0 & 0 \\ 0 & -1 & 0 \\ 0 & 0 & -1 \end{bmatrix}$$

TABLE 7.2 (*Cont.*)

System and Class	Generator Matrices

Tetragonal (*Cont.*)

4mm

$$4 \rightarrow \begin{bmatrix} 0 & 1 & 0 \\ -1 & 0 & 0 \\ 0 & 0 & 1 \end{bmatrix}, \quad m \rightarrow \begin{bmatrix} 1 & 0 & 0 \\ 0 & -1 & 0 \\ 0 & 0 & 1 \end{bmatrix}$$

$\overline{4}2m$

$$\overline{4} \rightarrow \begin{bmatrix} 0 & -1 & 0 \\ 1 & 0 & 0 \\ 0 & 0 & -1 \end{bmatrix}, \quad 2 \rightarrow \begin{bmatrix} 1 & 0 & 0 \\ 0 & -1 & 0 \\ 0 & 0 & -1 \end{bmatrix}$$

4/mmm

$$4 \rightarrow \begin{bmatrix} 0 & 1 & 0 \\ -1 & 0 & 0 \\ 0 & 0 & 1 \end{bmatrix}, \quad m \rightarrow \begin{bmatrix} 1 & 0 & 0 \\ 0 & 1 & 0 \\ 0 & 0 & -1 \end{bmatrix}, \quad m \rightarrow \begin{bmatrix} 1 & 0 & 0 \\ 0 & -1 & 0 \\ 0 & 0 & 1 \end{bmatrix}$$

Trigonal

3

$$3 \rightarrow \begin{bmatrix} -\dfrac{1}{2} & \dfrac{\sqrt{3}}{2} & 0 \\ -\dfrac{\sqrt{3}}{2} & -\dfrac{1}{2} & 0 \\ 0 & 0 & 1 \end{bmatrix}$$

$\overline{3}$

$$\overline{3} \rightarrow \begin{bmatrix} \dfrac{1}{2} & -\dfrac{\sqrt{3}}{2} & 0 \\ \dfrac{\sqrt{3}}{2} & \dfrac{1}{2} & 0 \\ 0 & 0 & -1 \end{bmatrix}$$

TABLE 7.2 (*Cont.*)

System and Class	Generator Matrices $[a_{ij}]$

Trigonal (*Cont.*)

32

$$3 \rightarrow \begin{bmatrix} -\dfrac{1}{2} & \dfrac{\sqrt{3}}{2} & 0 \\ -\dfrac{\sqrt{3}}{2} & -\dfrac{1}{2} & 0 \\ 0 & 0 & 1 \end{bmatrix}, \quad 2 \rightarrow \begin{bmatrix} 1 & 0 & 0 \\ 0 & -1 & 0 \\ 0 & 0 & -1 \end{bmatrix}$$

3m

$$3 \rightarrow \begin{bmatrix} -\dfrac{1}{2} & \dfrac{\sqrt{3}}{2} & 0 \\ -\dfrac{\sqrt{3}}{2} & -\dfrac{1}{2} & 0 \\ 0 & 0 & 1 \end{bmatrix}, \quad m \rightarrow \begin{bmatrix} -1 & 0 & 0 \\ 0 & 1 & 0 \\ 0 & 0 & 1 \end{bmatrix}$$

$\bar{3}m$

$$\bar{3} \rightarrow \begin{bmatrix} \dfrac{1}{2} & -\dfrac{\sqrt{3}}{2} & 0 \\ \dfrac{\sqrt{3}}{2} & \dfrac{1}{2} & 0 \\ 0 & 0 & -1 \end{bmatrix}, \quad m \rightarrow \begin{bmatrix} -1 & 0 \\ 0 & 1 \\ 0 & 0 \end{bmatrix}$$

Hexagonal

6

$$6 \rightarrow \begin{bmatrix} \dfrac{1}{2} & \dfrac{\sqrt{3}}{2} & 0 \\ -\dfrac{\sqrt{3}}{2} & \dfrac{1}{2} & 0 \\ 0 & 0 & 1 \end{bmatrix}$$

TABLE 7.2 (*Cont.*)

System and Class	Generator Matrices $[a_{ij}]$

Hexagonal (*Cont.*)

$\bar{6}$

$$\bar{6} \rightarrow \begin{bmatrix} -\dfrac{1}{2} & -\dfrac{\sqrt{3}}{2} & 0 \\[2ex] \dfrac{\sqrt{3}}{2} & -\dfrac{1}{2} & 0 \\[2ex] 0 & 0 & -1 \end{bmatrix}$$

$\dfrac{6}{m}$

$$6 \rightarrow \begin{bmatrix} \dfrac{1}{2} & \dfrac{\sqrt{3}}{2} & 0 \\[2ex] -\dfrac{\sqrt{3}}{2} & \dfrac{1}{2} & 0 \\[2ex] 0 & 0 & 1 \end{bmatrix}, \quad m \rightarrow \begin{bmatrix} 1 & 0 & 0 \\ 0 & 1 & 0 \\ 0 & 0 & -1 \end{bmatrix}$$

622

$$6 \rightarrow \begin{bmatrix} \dfrac{1}{2} & \dfrac{\sqrt{3}}{2} & 0 \\[2ex] -\dfrac{\sqrt{3}}{2} & \dfrac{1}{2} & 0 \\[2ex] 0 & 0 & 1 \end{bmatrix}, \quad 2 \rightarrow \begin{bmatrix} 1 & 0 & 0 \\ 0 & -1 & \\ 0 & 0 & -1 \end{bmatrix}$$

6mm

$$6 \rightarrow \begin{bmatrix} \dfrac{1}{2} & \dfrac{\sqrt{3}}{2} & 0 \\[2ex] -\dfrac{\sqrt{3}}{2} & \dfrac{1}{2} & 0 \\[2ex] 0 & 0 & 1 \end{bmatrix}, \quad m \rightarrow \begin{bmatrix} 1 & 0 & 0 \\ 0 & -1 & 0 \\ 0 & 0 & 1 \end{bmatrix}$$

$\bar{6}m2$

$$6 \rightarrow \begin{bmatrix} -\dfrac{1}{2} & -\dfrac{\sqrt{3}}{2} & 0 \\[2ex] \dfrac{\sqrt{3}}{2} & -\dfrac{1}{2} & 0 \\[2ex] 0 & 0 & -1 \end{bmatrix}, \quad 2 \rightarrow \begin{bmatrix} 1 & 0 & 0 \\ 0 & -1 & 0 \\ 0 & 0 & -1 \end{bmatrix}$$

TABLE 7.2 (*Cont.*)

System and Class	Generator Matrices $[a_{ij}]$

Hexagonal (*Cont.*)

$$6/mmm \quad 6 \rightarrow \begin{bmatrix} \dfrac{1}{2} & \dfrac{\sqrt{3}}{2} & 0 \\ -\dfrac{\sqrt{3}}{2} & \dfrac{1}{2} & 0 \\ 0 & 0 & 1 \end{bmatrix}, \quad m \rightarrow \begin{bmatrix} 1 & 0 & 0 \\ 0 & 1 & 0 \\ 0 & 0 & -1 \end{bmatrix}, \quad m \rightarrow \begin{bmatrix} 1 & 0 & 0 \\ 0 & -1 & 0 \\ 0 & 0 & 1 \end{bmatrix}$$

Cubic

$$23 \quad 2 \rightarrow \begin{bmatrix} -1 & 0 & 0 \\ 0 & -1 & 0 \\ 0 & 0 & 1 \end{bmatrix}, \quad 3 \rightarrow \begin{bmatrix} 0 & 1 & 0 \\ 0 & 0 & 1 \\ 1 & 0 & 0 \end{bmatrix}$$

$$m3 \quad 2 \rightarrow \begin{bmatrix} -1 & 0 & 0 \\ 0 & -1 & 0 \\ 0 & 0 & 1 \end{bmatrix}, \quad 3 \rightarrow \begin{bmatrix} 0 & 1 & 0 \\ 0 & 0 & 1 \\ 1 & 0 & 0 \end{bmatrix}, \quad m \rightarrow \begin{bmatrix} 1 & 0 & 0 \\ 0 & 1 & 0 \\ 0 & 0 & -1 \end{bmatrix}$$

$$432 \quad 4 \rightarrow \begin{bmatrix} 0 & 1 & 0 \\ -1 & 0 & 0 \\ 0 & 0 & 1 \end{bmatrix}, \quad 2 \rightarrow \begin{bmatrix} -1 & 0 & 0 \\ 0 & 0 & 1 \\ 0 & 1 & 0 \end{bmatrix}$$

$$m3m \quad 4 \rightarrow \begin{bmatrix} 0 & 1 & 0 \\ -1 & 0 & 0 \\ 0 & 0 & 1 \end{bmatrix}, \quad 2 \rightarrow \begin{bmatrix} -1 & 0 & 0 \\ 0 & 0 & 1 \\ 0 & 1 & 0 \end{bmatrix}, \quad m \rightarrow \begin{bmatrix} 1 & 0 & 0 \\ 0 & 1 & 0 \\ 0 & 0 & -1 \end{bmatrix}$$

$$43m \quad 4 \rightarrow \begin{bmatrix} 0 & -1 & 0 \\ 1 & 0 & 0 \\ 0 & 0 & -1 \end{bmatrix}, \quad m \rightarrow \begin{bmatrix} 1 & 0 & 0 \\ 0 & 0 & 1 \\ 0 & 1 & 0 \end{bmatrix}$$

† X, Y, Z in Table 7.1 are taken as the "old" axes in the sense of Fig. 1.13.

When (7.1) and (7.2) are multiplied on the right by $[M]$ and $[N]$ respectively, one obtains the *symmetry conditions*

$$[s][M] = [N][s] \tag{7.4}$$

and

$$[c][N] = [M][c]. \tag{7.5}$$

These must be satisfied for transformation matrices corresponding to *every* symmetry operation for the crystal lattice.

Symmetry restrictions on the elements of the compliance matrix $[s]$ are obtained by taking a general 6×6 matrix and requiring that it satisfy (7.4) for all operations of the symmetry group. In group $m3m$, for example, this means 48 different operations! Fortunately, it is necessary to apply symmetry conditions for the generator operations only, since (7.4) will be satisfied by the product of two operations whenever it is satisfied for the two operations individually. Suppose, for example, that

$$[N^{(a)}] \qquad [M^{(a)}]$$
$$[N^{(b)}] \qquad [M^{(b)}]$$

are transformation matrices corresponding to symmetry operations a and b. A successive application of the operation a and then the operation b (that is, the product operation ba) therefore has the transformation matrices

$$[N^{(ba)}] = [N^{(b)}][N^{(a)}] \tag{7.6}$$

and

$$[M^{(ba)}] = [M^{(b)}][M^{(a)}]. \tag{7.7}$$

If (7.4) is satisfied for a and b,

$$[s][M^{(a)}] = [N^{(a)}][s] \tag{7.8}$$

$$[s][M^{(b)}] = [N^{(b)}][s], \tag{7.9}$$

then multiplication of (7.8) on the left by $[N^{(b)}]$ and use of (7.9) gives

$$[s][M^{(b)}][M^{(a)}] = [N^{(b)}][N^{(a)}][s].$$

From (7.6) and (7.7) this is equivalent to

$$[s][M^{(ba)}] = [N^{(ba)}][s]. \tag{7.10}$$

EXAMPLE 1. Consider the symmetry group *432*. From Table 7.2, the coordinate transformation matrices $[a]$ for the group generators are

$$4 \rightarrow \begin{bmatrix} 0 & 1 & 0 \\ -1 & 0 & 0 \\ 0 & 0 & 1 \end{bmatrix} \qquad 2 \rightarrow \begin{bmatrix} -1 & 0 & 0 \\ 0 & 0 & 1 \\ 0 & 1 & 0 \end{bmatrix}$$

and transformation matrices $[M]$ and $[N]$ can be written down directly from (3.32) and (3.34). They are

$$[M^{(4)}] = [N^{(4)}] = \left[\begin{array}{ccc|ccc} 0 & 1 & 0 & 0 & 0 & 0 \\ 1 & 0 & 0 & 0 & 0 & 0 \\ 0 & 0 & 1 & 0 & 0 & 0 \\ \hline 0 & 0 & 0 & 0 & -1 & 0 \\ 0 & 0 & 0 & 1 & 0 & 0 \\ 0 & 0 & 0 & 0 & 0 & -1 \end{array}\right]$$

and

$$[M^{(2)}] = [N^{(2)}] = \left[\begin{array}{ccc|ccc} 1 & 0 & 0 & 0 & 0 & 0 \\ 0 & 0 & 1 & 0 & 0 & 0 \\ 0 & 1 & 0 & 0 & 0 & 0 \\ \hline 0 & 0 & 0 & 1 & 0 & 0 \\ 0 & 0 & 0 & 0 & 0 & -1 \\ 0 & 0 & 0 & 0 & -1 & 0 \end{array}\right]$$

If an arbitrary compliance matrix is assumed, the symmetry condition (7.4) for generator operation 4 leads to the matrix equation

$$\begin{bmatrix} s_{12} & s_{11} & s_{13} & s_{15} & -s_{14} & -s_{16} \\ s_{22} & s_{12} & s_{23} & s_{25} & -s_{24} & -s_{26} \\ s_{23} & s_{13} & s_{33} & s_{35} & -s_{34} & -s_{36} \\ s_{24} & s_{14} & s_{34} & s_{45} & -s_{44} & -s_{46} \\ s_{25} & s_{15} & s_{35} & s_{55} & -s_{45} & -s_{56} \\ s_{26} & s_{16} & s_{36} & s_{56} & -s_{46} & -s_{66} \end{bmatrix} = \begin{bmatrix} s_{12} & s_{22} & s_{23} & s_{24} & s_{25} & s_{26} \\ s_{11} & s_{12} & s_{13} & s_{14} & s_{15} & s_{16} \\ s_{13} & s_{23} & s_{33} & s_{34} & s_{35} & s_{36} \\ -s_{15} & -s_{25} & -s_{35} & -s_{45} & -s_{55} & -s_{56} \\ s_{14} & s_{24} & s_{34} & s_{44} & s_{45} & s_{46} \\ -s_{16} & -s_{26} & -s_{36} & -s_{46} & -s_{56} & -s_{66} \end{bmatrix}$$

Comparison term by term gives

$$s_{22} = s_{11} \qquad s_{33}\text{—no condition}$$
$$s_{23} = s_{13} \qquad s_{66}\text{—no condition}$$
$$s_{55} = s_{44} \qquad s_{12}\text{—no condition.}$$
$$s_{26} = -s_{16}$$

All other elements are zero from conditions such as $s_{24} = s_{15} = -s_{24} = 0$. The symmetry condition for generator 2

$$
\begin{bmatrix}
s_{11} & s_{13} & s_{12} & 0 & -s_{16} & 0 \\
s_{12} & s_{13} & s_{11} & 0 & s_{16} & 0 \\
s_{13} & s_{33} & s_{13} & 0 & 0 & 0 \\
0 & 0 & 0 & s_{44} & 0 & 0 \\
0 & 0 & 0 & 0 & 0 & -s_{44} \\
s_{16} & 0 & -s_{16} & 0 & -s_{66} & 0
\end{bmatrix}
=
\begin{bmatrix}
s_{11} & s_{12} & s_{13} & 0 & 0 & s_{16} \\
s_{13} & s_{13} & s_{33} & 0 & 0 & 0 \\
s_{12} & s_{11} & s_{13} & 0 & 0 & -s_{16} \\
0 & 0 & 0 & s_{44} & 0 & 0 \\
-s_{16} & s_{16} & 0 & 0 & 0 & -s_{66} \\
0 & 0 & 0 & 0 & -s_{44} & 0,
\end{bmatrix}
$$

leads to additional relations

$$s_{33} = s_{11}$$
$$s_{66} = s_{44}$$
$$s_{13} = s_{12}$$
$$s_{16} = 0.$$

The result of these restrictions is to reduce the compliance matrix to the form

$$
\begin{bmatrix}
s_{11} & s_{12} & s_{12} & 0 & 0 & 0 \\
s_{12} & s_{11} & s_{12} & 0 & 0 & 0 \\
s_{12} & s_{12} & s_{11} & 0 & 0 & 0 \\
0 & 0 & 0 & s_{44} & 0 & 0 \\
0 & 0 & 0 & 0 & s_{44} & 0 \\
0 & 0 & 0 & 0 & 0 & s_{44}
\end{bmatrix}
$$

Repetition of this procedure for the stiffness matrix gives the identical form,

$$
\begin{bmatrix}
c_{11} & c_{12} & c_{12} & 0 & 0 & 0 \\
c_{12} & c_{11} & c_{12} & 0 & 0 & 0 \\
c_{12} & c_{12} & c_{11} & 0 & 0 & 0 \\
0 & 0 & 0 & c_{44} & 0 & 0 \\
0 & 0 & 0 & 0 & c_{44} & 0 \\
0 & 0 & 0 & 0 & 0 & c_{44}
\end{bmatrix}
$$

The same result is obtained for all classes in the cubic system of Table 7.2, and is the cubic stiffness matrix given in Examples 1 and 3 of Chapter 3.

For group *432* in the above example, and for most other groups, the effect of matrices $[M]$ and $[N]$ is to rearrange the rows and columns of $[s]$ and to multiply some of them by -1. In these cases the results of symmetry can be written down without actually multiplying out the matrix products, and physical arguments such as those used for the isotropic case can often be used to obtain some of the symmetry restrictions. For trigonal and hexagonal classes, the coordinate transformations (Table 7.2) have more than one nonzero matrix element per row and column. This leads to the appearance of linear combinations of compliances in the symmetry equations. A technique for treating these cases is given in Reference 5 at the end of the chapter.

A summary of symmetry characteristics for compliance and stiffness matrices is given in Part A.2 of Appendix 2, and numerical values of the compliances and stiffnesses for commonly used materials are given in Parts A.4 and A.5 of the same appendix. Since the viscosity matrix $[\eta]$ in (3.57) transforms in the same way as the stiffness matrix in (7.2), it is subject to the same symmetry conditions. Numerical values of the viscosity coefficients, however, are available only for a very few materials. (See Reference 9 in Chapter 3.)

C. CHRISTOFFEL EQUATIONS FOR ANISOTROPIC SOLIDS

The most general kind of anisotropic solid is represented by the classes 1 and $\bar{1}$ in the triclinic system of Table 7.1. These have no effective symmetries whatsoever and the compliances and stiffnesss matrices have the full complement of 21 elastic constants identified in Section 3.A. When this general stiffness matrix is substituted into (6.11) and the matrix multiplications are performed, the Christoffel equation for *triclinic* crystals appears in the form

$$
k^2 \begin{bmatrix} \alpha & \delta & \epsilon \\ \delta & \beta & \zeta \\ \epsilon & \zeta & \gamma \end{bmatrix} \begin{bmatrix} v_x \\ v_y \\ v_z \end{bmatrix} = \rho \omega^2 \begin{bmatrix} v_x \\ v_y \\ v_z \end{bmatrix} \tag{7.11}
$$

where

$$
\begin{aligned}
\alpha &= c_{11}l_x^2 + c_{66}l_y^2 + c_{55}l_z^2 + 2c_{56}l_yl_z + 2c_{15}l_zl_x + 2c_{16}l_xl_y \\
\beta &= c_{66}l_x^2 + c_{22}l_y^2 + c_{44}l_z^2 + 2c_{24}l_yl_z + 2c_{46}l_zl_x + 2c_{26}l_xl_y \\
\gamma &= c_{55}l_x^2 + c_{44}l_y^2 + c_{33}l_z^2 + 2c_{34}l_yl_z + 2c_{35}l_zl_x + 2c_{45}l_xl_y \\
\delta &= c_{16}l_x^2 + c_{26}l_y^2 + c_{45}l_z^2 + (c_{46} + c_{25})l_yl_z + (c_{14} + c_{56})l_zl_x + (c_{12} + c_{66})l_xl_y \\
\epsilon &= c_{15}l_x^2 + c_{46}l_y^2 + c_{35}l_z^2 + (c_{45} + c_{36})l_yl_z + (c_{13} + c_{55})l_zl_x + (c_{14} + c_{56})l_xl_y \\
\zeta &= c_{56}l_x^2 + c_{24}l_y^2 + c_{34}l_z^2 + (c_{44} + c_{23})l_yl_z + (c_{36} + c_{45})l_zl_x + (c_{25} + c_{46})l_xl_y
\end{aligned} \tag{7.12}
$$

and

$$
\begin{aligned}
l_x &= k_x/k \\
l_y &= k_y/k \\
l_z &= k_z/k
\end{aligned}
$$

are the direction cosines of the propagation direction. This result can be specialized to other crystal systems by imposing the stiffness constraints given in Part A.2 of Appendix 2.

EXAMPLE 2.   In the case of crystals in the *cubic* system, the stiffness constants are subject to the following conditions

$$
\begin{aligned}
&c_{14} = c_{15} = c_{16} = 0 \qquad c_{12} = c_{13} = c_{23} \\
&c_{24} = c_{25} = c_{26} = 0 \\
&c_{34} = c_{35} = c_{36} = 0 \qquad c_{11} = c_{22} = c_{33} \\
&c_{45} = c_{46} = 0 \qquad\qquad\;\; c_{44} = c_{55} = c_{66} \\
&c_{56} = 0
\end{aligned}
$$

and substitution into (7.12) gives

$$
\begin{aligned}
\alpha &= c_{11}l_x^2 + c_{44}(l_y^2 + l_z^2) = c_{11}l_x^2 + c_{44}(1 - l_x^2) \\
\beta &= c_{11}l_y^2 + c_{44}(l_x^2 + l_z^2) = c_{11}l_y^2 + c_{44}(1 - l_y^2) \\
\gamma &= c_{11}l_z^2 + c_{44}(l_x^2 + l_y^2) = c_{11}l_z^2 + c_{44}(1 - l_z^2) \\
\delta &= (c_{12} + c_{44})l_xl_y \\
\epsilon &= (c_{12} + c_{44})l_zl_x \\
\zeta &= (c_{12} + c_{44})l_yl_z .
\end{aligned}
$$

This gives a Christoffel matrix having the same form as (6.14), but without the isotropy condition

$$c_{12} = c_{11} - 2c_{44}$$

Christoffel equations for all of the other crystal systems may be obtained in exactly the same way. These have been tabulated in Part A of Appendix 3. In all cases the coordinate axes x, y, z have been chosen to coincide with the crystal axes X, Y, Z defined in Table 7.1.

D. SLOWNESS (OR INVERSE VELOCITY) SURFACE

The propagation characteristics of plane waves in an anisotropic solid may be found by rewriting the Christoffel equation (6.10) as

$$[k^2 \Gamma_{ij} - \rho \omega^2 \delta_{ij}][v_j] = 0. \tag{7.13}$$

A dispersion relation is then obtained by setting the characteristic determinant of (7.13) equal to zero,

$$\Omega(\omega, k_x, k_y, k_z) = |k^2 \Gamma_{ij}(l_x, l_y, l_z) - \rho \omega^2 \delta_{ij}| = 0. \tag{7.14}$$

At fixed ω, (7.14) defines a surface in k-space that gives k as a function of its direction $\hat{\mathbf{l}}$. This is called the *wave vector surface*. Since the first term in the dispersion relation is proportional to k^2 and the second is proportional to ω^2, the relation can always be expressed entirely in terms of the variable k/ω. The wave vector $\mathbf{k}$ is therefore always proportional to ω and it is more convenient to consider the *slowness (or inverse velocity) surface*—which gives the inverse of the phase velocity $k/\omega = 1/V_p$ as a function of propagation direction and is independent of ω—rather than the wave vector surface, which scales with ω. Some important physical properties of the slowness surface will be examined in the following sections.

Once the characteristic equation (7.14) has been solved, the particle velocity polarization is obtained from (7.13). In general, numerical computation is required in carrying out this procedure; but, for special cases, the entire calculation can be performed analytically.

EXAMPLE 3.   From (7.11) and (7.12) the general form of the characteristic equation for a triclinic medium is as shown at the top of the facing page. Expansion of the determinant gives the dispersion relation

$$\left(\alpha \frac{k^2}{\omega^2} - \rho \right) \left(\beta \frac{k^2}{\omega^2} - \rho \right) \left(\gamma \frac{k^2}{\omega^2} - \rho \right)$$

$$= \frac{k^4}{\omega^4} \left\{ (\alpha \zeta^2 + \beta \epsilon^2 + \gamma \delta^2 - 2\delta \epsilon \zeta) \frac{k^2}{\omega^2} - \rho(\delta^2 + \epsilon^2 + \zeta^2) \right\}. \tag{7.15}$$

This is a third degree equation in k^2/ω^2 and is most conveniently solved by numerical computation.

$$\begin{vmatrix} \alpha k^2 - \rho\omega^2 & \delta k^2 & \epsilon k^2 \\ \\ \delta k^2 & \beta k^2 - \rho\omega^2 & \zeta k^2 \\ \\ \epsilon k^2 & \zeta k^2 & \gamma k^2 - \rho\omega^2 \end{vmatrix} = 0$$

Once k^2/ω^2 has been found from (7.15) the particle displacement vector can be obtained from (7.13). The first two lines are rearranged as

$$(\alpha - \rho\omega^2/k^2)v_x + \delta v_y = -\epsilon v_z$$
$$\delta v_x + (\beta - \rho\omega^2/k^2)v_y = -\zeta v_z . \tag{7.16}$$

Elimination of v_z gives

$$\frac{v_y}{v_x} = \frac{\zeta(\alpha - \rho\omega^2/k^2) - \epsilon\delta}{\epsilon(\beta - \rho\omega^2/k^2) - \zeta\delta} . \tag{7.17}$$

The z component of particle displacement is then found by substituting into one of the relations (7.16). That is,

$$v_z/v_x = \frac{(\alpha - \rho\omega^2/k^2)(\beta - \rho\omega^2/k^2) - \delta^2}{\epsilon(\beta - \rho\omega^2/k^2) - \zeta\delta} . \tag{7.18}$$

From (7.17) and (7.18), the particle velocity polarization can be calculated in terms of the phase velocity, the propagation direction, and the stiffness constants of the medium. The solutions will, in general, be quasilongitudinal and quasishear waves.

EXAMPLE 4. For a cubic crystal, with coordinate axes x, y, z as in Fig. 7.1, the Christoffel equation given in Part A of Appendix 3 leads to the characteristic equation

$$\begin{vmatrix} \left(\frac{k}{\omega}\right)^2 \{c_{11}l_x^2 + c_{44}(1-l_x^2)\} - \rho & \left(\frac{k}{\omega}\right)^2 \{c_{12}+c_{44}\}l_xl_y & \left(\frac{k}{\omega}\right)^2 \{c_{12}+c_{44}\}l_xl_z \\ \\ \left(\frac{k}{\omega}\right)^2 \{c_{12}+c_{44}\}l_yl_x & \left(\frac{k}{\omega}\right)^2 \{c_{11}l_y^2 + c_{44}(1-l_y^2)\} - \rho & \left(\frac{k}{\omega}\right)^2 \{c_{12}+c_{44}\}l_yl_z \\ \\ \left(\frac{k}{\omega}\right)^2 \{c_{12}+c_{44}\}l_zl_x & \left(\frac{k}{\omega}\right)^2 \{c_{12}+c_{44}\}l_zl_y & \left(\frac{k}{\omega}\right)^2 \{c_{11}l_z^2 + c_{44}(1-l_z^2)\} - \rho \end{vmatrix}$$
$$= 0$$
$$\tag{7.19}$$

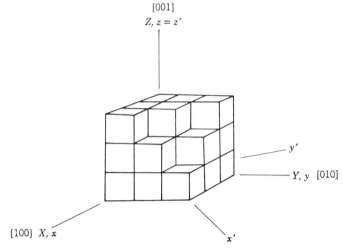

FIGURE 7.1.   Cubic crystal structure.

Numerical computation is required for a complete solution of (7.19). However, general features of the dispersion relation can be deduced by considering special propagation directions, for which the determinant factors. For example, if propagation is along any crystal axis.

$$\left[\left(\frac{k}{\omega}\right)^2 c_{11} - \rho\right]\left[\left(\frac{k}{\omega}\right)^2 c_{44} - \rho\right]\left[\left(\frac{k}{\omega}\right)^2 c_{44} - \rho\right] = 0. \qquad (7.20)$$

This factors into the dispersion relations (3.24) and (3.29) given in Chapter 3 for shear and compressional wave propagation along a cubic crystal axis. The wave solutions are pure shear and pure longitudinal.

The next simplest case is propagation in a cube face. For propagation in the X, Z plane, $l_y = 0$ and the Christoffel equation reduces to

$$\begin{bmatrix} \left(\frac{k}{\omega}\right)^2\{c_{11}l_x^2 + c_{44}l_z^2\} - \rho & 0 & \left(\frac{k}{\omega}\right)^2(c_{12} + c_{44})l_xl_z \\ 0 & \left(\frac{k}{\omega}\right)^2 c_{44} - \rho & 0 \\ \left(\frac{k}{\omega}\right)^2(c_{12} + c_{44})l_xl_z & 0 & \left(\frac{k}{\omega}\right)^2\{c_{11}l_z^2 + c_{44}l_x^2\} - \rho \end{bmatrix}\begin{bmatrix} v_x \\ v_y \\ v_z \end{bmatrix} = 0 \qquad (7.21)$$

From this, the dispersion relation separates into a linear factor

$$\left(\frac{k}{\omega}\right)^2 c_{44} - \rho = 0 \qquad (7.22)$$

and a quadratic factor

$$\left[\left(\frac{k}{\omega}\right)^2\{c_{11}l_x^2 + c_{44}l_z^2\} - \rho\right]\left[\left(\frac{k}{\omega}\right)^2\{c_{11}l_z^2 + c_{44}l_x^2\} - \rho\right] - \left(\frac{k}{\omega}\right)^4(c_{12} + c_{44})^2l_x^2l_z^2 = 0 \qquad (7.23)$$

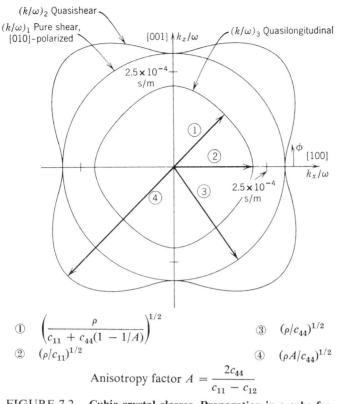

$$\textcircled{1}\quad \left(\frac{\rho}{c_{11} + c_{44}(1 - 1/A)}\right)^{1/2}$$

$$\textcircled{2}\quad (\rho/c_{11})^{1/2}$$

$$\textcircled{3}\quad (\rho/c_{44})^{1/2}$$

$$\textcircled{4}\quad (\rho A/c_{44})^{1/2}$$

Anisotropy factor $A = \dfrac{2c_{44}}{c_{11} - c_{12}}$

FIGURE 7.2. Cubic crystal classes. Propagation in a cube face.
Curves shown are for GaAs, with the piezoelectric effect ignored.
Sketches may be made for other materials, using the key dimensions
shown.

If (7.23) is multiplied by $(\omega/k)^4$, solutions to (7.22) and (7.23) are found to be

$$\left(\frac{k}{\omega}\right)_1 = \sqrt{\frac{\rho}{c_{44}}}$$

$$\left(\frac{k}{\omega}\right)_2 = (2\rho)^{1/2}\left\{c_{11} + c_{44} - \sqrt{(c_{11} - c_{44})^2 \cos^2 2\phi + (c_{12} + c_{44})^2 \sin^2 2\phi}\right\}^{-1/2}$$

$$\left(\frac{k}{\omega}\right)_3 = (2\rho)^{1/2}\left\{c_{11} + c_{44} + \sqrt{(c_{11} - c_{44})^2 \cos^2 2\phi + (c_{12} + c_{44})^2 \sin^2 2\phi}\right\}^{-1/2},$$

$$(7.24)$$

where $\cos\phi = l_x$. Inverse velocity curves are shown in Fig. 7.2 for gallium arsenide.
Representative dimensions are given in terms of material parameters. From these,
the general shape of the curves can be deduced for any cubic medium.

From (7.21) it is seen immediately that solution 1 has a particle velocity which
is normal to the XZ plane and is therefore normal to **k**. This is a *pure shear* wave

polarized along Y. The phase velocity is independent of the direction of propagation in the cube face and is equal to the phase velocity of a shear wave in an isotropic medium having the same c_{44} .

Solutions 2 and 3 are quasishear and quasilongitudinal waves, respectively, but they do reduce to pure modes for special propagation directions.† When $\phi = 0$ ($l_z = 0$) or $\phi = \pi/2$ ($l_x = 0$), the matrix in (7.21) becomes diagonal and all solutions are pure modes.

Pure mode solutions also occur at $\phi = \pi/4$ and $\phi = 3\pi/4$, where

$$l_x = \frac{1}{\sqrt{2}} \text{ at } \phi = \frac{\pi}{4} \qquad l_x = \frac{-1}{\sqrt{2}} \text{ at } \phi = \frac{3\pi}{4}$$

$$l_z = \frac{1}{\sqrt{2}} \text{ at } \phi = \frac{\pi}{4} \qquad l_z = \frac{1}{\sqrt{2}} \text{ at } \phi = \frac{3\pi}{4}. \qquad (7.25)$$

From the first row of (7.21) one has that

$$\frac{v_x}{v_z} = - \frac{\left(\dfrac{k}{\omega}\right)_2 \dfrac{(c_{12} + c_{44})}{2}}{\left(\dfrac{k}{\omega}\right)^2 \left\{ \dfrac{c_{11} + c_{44}}{2} \right\} - \rho} \qquad (7.26)$$

at $\phi = \pi/4$; and

$$\frac{v_x}{v_z} = + \frac{\left(\dfrac{k}{\omega}\right)^2 \dfrac{(c_{12} + c_{44})}{2}}{\left(\dfrac{k}{\omega}\right)^2 \left\{ \dfrac{c_{11} + c_{44}}{2} \right\} - \rho} \qquad (7.27)$$

at $\phi = 3\pi/4$. From (7.24), one can show that

$$\left(\frac{k}{\omega}\right)_2^2 \left\{ \frac{c_{11} + c_{44}}{2} \right\} - \rho = + \left(\frac{k}{\omega}\right)_2^2 \left\{ \frac{c_{12} + c_{44}}{2} \right\} \qquad (7.28)$$

and

$$\left(\frac{k}{\omega}\right)_3^2 \left\{ \frac{c_{11} + c_{44}}{2} \right\} - \rho = - \left(\frac{k}{\omega}\right)_3^2 \left\{ \frac{c_{12} + c_{44}}{2} \right\} \qquad (7.29)$$

for these propagation directions; and substitution into (7.26) and (7.27) therefore gives

$$\frac{v_x}{v_z} = -1, \qquad \text{Solution 2 at } \phi = \frac{\pi}{4}$$

$$\frac{v_x}{v_z} = +1, \qquad \text{Solution 3 at } \phi = \frac{\pi}{4}$$

$$\frac{v_x}{v_z} = +1, \qquad \text{Solution 2 at } \phi = \frac{3\pi}{4} \qquad (7.30)$$

$$\frac{v_x}{v_z} = -1, \qquad \text{Solution 3 at } \phi = \frac{3\pi}{4}.$$

† Conditions for pure mode propagation will be reviewed in Section 7.I.

Calculation of the quantities $\hat{\mathbf{I}} \cdot \mathbf{v}$ and $\hat{\mathbf{I}} \times \mathbf{v}$ from (7.25) and (7.30) shows that

$$(\hat{\mathbf{I}} \cdot \mathbf{v})_{\text{Solution 2}} = 0$$

and

$$(\hat{\mathbf{I}} \times \mathbf{v})_{\text{Solution 3}} = 0.$$

Solution 2 is thus a pure shear wave and solution 3 is a pure longitudinal wave at $\phi = \pi/4, 3\pi/4$. These results are in agreement with the solutions obtained in Examples 8 and 9 of Chapter 3.

To summarize, the dispersion relations (7.24) reduce to

$$\left(\frac{k}{\omega}\right)_1 = \sqrt{\frac{\rho}{c_{44}}} \qquad \text{Pure shear}$$

$$\left(\frac{k}{\omega}\right)_2 = \sqrt{\frac{2\rho}{c_{11} - c_{12}}} \qquad \text{Pure shear} \qquad (7.31)$$

$$\left(\frac{k}{\omega}\right)_3 = \sqrt{\frac{2\rho}{c_{11} + c_{12} + 2c_{44}}} \qquad \text{Pure longitudinal}$$

for propagation at $\phi = \pi/4, 3\pi/4$.

The ratio of the squares of the shear wave phase velocities at $\phi = \pi/4, 3\pi/4$ is defined as the *anisotropy factor* for a cubic crystal,

$$A = \frac{2c_{44}}{c_{11} - c_{12}}, \qquad (7.32)$$

and the longitudinal wave inverse velocity may be written in terms of this parameter as

$$\left(\frac{k}{\omega}\right)_3 = \sqrt{\frac{\rho}{c_{11} + c_{44}\left(1 - \frac{1}{A}\right)}}. \qquad (7.33)$$

The anisotropy factor A may be either greater or less than unity. When $A > 1$ (Fig. 7.2) the quasishear curve bulges out and the quasilongitudinal curve bulges in. For $A < 1$ the distortions are reversed.

EXAMPLE 5. The general Christoffel equation given by (7.11) and (7.12) is not restricted to any particular coordinate system. It is necessary only that the propagation direction $\hat{\mathbf{I}}$ and the stiffness constants c_{IJ} be referred to the *same* coordinate system. For some problems the analysis is considerably simplified by choosing coordinate axes that differ from the crystal axes. Suppose, for example, that one wishes to find the shape of the slowness curves for propagation in a plane $(X = Y, Z)$ passing through a cube face diagonal in a cubic crystal. In this case it is desirable to choose coordinate axes that are rotated by $45°$ about the Z crystal axis (Fig. 7.1). The transformed stiffness matrix is then given by (3.44). Substitution into (7.12) gives

$$\alpha = c'_{11} l^2_{x'} + c'_{66} l^2_{y'} + c_{44} l^2_{z'}$$

$$\beta = c'_{11} l^2_{y'} + c'_{66} l^2_{x'} + c_{44} l^2_{z'}$$

$$\gamma = c_{11} l^2_{z'} + c_{44}(1 - l^2_{z'})$$

$$\delta = (c'_{12} + c'_{66}) l_{x'} l_{y'}$$

$$\epsilon = (c_{12} + c_{44}) l_{z'} l_{x'}$$

$$\zeta = (c_{12} + c_{44}) l_{y'} l_{z'}$$

and the Christoffel equation is obtained from (7.11). For propagation in the diagonal (x', z') plane, $l_{y'} = 0$. The characteristic determinant then factors into a linear term which defines a pure shear wave polarized normal to the plane of propagation

$$\left(\frac{k}{\omega}\right)_2 = \left(\frac{\rho}{\dfrac{c_{11} - c_{12}}{2} \cos^2 \theta + c_{44} \sin^2 \theta}\right)^{1/2}, \tag{7.34}$$

and a quadratic term, which defines a quasishear wave

$$\left(\frac{k}{\omega}\right)_1 = \left(\frac{2\rho}{B - \sqrt{B^2 - C}}\right)^{1/2} \tag{7.35}$$

and a quasilongitudinal wave

$$\left(\frac{k}{\omega}\right)_3 = \left(\frac{2\rho}{B + \sqrt{B^2 - C}}\right)^{1/2}. \tag{7.36}$$

Here

$$B = (c_{11} + c_{12} + 4c_{44}) \frac{\cos^2 \theta}{2} + (c_{11} + c_{44}) \sin^2 \theta,$$

and

$$C = (c_{11} c'_{11} - c^2_{12} - 2c_{12} c_{44}) \sin^2 2\theta + 4c_{44} (c'_{11} \cos^4 \theta + c_{11} \sin^4 \theta),$$

with

$$c'_{11} = \frac{c_{11} + c_{12} + 2c_{44}}{2}$$

The direction angle θ is defined in Fig. 7.3, which gives curves for gallium arsenide and also some key dimensions.

Similar calculations for other crystal systems are summarized in Part B of Appendix 3. The results are presented in the form of curves like Fig. 7.2 and 7.3, with key dimensions specified in terms of material parameters. Using this information, the general shape of the slowness curves for any medium can be easily sketched.

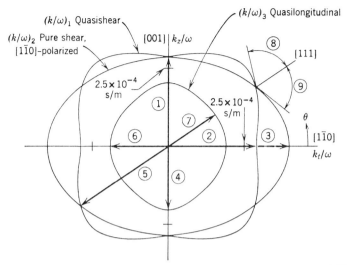

$$①② \quad (\rho/c_{44})^{1/2}$$

$$③ \quad (\rho A/c_{44})^{1/2}$$

$$④ \quad (\rho/c_{11})^{1/2}$$

$$⑤ \quad \left(\frac{3\rho A}{(2+A)c_{44}}\right)^{1/2}$$

$$⑥ \quad \left(\frac{\rho}{c_{11}+c_{44}\left(\dfrac{A-1}{A}\right)}\right)^{1/2}$$

$$⑦ \quad \left(\frac{3\rho}{3c_{11}+4c_{44}\left(\dfrac{A-1}{A}\right)}\right)^{1/2}$$

$$⑧⑨ \quad \cot^{-1}\sqrt{2}\left(\frac{A-1}{A+2}\right)$$

Anisotropy factor

$$A = \frac{2c_{44}}{c_{11}-c_{12}}$$

FIGURE 7.3.   Cubic crystal classes. Propagation in a cube diagonal plane. Curves shown are for GaAs, with the piezoelectric effect ignored.

E.   ORTHOGONALITY OF PARTICLE VELOCITY POLARIZATIONS

In all of the specific wave solutions obtained in Chapters 3 and 4, it was seen that the three waves propagating along a particular direction always had particle velocity vectors that were mutually orthogonal. It can now be proved from the symmetry of the Christoffel matrix

$$[\Gamma] = \begin{bmatrix} \alpha & \delta & \epsilon \\ \delta & \beta & \zeta \\ \epsilon & \zeta & \gamma \end{bmatrix}$$

in (7.11) that this must always be so. First, (7.11) is written as

$$\Gamma_{ij}v_j = \rho\left(\frac{\omega}{k}\right)^2 v_i, \tag{7.37}$$

and the three solutions for a given propagation direction $\hat{\mathbf{l}}$ are denoted by

$$(v_i)_1 \quad \left(\frac{\omega}{k}\right)_1$$

$$(v_i)_2 \quad \left(\frac{\omega}{k}\right)_2$$

$$(v_i)_3 \quad \left(\frac{\omega}{k}\right)_3.$$

The proof is simply a generalization of the argument used in deriving (4.60) in Example 2(b) of Chapter 4. One of the solutions "a" is written into (7.37) and the scalar product of both sides is taken with another solution "b",

$$(v_i)_b\Gamma_{ij}(v_j)_a = \rho\left(\frac{\omega}{k}\right)_a^2 (v_i)_b(v_i)_a. \tag{7.38}$$

Repetition of the same procedure, with the roles of solutions "a" and "b" interchanged, gives

$$(v_i)_a\Gamma_{ij}(v_j)_b = \rho\left(\frac{\omega}{k}\right)_b^2 (v_i)_a(v_i)_b. \tag{7.39}$$

Since Γ is a symmetric matrix ($\Gamma_{ij} = \Gamma_{ji}$), the subscripts i and j on the left-hand side of (7.39) may be interchanged. Also, the order of the factors may be reversed in the scalar product on the right-hand side. The result is

$$(v_i)_b\Gamma_{ij}(v_j)_a = \rho\left(\frac{\omega}{k}\right)_b^2 (v_i)_b(v_i)_a, \tag{7.40}$$

and subtraction of (7.38) from (7.40) gives

$$\rho\left(\left(\frac{\omega}{k}\right)_b^2 - \left(\frac{\omega}{k}\right)_a^2\right)(v_i)_b(v_i)_a = \rho\left(\left(\frac{\omega}{k}\right)_b^2 - \left(\frac{\omega}{k}\right)_a^2\right)\mathbf{v}_b \cdot \mathbf{v}_a = 0. \tag{7.41}$$

If the phase velocities $(\omega/k)_b$ and $(\omega/k)_a$ are different, this means that

$$\mathbf{v}_b \cdot \mathbf{v}_a = 0. \tag{7.42}$$

In other words, the particle velocities are orthogonal. When

$$\left(\frac{\omega}{k}\right)_b = \left(\frac{\omega}{k}\right)_a$$

solutions "a" and "b" can always be combined into solutions that are orthogonal, if they are not already so.

F. ENERGY VELOCITY

An important relation between the time-average power flow density and the time-average stored energy density of a plane acoustic wave can be obtained by manipulating the acoustic field equations in the same manner used to derive the complex Poynting's Theorems of Chapter 5. The dot product of (6.1), with $\mathbf{F} = 0$, is taken with the complex conjugate of $\mathbf{v}$; and the double dot product of the complex conjugate of (6.2) is taken with $\mathbf{T}$,

$$\mathbf{v}^* \cdot (\nabla \cdot \mathbf{T}) = \mathbf{v}^* \cdot \rho \frac{\partial \mathbf{v}}{\partial t} \tag{7.43}$$

$$\mathbf{T} : \nabla_s \mathbf{v}^* = \mathbf{T} : \mathbf{s} : \frac{\partial \mathbf{T}^*}{\partial t}, \tag{7.44}$$

noting that $\mathbf{s}$ is a real quantity in the case of a lossless medium. From (2.23), expansion of (7.43) into rectangular components gives

$$v_i^* \frac{\partial}{\partial r_j} T_{ij} = \rho v_i^* \frac{\partial v_i}{\partial t}. \tag{7.45}$$

If the stress tensor is symmetric ($T_{ij} = T_{ji}$), (7.44) may be expanded as

$$\tfrac{1}{2} T_{ij} \left(\frac{\partial v_i^*}{\partial r_j} + \frac{\partial v_j^*}{\partial r_i} \right) = T_{ij} \frac{\partial v_i^*}{\partial r_j} = T_I s_{IJ} \frac{\partial T_J^*}{\partial t}. \tag{7.46}$$

Complex exponential wave solutions are now assumed,

$$\mathbf{v} \rightarrow \mathbf{v} e^{i(\omega t - \mathbf{k} \cdot \mathbf{r})}$$
$$\mathbf{T} \rightarrow \mathbf{T} e^{i(\omega t - \mathbf{k} \cdot \mathbf{r})}, \tag{7.47}$$

and this converts (7.45) and (7.46) into

$$-v_i^* k_j T_{ij} = \omega \rho v_i^* v_i \tag{7.48}$$
$$-v_i^* k_j T_{ij} = \omega T_I s_{IJ} T_J^*. \tag{7.49}$$

The left-hand sides of (7.48) and (7.49) are both equal to

$$2\mathbf{k} \cdot \mathbf{P},$$

where

$$\mathbf{P} = -\frac{\mathbf{v}^* \cdot \mathbf{T}}{2} = -\left(\frac{\hat{x} v_i^* T_{ix} + \hat{y} v_i^* T_{iy} + \hat{z} v_i^* T_{iz}}{2} \right) \tag{7.50}$$

is the complex acoustic Poynting vector defined by (5.77). From (5.73), the right-hand side of (7.48) is

$$\omega \rho v_i^* v_i = \omega \rho |v|^2 = 2\omega (u_v)_{\text{peak}},$$

where $(u_v)_{peak}$ is the peak kinetic energy density. If the stresses are eliminated by using Hooke's Law $(T_I = c_{IJ}S_J)$, the right-hand side of (7.49) is

$$\omega S_I c_{IJ} S_J^* = \omega \mathbf{S}:\mathbf{c}:\mathbf{S}^* = 2\omega(u_s)_{peak} ,$$

where $(u_v)_{peak}$ is the peak strain energy density from (5.74). Relations (7.48) and (7.49) thus become

$$2\mathbf{k} \cdot \mathbf{P} = 2\omega(u_v)_{peak} \tag{7.51}$$

$$2\mathbf{k} \cdot \mathbf{P} = 2\omega(u_s)_{peak} , \tag{7.52}$$

which show that

$$(u_v)_{peak} = (u_s)_{peak} = u_{AV} , \tag{7.53}$$

as in (5.81) and (5.82). Since the right-hand sides of (7.51) and (7.52) are pure real, the Poynting vector on the left-hand side is also pure real and therefore represents *average* power flow density. One then has

$$\mathbf{k} \cdot \mathbf{P}_{AV} = \omega u_{AV} .$$

Division on both sides by k converts this to

$$\hat{\mathbf{k}} \cdot \mathbf{P}_{AV} = \frac{\omega}{k} u_{AV} = V_p u_{AV} , \tag{7.54}$$

which is a generalization of (5.83) when $\mathbf{P}_{AV}$ and $\mathbf{k}$ are not colinear. A simple rearrangement of (7.54) gives

$$\hat{\mathbf{k}} \cdot \frac{\mathbf{P}_{AV}}{u_{AV}} = V_p . \tag{7.55}$$

If the *energy velocity* is defined as

$$\mathbf{V}_e = \frac{\mathbf{P}_{AV}}{u_{AV}} , \tag{7.56}$$

(7.55) takes the form

$$\hat{\mathbf{k}} \cdot \mathbf{V}_e = V_p . \tag{7.57}$$

The physical significance of this important equation will be explored in the next section.

EXAMPLE 6.   The Poynting vector evaluated in Example 4 of Chapter 5 provides a starting point for illustrating the calculation of energy velocity. The wave vector **k** lies at an angle η in the XZ plane of a hexagonal crystal, and the coordinate axes have been rotated so that x' is along the direction of **k**. For the *quasishear* and *quasilongitudinal* solutions, it follows from (5.50) that the time-average Poynting vector is

$$\mathbf{P}_{AV} = \hat{\mathbf{x}}' \frac{\omega}{k} \rho((v_{x'})^2 + (v_{z'})^2)_{AV}$$

$$+ \hat{\mathbf{z}}' \left\{ \left(\frac{\omega}{k} \rho + \frac{k}{\omega} c'_{13} \right) v_{x'} v_{z'} + \frac{k}{\omega} c'_{35}(v_{z'})^2 \right\}_{AV} \tag{7.58}$$

where the field components are real functions of t and x'. According to (7.53), the time-average stored energy density is

$$u_{\mathrm{AV}} = \frac{\rho}{2}((v_{x'})^2 + (v_{z'})^2)_{\mathrm{peak}} = \rho((v_{x'})^2 + (v_{z'})^2)_{\mathrm{AV}} \qquad (7.59)$$

and the energy velocity is therefore

$$\mathbf{V}_e = \frac{\mathbf{P}_{\mathrm{AV}}}{u_{\mathrm{AV}}} = \hat{\mathbf{x}}' V_p + \hat{\mathbf{z}}' V_p \frac{\left\{ \left(\rho + \left(\frac{k}{\omega}\right)^2 c'_{13} \right) v_{x'} v_{y'} + \left(\frac{k}{\omega}\right)^2 c'_{35}(v_{z'})^2 \right\}_{\mathrm{AV}}}{\rho((v_{x'})^2 + (v_{z'})^2)_{\mathrm{AV}}} . \qquad (7.60)$$

The first term on the right-hand side confirms relation (7.57), since $\hat{\mathbf{k}} = \hat{\mathbf{x}}'$ in this example. To proceed further with the evaluation, it is necessary to find the field components $v_{x'}$ and $v_{z'}$ from Example 2(b) of Chapter 4 and then take the time average indicated in (7.60).

G. RAY AND NORMAL SURFACES

For anisotropic media, $\mathbf{k}$ and $\mathbf{V}_e$ in (7.57) do not always have the same direction. Consequently a plot of V_e as a function of its direction, called the *ray surface*, cannot be obtained by simply inverting the slowness surface of Section 7.D. There is, however, an interesting and useful relationship between these surfaces.

Consider two plane waves having infinitesimally different directions of propagation but the same frequency. These waves, shown on the slowness surface in Fig. 7.4(a), are defined as

$$\frac{\mathbf{k}}{\omega}, \mathbf{T}, \mathbf{v}, \mathbf{P}, \mathbf{V}_e$$

and

$$\frac{\mathbf{k}'}{\omega} = \frac{\mathbf{k} + \delta\mathbf{k}}{\omega}, \ \mathbf{T}' = \mathbf{T} + \delta\mathbf{T}, \ \mathbf{v}' = \mathbf{v} + \delta\mathbf{v}, \ \mathbf{P}' = \mathbf{P} + \delta\mathbf{P}, \ \mathbf{V}'_e = \mathbf{V}_e + \delta\mathbf{V}_e,$$

respectively. For complex exponential wave solutions of the form (7.47), where ω and k are real quantities, the acoustic field equations (6.1) and (6.2), with $\mathbf{F} = 0$, reduce to

$$-k_{iJ} T_J = \omega \rho v_i \qquad (7.61)$$

$$-k_{Ij} v_j = \omega s_{IJ} T_J , \qquad (7.62)$$

and

$$-k'_{iJ} T'_J = \omega \rho v'_i \qquad (7.63)$$

$$-k'_{Ij} v'_j = \omega s_{IJ} T'_J \qquad (7.64)$$

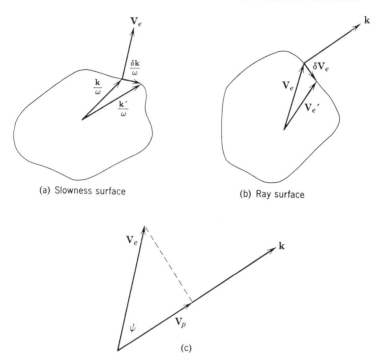

(a) Slowness surface   (b) Ray surface

(c)

FIGURE 7.4. (a) and (b) Relationships between the slowness surface and the ray surface. (c) Relation between the phase velocity V_p and the energy velocity V_e.

for these two waves. The matrices k_{iJ} and k_{Ij} are defined by (6.8) and (6.9). Subtraction of (7.61) from (7.63) and (7.62) from (7.64) gives

$$-(\delta k_{iJ} T_J + k_{iJ} \delta T_J) = \omega \rho \, \delta v_i \tag{7.65}$$

$$-(\delta k_{Ij} v_j + k_{Ij} \delta v_j) = \omega s_{IJ} \delta T_J , \tag{7.66}$$

when second order terms in the δ's are ignored. The dot product of (7.65) is then taken with $\mathbf{v}^*$ and the double dot product of (7.66) with $\mathbf{T}^*$,

$$-v_i^*(\delta k_{iJ} T_J + k_{iJ} \delta T_J) = \omega \rho v_i^* \delta v_i$$
$$-T_I^*(\delta k_{Ij} v_j + k_{Ij} \delta v_j) = \omega T_I^* s_{IJ} \delta T_J . \tag{7.67}$$

These equations are added, giving

$$-v_i^* \delta k_{iJ} T_J - T_I^* \delta k_{Ij} v_j = (v_i^* k_{iJ} + \omega T_I^* s_{IJ}) \, \delta T_J + (T_I^* k_{Ii} + \omega \rho v_i^*) \, \delta v_i \tag{7.68}$$

after rearrangement of terms and relabelling of some subscripts.

In (7.68), terms such as

$$v_i^* k_{iJ} \, \delta T_J$$

represent the product of a 3-element column matrix, a 3×6 matrix, and a 6-element column matrix, and are therefore unchanged by transposing subscripts,

$$v_i^* k_{iJ} \delta T_J = \delta T_J k_{Ji} v_i^*.$$

All terms in (7.68), except the third and fourth on the right-hand side are of this same general form. With some relabeling of the subscripts, the equation can thus be rearranged as

$$-v_i^* \delta k_{iJ} T_J - v_i \delta k_{iJ} T_J^* = (k_{Ji} v_i^* + \omega s_{JI} T_I^*) \, \delta T_J + (k_{iI} T_I^* + \omega \rho v_i^*) \, \delta v_i . \tag{7.69}$$

Here, the symmetry of the compliance matrix ($s_{IJ} = s_{JI}$) has also been used. From the complex conjugates of (7.61) and (7.62), the factors multiplying δT_J and δv_i on the right-hand side of (7.69) are both zero.† Because δk_{iJ} is real, the left-hand side is simply two times the real part of $-v_i^* \, \delta k_{iJ} T_J$; and (7.69) therefore reduces to

$$2 \, \mathscr{R}e \, (-v_i^* \delta k_{iJ} T_J) = 0. \tag{7.70}$$

The physical meaning of this result is made apparent by converting the term in brackets to full subscript notation. This gives

$$2 \, \mathscr{R}e \, (-v_i^* \delta k_j T_{ij}) = 2\delta k_j \, \{ \mathscr{R}e \, (-v_i^* T_{ij}) \} = 4 \, \delta\mathbf{k} \cdot \mathscr{R}e \, \mathbf{P},$$

where $\mathbf{P}$ is the complex acoustic Poynting vector (7.50) and the real part represents average power flow density $\mathbf{P}_{AV}$. It follows, then, that (7.70) is equivalent to

$$4\delta\mathbf{k} \cdot \mathbf{P}_{AV} = 0;$$

and, after division by $4u_{AV}$,

$$\delta\mathbf{k} \cdot \frac{\mathbf{P}_{AV}}{u_{AV}} = \delta\mathbf{k} \cdot \mathbf{V}_e = 0. \tag{7.71}$$

Since $\delta\mathbf{k}$ must always lie in the slowness surface and (7.71) applies to all $\delta\mathbf{k}$'s at the point where $\mathbf{k}$ touches the slowness surface (Fig. 7.4a), the *energy velocity* $\mathbf{V}_e$ must always be normal to the *slowness surface*. As will be seen in the following example, this property provides a simple and useful means for finding the direction of $\mathbf{V}_e$ without carrying out the detailed Poynting vector calculation used in Example 6.

---

† Note that s_{IJ} is a real quantity in the case of a lossless medium.

An analogous property applies to the ray surface. Since (7.57) applies to the wave with wave vector $\mathbf{k}' = \mathbf{k} + \delta\mathbf{k}$ in Fig. 7.4a, one can write

$$\frac{\hat{\mathbf{k}}' \cdot \mathbf{V}'_e}{V'_p} = \frac{\mathbf{k}'}{\omega} \cdot \mathbf{V}'_e = 1.$$

That is,

$$\frac{(\mathbf{k} + \delta\mathbf{k})}{\omega} \cdot (\mathbf{V}_e + \delta\mathbf{V}_e) = \frac{\mathbf{k} \cdot \mathbf{V}_e}{\omega} + \frac{\delta\mathbf{k} \cdot \mathbf{V}_e}{\omega} + \frac{\mathbf{k} \cdot \delta\mathbf{V}_e}{\omega} = 1, \qquad (7.72)$$

where second order terms have been ignored. From (7.57),

$$\frac{\mathbf{k} \cdot \mathbf{V}_e}{\omega} = 1$$

and (7.72) is therefore equivalent to

$$\frac{\delta\mathbf{k} \cdot \mathbf{V}_e}{\omega} + \frac{\mathbf{k} \cdot \delta\mathbf{V}_e}{\omega} = 0. \qquad (7.73)$$

Consequently, it follows from (7.71) that

$$\frac{\mathbf{k} \cdot \delta\mathbf{V}_e}{\omega} = 0. \qquad (7.74)$$

This means that the *wave vector* $\mathbf{k}$ must always be normal to the *ray surface* (Fig. 7.4b). An important physical interpretation of this result is made by

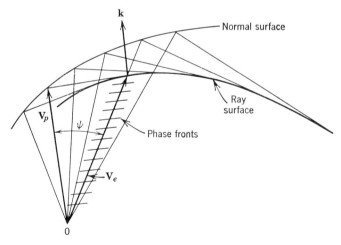

FIGURE 7.5.   **Definition of the normal (or phase velocity) surface and its relationship with the ray surface.**

writing (7.57) as

$$V_p = V_e \cos \psi \qquad (7.75)$$

where ψ is the angle between the energy velocity $\mathbf{V}_e$ and the wave vector $\mathbf{k}$ (Fig. 7.4c). This leads to a useful relation between the *normal surface* (a plot of the phase velocity V_p versus the wave vector direction) and the *ray surface* (a plot of energy velocity V_e versus the energy flow direction). Equation (7.75) and Fig. 7.4c apply at every point on the normal surface. The ray surface must therefore be the envelope of planes normal to $\mathbf{V}_p$, as illustrated in Fig. 7.5. Since the phase fronts of a plane wave are normal to $\mathbf{k}$ it is seen from Fig. 7.4b that each portion of the ray surface corresponds to the phase front for a plane wave with energy traveling in that direction.

EXAMPLE 7.   The problem treated in Example 6 will also serve to demonstrate the usefulness of the construction shown in Fig. 7.4a. Figure 7.6 illustrates a particular set of slowness curves for the meridian plane in a hexagonal crystal. The direction in which $\mathbf{V}_e$ is deflected away from $\mathbf{k}$ cannot be seen easily from (7.60), but this information is immediately obvious from examination of the normal directions to the slowness curves in the figure. For the quasishear wave in cadmium sulfide $\mathbf{V}_e$ is deflected toward the 45° line as $\mathbf{k}$ rotates from the t direction to the z direction. Deflection of $\mathbf{V}_e$ for the quasilongitudinal wave is away from the 45° line. The figure also shows that the quasilongitudinal deflection is very small in cadmium sulfide. Since the general shape of the slowness curve for any material is easily sketched from the information given in Part B of Appendix 3, it is a simple matter to deduce the energy velocity directions for many cases of interest.

H.   GROUP VELOCITY†

The velocity of the modulation on a wave is called the group velocity. For a wave with a one-dimensional modulation envelope, this is defined in the following manner. A modulated wave is constructed by taking two waves with slightly differing values of ω and k.

$$\cos(\omega t - kz) + \cos[(\omega + \delta\omega)t - (k + \delta k)z]$$

$$= 2 \underbrace{\cos\left\{\left(\omega + \frac{\delta\omega}{2}\right)t - \left(k + \frac{\delta k}{2}\right)z\right\}}_{\text{Carrier wave}} \underbrace{\cos\left(\frac{\delta\omega t}{2} - \frac{\delta k z}{2}\right)}_{\substack{\text{Modulation}\\\text{envelope}}} .$$

---

† Note that the results obtained in this section apply rigorously only to lossless media. For a discussion of the effect of losses in the *electromagnetic* case, see C. O. Hines, *J. Geophys. Res.* **56**, pp. 197–220 (1951).

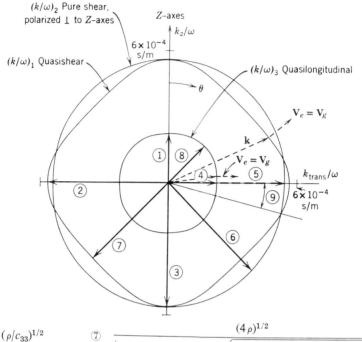

$$① \quad (\rho/c_{33})^{1/2}$$

$$⑦ \quad \frac{(4\rho)^{1/2}}{\left\{c_{11} + c_{33} + 2c_{44} - \sqrt{(c_{11} - c_{33})^2 + 4(c_{13} + c_{44})^2}\right\}^{1/2}}$$

$$② \quad (\rho/c_{44})^{1/2}$$
$$③ \quad (\rho/c_{44})^{1/2}$$

$$④ \quad (\rho/c_{11})^{1/2}$$

$$⑧ \quad \frac{(4\rho)^{1/2}}{\left\{c_{11} + c_{33} + 2c_{44} + \sqrt{(c_{11} - c_{33})^2 + 4(c_{13} + c_{44})^2}\right\}^{1/2}}$$

$$⑤ \quad (\rho/c_{66})^{1/2}$$

$$⑥ \quad \left(\frac{2\rho}{c_{66} + c_{44}}\right)^{1/2}$$

$$⑨ \quad \cot^{-1}\sqrt{\frac{(c_{13} + c_{44})^2 - (c_{33} - c_{66})(c_{11} - c_{44})}{(c_{44} - c_{66})(c_{11} - c_{66})}}$$

FIGURE 7.6. Hexagonal crystal classes. Propagation in a meridian plane. Curves shown are for CdS, with the piezoelectric effect ignored.

This shows that the propagation velocity of the carrier is the phase velocity

$$V_p = \frac{\omega + \delta\omega/2}{k + \delta k/2} \to \frac{\omega}{k},$$

and the propagation velocity of the modulation envelope is the group velocity

$$V_g = \frac{\delta\omega}{\delta k} \to \frac{\partial\omega}{\partial k}.$$

For a 3-dimensional modulation envelope, the procedure given in Problem 13 at the end of the chapter can be used to prove that

$$\mathbf{V}_g = \hat{\mathbf{x}}\frac{\partial\omega}{\partial k_x} + \hat{\mathbf{y}}\frac{\partial\omega}{\partial k_y} + \hat{\mathbf{z}}\frac{\partial\omega}{\partial k_z}. \tag{7.76}$$

This expression is convenient to use only if the dispersion relation is given explicitly as

$$\omega = f(k_x, k_y, k_z).$$

In Section 7.D it was seen that the dispersion relation for plane acoustic waves is obtained in the implicit form

$$|k^2\Gamma_{ij}(\hat{\mathbf{l}}) - \rho\omega^2\delta_{ij}| = \Omega(\omega, k_x, k_y, k_z) = 0; \tag{7.77}$$

and this cannot always be transformed into an explicit equation for ω. In such cases the required derivatives in (7.76) are obtained by implicit differentiation. For example,

$$\left(\frac{\partial\Omega}{\partial\omega}\delta\omega + \frac{\partial\Omega}{\partial k_x}\delta k_x\right)_{k_y, k_z} = 0,$$

or

$$\left(\frac{\partial\omega}{\partial k_x}\right)_{k_y, k_z} = -\frac{\partial\Omega/\partial k_x}{\partial\Omega/\partial\omega}, \text{ etc.}$$

The group velocity (7.76) can therefore always be evaluated as

$$\mathbf{V}_g = -\frac{\nabla_k\Omega}{\partial\Omega/\partial\omega}, \tag{7.78}$$

where the gradient of Ω is taken with respect to variables k_x, k_y, k_z.

Since $\mathbf{V}_g$ and $\mathbf{V}_e$ are identical for acoustic waves in a lossless medium (Problem 14), (7.78) provides an alternative to the Poynting vector calculation needed for evaluation of (7.56). As will be seen in the following example, this provides a significant computational advantage. The group velocity also has a directly measurable physical meaning that is not apparent in the definition of energy velocity. If a pulse of acoustic energy is radiated by a plane wave transducer, the wavepacket (which is limited in two dimensions by the size of the transducer and in the third dimension by the pulse length) travels along the trajectory shown in Fig. 7.7. The wave fronts travel along the direction of $\mathbf{k}$, which is normal to the transducer surface; but the wavepacket modulation envelope travels in the direction $\mathbf{V}_g = \mathbf{V}_e$, which may be canted at an angle to $\mathbf{k}$. This means that the receiving transducer must be offset in order to intercept the acoustic pulse. Experimental evidence of this effect in single crystal quartz is shown by Fig. 7.8, where the path of the acoustic

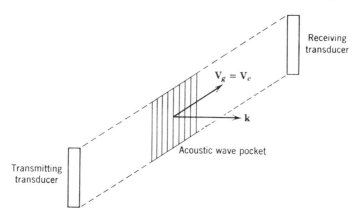

FIGURE 7.7. **Deflected acoustic beam trajectory in an anisotropic solid.**

beam has been made visible by means of optical scattering from changes in ϵ produced by the acoustic deformations. The vertical striations indicate the orientation of the phase fronts, normal to $\mathbf{k}$, and the quartz crystal is oriented with $\mathbf{k}$ parallel to the Y crystal axis. Reference to Fig. 3.7 in Part B.4 of Appendix 3 and use of Fig. 7.4a to graphically determine the directions of $\mathbf{V}_e = \mathbf{V}_g$ confirms that the quasilongitudinal wave should be deflected upward and the quasishear waves deflected downward by an angle of approximately $25°$, as illustrated in Fig. 7.8.

EXAMPLE 8. *Energy Velocity and Group Velocity Calculations for the XZ Plane of a Cubic Crystal.*

(a) Energy Velocity. In the case of Example 4 the quasishear and quasilongitudinal waves were polarized in the xz plane and had particle velocity components that may be written as

$$v_x = A_x e^{i(\omega t - k_x x - k_z z)}$$
$$v_z = A_z e^{i(\omega t - k_x x - k_z z)}. \tag{7.79}$$

The associated strain components are

$$S_1 = \frac{1}{i\omega}\frac{\partial v_x}{\partial x} = -\frac{k_x}{\omega}A_x e^{i(\omega t - \mathbf{k}\cdot\mathbf{r})}$$

$$S_3 = \frac{1}{i\omega}\frac{\partial v_z}{\partial z} = -\frac{k_z}{\omega}A_z e^{i(\omega t - \mathbf{k}\cdot\mathbf{r})}$$

$$S_5 = \frac{1}{i\omega}\left(\frac{\partial v_x}{\partial z} + \frac{\partial v_z}{\partial x}\right) = -\frac{1}{\omega}(k_z A_x + k_x A_z)e^{i(\omega t - \mathbf{k}\cdot\mathbf{r})}.$$

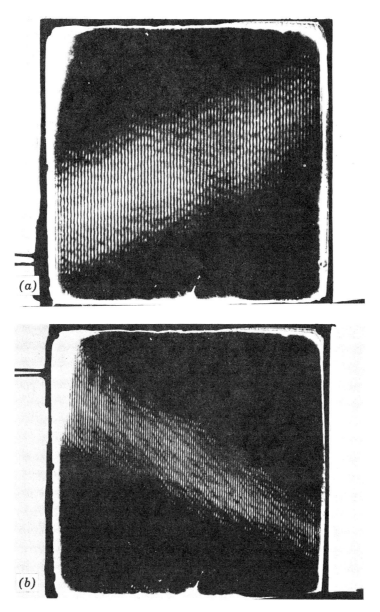

FIGURE 7.8. Deflection of (a) quasilongitudinal and (b) quasishear acoustic beams with k parallel to the Y crystal axis in quartz. (After Staudt and Cook.)

Using the stiffness matrix from Part A.2 of Appendix 2, one calculates the stress components

$$T_1 = c_{11}S_1 + c_{12}S_3 = -\frac{1}{\omega}(c_{11}k_xA_x + c_{12}k_zA_z)e^{i(\omega t - \mathbf{k} \cdot \mathbf{r})}$$

$$T_3 = c_{12}S_1 + c_{11}S_3 = -\frac{1}{\omega}(c_{12}k_xA_x + c_{11}k_zA_z)e^{i(\omega t - \mathbf{k} \cdot \mathbf{r})}$$

$$T_5 = c_{44}S_5 = -\frac{1}{\omega}(c_{44}k_zA_x + c_{44}k_xA_z)e^{i(\omega t - \mathbf{k} \cdot \mathbf{r})} . \qquad (7.80)$$

From (7.50) the complex Poynting vector is

$$\mathbf{P} = \frac{\hat{\mathbf{x}}}{2}(-v_x^*T_{xx} - v_z^*T_{zx}) + \frac{\hat{\mathbf{z}}}{2}(-v_x^*T_{xz} - v_z^*T_{zz})$$

$$= \frac{\hat{\mathbf{x}}}{2}(-v_x^*T_1 - v_z^*T_5) + \frac{\hat{\mathbf{z}}}{2}(-v_x^*T_5 - v_z^*T_3),$$

and substitution from (7.79) and (7.80) gives

$$P_x = \frac{k}{2\omega}[A_x^*A_xc_{11}l_x + A_x^*A_zc_{12}l_z + A_xA_z^*c_{44}l_z + A_zA_x^*c_{44}l_x]$$

$$P_z = \frac{k}{2\omega}[A_x^*A_xc_{44}l_z + A_x^*A_zc_{44}l_x + A_xA_z^*c_{12}l_x + A_zA_z^*c_{11}l_z] , \qquad (7.81)$$

where

$$k_x = kl_x$$
$$k_y = kl_y .$$

The ratio of A_x to A_z can be determined by substituting (7.79) into (7.21). From the first line

$$\frac{A_z}{A_x} = -\frac{\left(\dfrac{k}{\omega}\right)^2\{c_{11}l_x^2 + c_{44}l_z^2\} - \rho}{\left(\dfrac{k}{\omega}\right)^2(c_{12} + c_{44})l_xl_z}, \qquad (7.82)$$

and from the third line

$$\frac{A_z}{A_x} = -\frac{\left(\dfrac{k}{\omega}\right)^2(c_{12} + c_{44})l_xl_z}{\left(\dfrac{k}{\omega}\right)^2\{c_{11}l_z^2 + c_{44}l_x^2\} - \rho}. \qquad (7.83)$$

These two expressions are identical by virtue of the fact that $(k/\omega)^2$ satisfies the dispersion relation (7.23). They are also pure real. The amplitudes A_x and A_z may therefore be taken to be pure real in (7.81), which is then written as

$$P_x = \frac{k}{2\omega}A_xA_z\left[\frac{A_x}{A_z}c_{11}l_x + \frac{A_z}{A_x}c_{44}l_x + (c_{12} + c_{44})l_z\right]$$

$$P_z = \frac{k}{2\omega}A_xA_z\left[\frac{A_x}{A_z}c_{44}l_z + \frac{A_z}{A_x}c_{11}l_z + (c_{12} + c_{44})l_x\right].$$

If (7.82) is used to evaluate the second term in the square brackets and (7.83) to evaluate the first term, the Poynting vector components reduce to

$$
P_x = \frac{k}{2\omega} \frac{A_x A_z l_x}{(c_{12} + c_{44}) l_x l_z} \left[-c_{11} \left\{ c_{11} l_z^2 + c_{44} l_x^2 - \rho \left(\frac{\omega}{k} \right)^2 \right\} \right.
$$
$$
\left. - c_{44} \left\{ c_{11} l_x^2 + c_{44} l_z^2 - \rho \left(\frac{\omega}{k} \right)^2 \right\} + (c_{12} + c_{44})^2 l_z^2 \right]
$$

$$
P_z = \frac{k}{2\omega} \frac{A_x A_z l_z}{(c_{12} + c_{44}) l_x l_z} \left[-c_{44} \left\{ c_{11} l_z^2 + c_{44} l_x^2 - \rho \left(\frac{\omega}{k} \right)^2 \right\} \right.
$$
$$
\left. - c_{11} \left\{ c_{11} l_x^2 + c_{44} l_z^2 - \rho \left(\frac{\omega}{k} \right)^2 \right\} + (c_{12} + c_{44})^2 l_x^2 \right].
$$

$$(7.84)$$

These are pure real and therefore represent $\mathbf{P}_{AV}$.

The average stored energy density is

$$
u_{AV} = \tfrac{1}{2} \rho (v_x^* v_x + v_z^* v_z)
$$
$$
= \frac{\rho}{2} A_x A_z \left(\frac{A_x}{A_z} + \frac{A_z}{A_x} \right)
$$

$$(7.85)$$

from (7.53) and (7.79). If the amplitude ratios are evaluated by means of (7.82) and (7.83), this becomes

$$
u_{AV} = -\frac{\rho}{2} \frac{A_x A_z}{(c_{12} + c_{44}) l_x l_z} \left((c_{11} + c_{44}) - 2\rho \left(\frac{\omega}{k} \right)^2 \right);
$$

and the energy velocity components are

$$
(V_e)_x =
$$
$$
\frac{k}{\omega} l_x \frac{\left[c_{11} \left\{ c_{11} l_z^2 + c_{44} l_x^2 - \rho \left(\frac{\omega}{k} \right)^2 \right\} + c_{44} \left\{ c_{11} l_x^2 + c_{44} l_z^2 - \rho \left(\frac{\omega}{k} \right)^2 \right\} - (c_{12} + c_{44})^2 l_z^2 \right]}{\rho \left[(c_{11} + c_{44}) - 2\rho \left(\frac{\omega}{k} \right)^2 \right]}
$$

$$
(V_e)_z =
$$
$$
\frac{k}{\omega} l_z \frac{\left[c_{44} \left\{ c_{11} l_z^2 + c_{44} l_x^2 - \rho \left(\frac{\omega}{k} \right)^2 \right\} + c_{11} \left\{ c_{11} l_x^2 + c_{44} l_z^2 - \rho \left(\frac{\omega}{k} \right)^2 \right\} - (c_{12} + c_{44})^2 l_x^2 \right]}{\rho \left[(c_{11} + c_{44}) - 2\rho \left(\frac{\omega}{k} \right)^2 \right]}
$$

$$(7.86)$$

from (7.56).

(b) Group Velocity. To evaluate the group velocity from (7.78), any convenient form of the dispersion relation

$$
\Omega = 0 \tag{7.14}
$$

may be used. For the quasishear and quasilongitudinal waves in the present problem this is given by (7.23). Therefore, Ω may be chosen as

$$\Omega = [c_{11}k_x^2 + c_{44}k_z^2 - \rho\omega^2][c_{11}k_z^2 + c_{44}k_x^2 - \rho\omega^2] - (c_{12} + c_{44})^2 k_x^2 k_z^2 = 0,$$

and

$$-\frac{\partial\Omega}{\partial k_x} = -2c_{11}k_x(c_{11}k_z^2 + c_{44}k_x^2 - \rho\omega^2) - 2c_{44}k_x(c_{11}k_x^2 + c_{44}k_z^2 - \rho\omega^2)$$
$$+ 2(c_{12} + c_{44})^2 k_x k_z^2$$

$$-\frac{\partial\Omega}{\partial k_y} = 0$$

$$-\frac{\partial\Omega}{\partial k_z} = -2c_{44}k_z(c_{11}k_z^2 + c_{44}k_x^2 - \rho\omega^2) - 2c_{11}k_z(c_{11}k_x^2 + c_{44}k_z^2 - \rho\omega^2)$$
$$+ 2(c_{12} + c_{44})^2 k_x^2 k_z$$

$$\frac{\partial\Omega}{\partial\omega} = -2\rho\omega\{(c_{11} + c_{44})k^2 - 2\,\rho\omega^2\}.$$

Substitution of these derivatives into (7.78) gives

$$(V_g)_x = (V_e)_x$$
$$(V_g)_y = 0$$
$$(V_g)_z = (V_e)_z$$

in (7.86). This is obviously a much simpler way to perform the calculation.

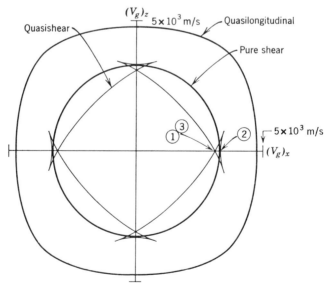

FIGURE 7.9. Section of the ray surface in the XZ plane of a cubic crystal. Curves are for gallium arsenide, with the piezoelectric effect ignored.

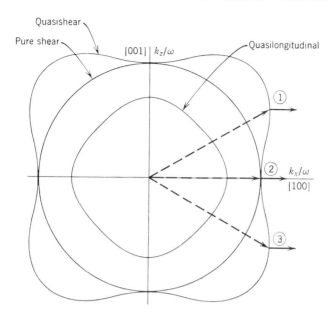

FIGURE 7.10.   **Section of the slowness surface in the** XZ **plane of a cubic medium, showing the origin of the cusps in Fig. 7.9.**

The energy velocity components evaluated in Example 8 are so complicated that it is difficult to visualize the shape of the ray surface they describe. Curves numerically computed from (7.86), using material constants for gallium arsenide, are shown in Fig. 7.9. These are marked pure shear, quasishear, and quasilongitudinal, and are obtained by substituting the appropriate values of k/ω into (7.86). The cusps on the quasishear curve arise from the convoluted form of the quasishear slowness curve in the vicinity of the cube axes (Fig. 7.10). Because of this, there are three wave vectors (marked ①, ②, and ③) that correspond to energy flow along the X axis. From symmetry, two of these (① and ③) must have the same energy velocity, as indicated by the numbering on Fig. 7.9. Similar behavior appears near the Z axis. The ray surface for the pure shear mode is also given in Fig. 7.9. Because the slowness surface is a sphere in this case, V_e is parallel to $\mathbf{k}$ and the ray surface is simply another sphere inverse to the slowness surface. In Fig. 7.11 the ray (or energy velocity) and the normal (or phase velocity) surfaces for the quasishear wave are superimposed. This demonstrates the geometric relationship of Fig. 7.4c for a ray surface with cusps.

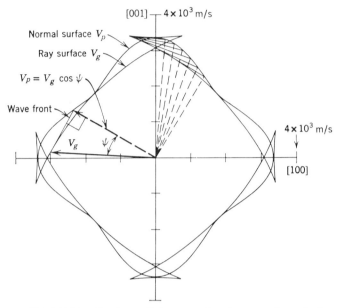

FIGURE 7.11.   Sections of the normal and ray surfaces in the *XZ* plane of a cubic crystal. Curves are for gallium arsenide, with the piezoelectric effect ignored.

I.   PURE MODE DIRECTIONS

In Example 1 of Chapter 6 it was seen that acoustic waves in an isotropic medium are always pure modes,† in the sense that the particle velocity is either parallel or normal to **k**. Example 2(b) of Chapter 4 and Example 4 of this chapter show that this is true only for special propagation directions in an anisotropic medium. Pure modes are so much easier to handle in scattering and boundary value problems that it is desirable to have a method for identifying the pure mode propagation directions without solving the entire anisotropic problem. Symmetry arguments provide a powerful, yet simple, tool for this purpose.

For any direction of propagation **k** there are three independent plane wave solutions,

$$\mathbf{v}_a e^{-i k a \cdot r} e^{i\omega t}$$
$$\mathbf{v}_b e^{-i k b \cdot r} e^{i\omega t} \qquad (7.87)$$
$$\mathbf{v}_c e^{-i k c \cdot r} e^{i\omega t},$$

† Pure modes are sometimes defined in the literature as waves that have the Poynting vector **P** parallel to **k**. This is an entirely different physical condition and has no relation to the definition used here.

where $\mathbf{v}_a$, $\mathbf{v}_b$, $\mathbf{v}_c$ are mutually perpendicular (from Section 7.E) and $\mathbf{k}_a$, $\mathbf{k}_b$, $\mathbf{k}_c$ have the same direction $\hat{\mathbf{k}}$ but have different magnitudes. If these three waves were degenerate $(k_a = k_b = k_c)$ for some direction of propagation, this would necessarily be a pure mode direction because linear combinations of the solutions could always be chosen to be mutually orthogonal and to have one polarization parallel to $\hat{\mathbf{k}}$. However, this situation does not occur in practice.

When a wave solution $\mathbf{v}_a e^{-i\mathbf{k}_a \cdot \mathbf{r}}$ is transformed by a symmetry operation $\mathcal{O}$ of the crystal, the resulting field $\mathcal{O}\mathbf{v}_a e^{-i\mathbf{k}_a \cdot \mathbf{r}}$ (which may be visualized by applying the rotation, reflection, etc., to a field plot of the original solution, as in Fig. 7.12) is also a solution. The reason is that $\mathcal{O}$ is a crystal symmetry operation, and the transformed solution therefore bears the same relationship to the crystal lattice that the original solution did. Suppose the symmetry operation $\mathcal{O}$, as Fig. 7.12, leaves the propagation direction $\hat{\mathbf{k}}$ unchanged. Then the transformed solution must be either (a) a scalar multiple of the original solution

$$\mathcal{O}\mathbf{v}_a e^{-i\mathbf{k}_a \cdot \mathbf{r}} = C\mathbf{v}_a e^{-i\mathbf{k}_a \cdot \mathbf{r}}, \tag{7.88}$$

or (b) a linear combination of solutions degenerate with $\mathbf{v}_a e^{-i\mathbf{k}_a \cdot \mathbf{r}}$,

$$\mathcal{O}\mathbf{v}_a e^{-i\mathbf{k}_a \cdot \mathbf{r}} = c_a\mathbf{v}_a e^{-i\mathbf{k}_a \cdot \mathbf{r}} + c_b\mathbf{v}_b e^{-i\mathbf{k}_b \cdot \mathbf{r}}$$
$$\mathbf{k}_b = \mathbf{k}_a. \tag{7.89}$$

Only two terms are allowed in (7.89), since acoustic plane wave solutions are degenerate only in pairs.

I.1 Propagation in a Symmetry Plane

Consider a quasilongitudinal wave propagating in a symmetry plane. From the above discussion, this mode must transform into a scalar multiple of

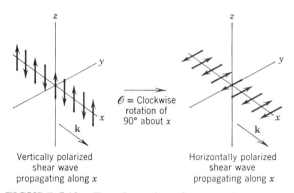

FIGURE 7.12. **Transformation of a pure shear wave by a 90° rotation about the propagation direction.**

itself under the reflection operation. It must therefore be polarized either normal or parallel to the plane. If it were normal to the plane it would be a pure shear mode, contrary to the original assumption. It must therefore be polarized in the symmetry plane. Since the polarizations are mutually orthogonal, this means there is one pure shear wave, with its polarization normal to the symmetry plane.

I.2 Propagation Normal to a Rotation Axis

Stiffness and compliance matrices are always invariant under the inversion operation, even when the crystal lattice does not possess this symmetry. This is easily confirmed by calculating the transformation matrices $[M]$ and $[N]$ in (3.32) and (3.34), which are both unity for the inversion operation. An important consequence of this is that any plane normal to a twofold crystal rotation axis is also an elastic reflection symmetry plane. The reason is that the combination of twofold rotation and inversion is identical with reflection in the normal plane. Planes normal to fourfold and sixfold crystal rotation axes are also elastic reflection symmetry planes, since fourfold and sixfold symmetries include twofold symmetries. In all these cases the pure mode characteristics are the same as for a crystal reflection symmetry plane, discussed above.

I.3 Propagation Along a Rotation Axis

Two Fold Axis. In this case the nondegenerate mode must be polarized along the axis in order to transform into a multiple of itself. It is therefore a pure longitudinal mode and the other two are pure shear. Since a 2-fold rotation transforms each shear wave into a scalar multiple of itself, shear wave degeneracy is not required.

Three Fold Axis. The nondegenerate mode is again polarized along the axis, and all modes are pure. The threefold rotation transforms the pure shear modes as shown in Fig. 7.13. Since the transformed shear modes can only be linear combinations of the original modes, the latter must be degenerate. Therefore a 3-fold rotation axis requires shear wave degeneracy.

Four and Six Fold Axes. The same kind of arguments show that all three modes are pure and that the shear waves are degenerate.

The above results are summarized in Part C of Appendix 3. Since the slowness surface must be invariant under any crystal symmetry transformation, information about its shape can also be obtained from symmetry arguments. At a twofold axis the surface is invariant with respect to 180°

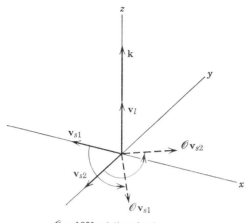

$\mathcal{O} = 120°$ rotation about z

FIGURE 7.13. **Pure mode propagation along a 3-fold symmetry axis.**

rotation and must therefore cross the axis at right angles. This also applies to four and sixfold axes, both of which possess a 180° rotation symmetry. A threefold rotation axis does not have this symmetry and therefore does not require a normal crossing of the axis by the slowness surface. According to the arguments given above, the pure longitudinal wave along the axis must transform into itself under the threefold rotation operation. This requires that the slowness surface for this wave be normal to the rotation axis. Otherwise, the energy velocity (which is normal to the surface) would not be invariant with respect to the threefold rotation. These general features of the slowness surface are summarized in Fig. 7.14.

This section has shown how simple symmetry arguments can be used to deduce many of the pure mode directions. Pure modes may, however, also appear in nonsymmetry directions. To find these directions, a detailed analysis is required. The results are summarized in Part C of Appendix 3.

EXAMPLE 9.   The symmetry conditions derived above are easily illustrated by referring to the table of crystal symmetries (Table 7.1) and the wave solutions in Part B of Appendix 3.

(*a*) *Propagation Normal to a Rotation Axis.*   According to Table 7.1 the *XZ* plane of any cubic crystal is normal to a rotation axis and there is therefore one pure shear mode. This is shown to be so in the "Cubic Section" of Part B, Appendix 3.

(*b*) *Propagation Along a Twofold Axis.*   For crystal classes *432* and *m3m* of the cubic system (Table 7.1) the direction $\phi = \pi/4$ in Fig. 3.2 of Part B, Appendix 3 is

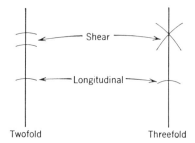

Twofold   Threefold

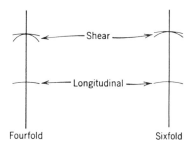

Fourfold   Sixfold

FIGURE 7.14. **Characteristics of the slowness surface at rotational symmetry axes.**

a twofold axis. The figure shows that all the slowness curves cross this axis at right angles and that the shear curves are nondegenerate. All modes are pure in this case. Although the other cubic classes *23*, *m3*, and *$\bar{4}3m$* do not have a twofold rotation axis along this direction, they do have the same wave propagation characteristics because the stiffness matrix is the same for all cubic classes.

(c) *Propagation Along a Threefold Axis.* Crystals in the trigonal system all have a threefold axis along *Z*. Figure 3.7 of Part B, Appendix 3 shows that the quasilongitudinal slowness curve crosses the axis at right angles. The other two curves are degenerate but cross the axis at an angle. All modes are pure.

(d) *Propagation Along a Fourfold Axis.* All classes in the tetragonal system have a fourfold axis along *Z*. In Fig. 3.10 of Part B, Appendix 3 it is seen that all slowness curves cross the axis at right angles and two of them are degenerate. All modes are pure.

(e) *Propagation Along a Sixfold Axis.* For crystals in the hexagonal system, the *Z* axis is always sixfold. According to Fig. 3.5 of Part B, Appendix 3 the slowness curves again cross at right angles and two are degenerate. All modes are pure.

J.   TRANSMISSION LINE MODEL FOR ANISOTROPIC SOLIDS

In Section 6.D a transmission line model was obtained for acoustic plane wave propagation in an isotropic medium. Some modification of this model is required for anisotropic solids. Assume a cubic medium, with propagation restricted to a cube face (Fig. 7.15). The wave fields are then functions of x, z, and t only, and matrix representations of the acoustic field equations are therefore

$$
\begin{bmatrix}
\dfrac{\partial}{\partial x} & 0 & 0 & 0 & \dfrac{\partial}{\partial z} & 0 \\[2ex]
0 & 0 & 0 & \dfrac{\partial}{\partial z} & 0 & \dfrac{\partial}{\partial x} \\[2ex]
0 & 0 & \dfrac{\partial}{\partial z} & 0 & \dfrac{\partial}{\partial x} & 0
\end{bmatrix}
\begin{bmatrix}
T_1 \\[1ex] T_2 \\[1ex] T_3 \\[1ex] T_4 \\[1ex] T_5 \\[1ex] T_6
\end{bmatrix}
= \rho \dfrac{\partial}{\partial t}
\begin{bmatrix}
v_x \\[1ex] v_y \\[1ex] v_z
\end{bmatrix}
-
\begin{bmatrix}
F_x \\[1ex] F_y \\[1ex] F_z
\end{bmatrix}
$$

and

$$
\begin{bmatrix}
c_{11} & c_{12} & c_{12} & 0 & 0 & 0 \\[1ex]
c_{12} & c_{11} & c_{12} & 0 & 0 & 0 \\[1ex]
c_{12} & c_{12} & c_{11} & 0 & 0 & 0 \\[1ex]
0 & 0 & 0 & c_{44} & 0 & 0 \\[1ex]
0 & 0 & 0 & 0 & c_{44} & 0 \\[1ex]
0 & 0 & 0 & 0 & 0 & c_{44}
\end{bmatrix}
\begin{bmatrix}
\dfrac{\partial v_x}{\partial x} \\[2ex]
0 \\[2ex]
\dfrac{\partial v_z}{\partial z} \\[2ex]
\dfrac{\partial v_y}{\partial z} \\[2ex]
\dfrac{\partial v_z}{\partial x} + \dfrac{\partial v_x}{\partial z} \\[2ex]
\dfrac{\partial v_y}{\partial x}
\end{bmatrix}
= \dfrac{\partial}{\partial t}
\begin{bmatrix}
T_1 \\[1ex] T_2 \\[1ex] T_3 \\[1ex] T_4 \\[1ex] T_5 \\[1ex] T_6
\end{bmatrix}
$$

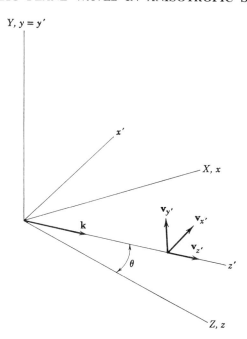

FIGURE 7.15. **Plane wave propagation in the** XZ **plane of a cubic medium.**

In the isotropic case these were found to separate into three sets of independent equations. Here there are only two independent sets,

$$\frac{\partial}{\partial x} T_1 + \frac{\partial}{\partial z} T_5 = \rho \frac{\partial v_x}{\partial t} - F_x$$

$$\frac{\partial}{\partial z} T_3 + \frac{\partial}{\partial x} T_5 = \rho \frac{\partial v_z}{\partial t} - F_z$$

$$c_{11} \frac{\partial v_x}{\partial x} + c_{12} \frac{\partial v_z}{\partial z} = \frac{\partial}{\partial t} T_1$$

$$c_{12} \frac{\partial v_x}{\partial x} + c_{11} \frac{\partial v_z}{\partial z} = \frac{\partial}{\partial t} T_3$$

$$c_{44} \left(\frac{\partial v_z}{\partial x} + \frac{\partial v_x}{\partial z} \right) = \frac{\partial}{\partial t} T_5 \tag{7.90}$$

and

$$\frac{\partial}{\partial z} T_4 + \frac{\partial}{\partial x} T_6 = \rho \frac{\partial}{\partial t} v_y - F_y$$

$$c_{44} \frac{\partial v_y}{\partial z} = \frac{\partial}{\partial t} T_4$$

$$c_{44} \frac{\partial v_y}{\partial x} = \frac{\partial}{\partial t} T_6 . \tag{7.91}$$

In addition there is an equation

$$c_{12} \frac{\partial v_x}{\partial x} + c_{12} \frac{\partial v_z}{\partial z} = \frac{\partial}{\partial t} T_2 ,$$

which determines the remaining stress component T_2.

To obtain a transmission line representation of this system, transformation is made to the rotated coordinates in Fig. 7.15. Since the fields of a *uniform* plane wave traveling along z' are independent of the coordinate x',

$$\frac{\partial}{\partial x} = \frac{\partial}{\partial z'} \sin \theta$$

$$\frac{\partial}{\partial z} = \frac{\partial}{\partial z'} \cos \theta \tag{7.92}$$

and the first equation of the set (7.91) becomes

$$\frac{\partial}{\partial z'} (T_4 \cos \theta + T_6 \sin \theta) = \rho \frac{\partial}{\partial t} v_{y'} - F_{y'} .$$

This can be rewritten as

$$\frac{\partial}{\partial z'} T_{4'} = \rho \frac{\partial}{\partial t} v_{y'} - F_{y'} , \tag{7.93}$$

where

$$T_{4'} = T_4 \cos \theta + T_6 \sin \theta = T_{y'z'}$$

is the y' component of traction force on a surface normal to z'. The same combination of stress components is obtained by multiplying the second and third equations in (7.91) by $\cos \theta$ and $\sin \theta$, respectively, and adding. After the partial derivatives have been expressed in terms of $\partial/\partial z'$, this gives

$$c_{44} \frac{\partial}{\partial z'} v_{y'} = \frac{\partial}{\partial t} T_{4'} . \tag{7.94}$$

When $\theta = 0$, (7.93) and (7.94) reduce to (6.36). Accordingly, $-T_{4'}$ and $v_{y'}$ can again be identified with voltage and current in the transmission line model (Fig. 7.16a). This is the pure shear wave that propagates isotropically in the xz plane and is polarized along $y = y'$.

The set of equations (7.90) relates to the quasilongitudinal and quasishear waves polarized in the xz plane. The first and second equations of the set are expressed in terms of the particle velocity components

$$v_{z'} = v_z \cos \theta + v_x \sin \theta$$

$$v_{x'} = -v_z \sin \theta + v_x \cos \theta$$

by multiplying with appropriate trigonometric functions and combining. This gives

$$\frac{\partial}{\partial z'} T_{3'} = \rho \frac{\partial}{\partial t} v_{z'} - F_{z'} \tag{7.95}$$

$$\frac{\partial}{\partial z'} T_{5'} = \rho \frac{\partial}{\partial t} v_{x'} - F_{x'} \tag{7.96}$$

where

$$T_{3'} = T_1 \sin^2 \theta + T_5 \sin 2\theta + T_3 \cos^2 \theta = T_{z'z'}$$

and

$$T_{5'} = T_1 \sin \theta \cos \theta + T_5 \cos 2\theta - T_3 \cos \theta \sin \theta = T_{x'z'}$$

are the z' and x' components, respectively, of the traction force on a surface normal to z'. Two other equations in the same variables are obtained by suitably combining the remaining equations in (7.90),

$$\left(c_{11}(\cos^4 \theta + \sin^4 \theta) + c_{12} \frac{\sin^2 2\theta}{2} + c_{44} \sin^2 2\theta \right) \frac{\partial v_{z'}}{\partial z'}$$

$$- \frac{\sin 4\theta}{4} (c_{11} - c_{12} - 2c_{44}) \frac{\partial v_{x'}}{\partial z'} = \frac{\partial T_{3'}}{\partial t} \tag{7.97}$$

and

$$\left(c_{44} \cos^2 2\theta + c_{11} \frac{\sin^2 2\theta}{2} - c_{12} \frac{\sin^2 2\theta}{2} \right) \frac{\partial v_{x'}}{\partial z'}$$

$$- \frac{\sin 4\theta}{4} (c_{11} - c_{12} - 2c_{44}) \frac{\partial v_{z'}}{\partial z'} = \frac{\partial T_{5'}}{\partial t}. \tag{7.98}$$

Equations (7.97) and (7.98) are coupled together through the stiffness matrix. If $\theta = 0$, or the isotropy condition ($c_{12} = c_{11} - 2c_{44}$) is satisfied, this coupling between longitudinal and horizontal shear motions vanishes. Equations (7.95) and (7.97) then reduce to (6.37), for a pure longitudinal wave, and Eqs. (7.96) and (7.98) reduce to (6.35), for a pure shear wave. If the same transmission line analogy is used for arbitrary θ, the model is a pair

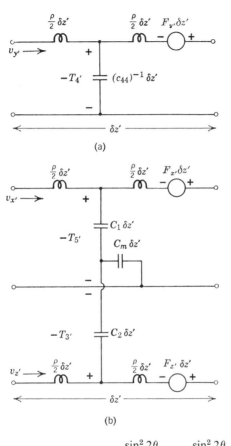

(a)

(b)

$$C_1^{-1} + C_m^{-1} = c_{44} \cos^2 2\theta + c_{11} \frac{\sin^2 2\theta}{2} - c_{12} \frac{\sin^2 2\theta}{2}$$

$$C_2^{-1} + C_m^{-1} = c_{11}(\cos^4 \theta + \sin^4 \theta) + c_{12} \frac{\sin^2 2\theta}{2} + c_{44} \sin^2 2\theta$$

$$C_m^{-1} = -(c_{11} - c_{12} - 2c_{44}) \frac{\sin 4\theta}{4}$$

FIGURE 7.16. **Transmission line model for uniform plane wave propagation in the** XZ **plane of a cubic crystal (Fig. 7.15).**

of capacitively coupled transmission lines (Fig. 7.16b). The coupling vanishes for the pure mode directions $\theta = 0$, $\pi/4$, $\pi/2$, $3\pi/4$.

In this example, the transmission line representing one of the waves is completely decoupled from the other two. For propagation in an arbitrary crystal direction the transmission line model generally consists of three mutually coupled lines.

EXAMPLE 10.   Examples 3 and 4 of Chapter 6 demonstrated the convenience of the normal mode formulation for solving problems of acoustic wave excitation by body force source distributions. It is instructive to consider how this approach may be applied to the coupled transmission line model that occurs in anisotropic problems. After introduction of minus signs as in (6.40), the set of equations (7.95)–(7.98) may be written as

$$\frac{\partial}{\partial z'}(-T_{3'}) = -\rho\frac{\partial}{\partial t}v_{z'} + F_{z'} \tag{7.99}$$

$$\frac{\partial}{\partial z'}(-T_{5'}) = -\rho\frac{\partial}{\partial t}v_{x'} + F_{x'} \tag{7.100}$$

$$c'_{11}\frac{\partial v_{z'}}{\partial z'} + c'_{15}\frac{\partial v_{x'}}{\partial z'} = -\frac{\partial}{\partial t}(-T_{3'}) \tag{7.101}$$

$$c'_{15}\frac{\partial v_{z'}}{\partial z'} + c'_{55}\frac{\partial v_{x'}}{\partial z'} = -\frac{\partial}{\partial t}(-T_{5'}). \tag{7.102}$$

Before applying the normal mode method it is necessary to decouple these equations. This may be accomplished by finding a transformation of variables that diagonalizes the matrix

$$\begin{bmatrix} c'_{11} & c'_{15} \\ c'_{15} & c'_{55} \end{bmatrix} \tag{7.103}$$

Application of this transformation to $(v_{z'}, v_{x'})$, $(T_{3'}, T_{5'})$ and $(F_{z'}, F_{x'})$ in (7.99)–(7.102) produces two pairs of decoupled equations, and the normal mode formalism can then be applied as in Examples 3 and 4 of Chapter 6.

K.   ACOUSTIC IMPEDANCE IN ANISOTROPIC SOLIDS

From (6.45) the characteristic acoustic impedances for pure longitudinal and shear waves propagating along the z axis in an isotropic medium are defined as

$$(Z_a)_l = -\frac{T_3}{v_z} = (\rho c_{11})^{1/2}$$

$$(Z_a)_s = -\frac{T_5}{v_x} = (\rho c_{44})^{1/2}$$

$$(Z_a)_{s'} = -\frac{T_4}{v_y} = (\rho c_{44})^{1/2}. \tag{7.104}$$

In an anisotropic medium, pure modes can exist for certain propagation directions (Section 7.I), and for these directions the characteristic impedance is completely analogous to (7.104). For example, the pure shear wave propagating in the XZ plane of a cubic crystal (Fig. 7.15) has a characteristic impedance

$$(Z_a)_{s'} = -\frac{T_{4'}}{v_{y'}} = (\rho c_{44})^{1/2},$$

from (7.93) and (7.94). This is in accord with the transmission line model for this wave (Fig. 7.16a). The quasilongitudinal and quasishear waves, on the other hand, are represented by a pair of coupled transmission lines (Fig. 7.16b). In this case, the stresses and particle velocities at any cross section are related by a 2×2 characteristic impedance matrix,

$$\begin{bmatrix} -T_{5'} \\ -T_{3'} \end{bmatrix} = \begin{bmatrix} (Z_a)_{5'x'} & (Z_a)_{5'y'} \\ (Z_a)_{3'x'} & (Z_a)_{3'y'} \end{bmatrix} \begin{bmatrix} v_{x'} \\ v_{y'} \end{bmatrix}$$

for each of the coupled wave solutions.

For a general propagation direction, where the transmission line model consists of three coupled lines, each plane wave solution has three stress components and three velocity components. In the transmission line model the stress components equivalent to transmission line voltages are components of the traction force on surfaces normal to the propagation direction $\hat{\mathbf{k}}$. For a general formulation of the characteristic impedance, it is useful to identify these stress components as

$$(\mathbf{T}_k)_i = T_{ik}$$

$$i = x, y, z$$

$$\hat{\mathbf{k}} = \text{propagation direction.} \tag{7.105}$$

The characteristic impedance for a plane wave propagating along $\hat{\mathbf{k}}$ is then defined by

$$\begin{bmatrix} -T_{xk} \\ -T_{yk} \\ -T_{zk} \end{bmatrix} = \begin{bmatrix} (Z_a^k)_{xx} & (Z_a^k)_{xy} & (Z_a^k)_{xz} \\ (Z_a^k)_{yx} & (Z_a^k)_{yy} & (Z_a^k)_{yz} \\ (Z_a^k)_{zx} & (Z_a^k)_{zy} & (Z_a^k)_{zz} \end{bmatrix} \begin{bmatrix} v_x \\ v_y \\ v_z \end{bmatrix}$$

or

$$-T_{ik} = (Z_a^k)_{ij} v_j$$

$$i, j = x, y, z$$

$$\hat{\mathbf{k}} = \text{propagation direction.} \tag{7.106}$$

For a given propagation direction $\hat{\mathbf{k}}$ there are three different impedance matrices, corresponding to one quasilongitudinal and two quasishear wave solutions.

Elements of the impedance matrix in (7.106) are easily found from the acoustic field equations. For complex exponential wave solutions (6.1), with $\mathbf{F} = 0$, becomes

$$-\mathbf{k} \cdot \mathbf{T} = \rho \omega \mathbf{v}. \tag{7.107}$$

Since the stress tensor is symmetric, the traction force normal to $\hat{\mathbf{k}}$ is

$$\mathbf{T}_k = \mathbf{T} \cdot \hat{\mathbf{k}} = \hat{\mathbf{k}} \cdot \mathbf{T},$$

and (7.107) is equivalent to

$$-\mathbf{T}_k = \frac{\rho \omega}{k} \mathbf{v}. \tag{7.108}$$

This shows that the acoustic characteristic impedance matrix in (7.106) is equal to the identity matrix multiplied by

$$Z_a = \frac{\rho \omega}{k} = \rho V_p, \tag{7.109}$$

where V_p is the phase velocity.† For an isotropic medium it is easily confirmed that (7.109) is equivalent to (7.104).

EXAMPLE 11. Because of the extremely simple form of (7.108), the characteristic impedance of a plane wave propagating in an arbitrary direction is easily obtained from inspection of the slowness curves for the wave in question. Consider, for instance, the quasilongitudinal wave in Fig. 7.8a. This has its wave vector directed along the Y axis in quartz. Scaling from Fig. 3.7 in Part B.4 of Appendix 3, the phase velocity is found to be

$$V_p = 1/(k_Y/\omega) \cong 6.08 \times 10^3 \text{ m/s}$$

From Part A.1 of Appendix 2 the density of quartz is

$$\rho = 2651 \text{ kg/m}^3,$$

and the acoustic characteristic impedance is therefore

$$Z_a = 16.1 \times 10^6 \frac{\text{watts/m}^2}{(\text{m/s})^2}. \tag{7.110}$$

---

† See Problem 22 at the end of the chapter for an alternative form of the impedence matrix.

The characteristic impedance Z_a is of great practical importance because it allows one to calculate the magnitude of the particle velocity directly from the power flow density. From (5.77) the component of the complex Poynting vector along $\mathbf{k}$ is

$$\mathbf{P} \cdot \hat{\mathbf{k}} = \left(-\frac{\mathbf{v}^* \cdot \mathbf{T}}{2}\right) \cdot \hat{\mathbf{k}} = -\frac{\mathbf{v}^* \cdot \mathbf{T} \cdot \hat{\mathbf{k}}}{2},$$

which is the same as

$$-\frac{\mathbf{v}^* \cdot \mathbf{T}_k}{2}.$$

Accordingly,

$$\mathbf{P} \cdot \hat{\mathbf{k}} = \tfrac{1}{2} Z_a |v|^2 = (P_{AV})_k \tag{7.111}$$

is equal to the average power density at the surface of the transducer in Fig. 7.8. The particle velocity in the beam is therefore

$$|v| = \left(\frac{2}{Z_a} (P_{AV})_k\right)^{1/2}. \tag{7.112}$$

From (7.110) this gives

$$|v| = 3.52 \times 10^{-2} \text{ m/s}$$

at a power density of 1 W/cm² $= 10^4$ W/m².

For calculations of plane wave scattering at plane boundaries, to be considered in Chapter 9, it is desirable to use an impedance measured normal to the boundary, rather than in the direction of wave propagation. The appropriate generalization of (7.106) is then

$$-T_{in} = (Z_a^n)_{ij} v_j, \tag{7.113}$$

where $\hat{\mathbf{n}}$ is the direction in which the impedance is measured. From (6.4),

$$\mathbf{T}_n = \hat{\mathbf{n}} \cdot \mathbf{T} = \frac{\hat{\mathbf{n}} \cdot \mathbf{c} : \nabla_s \mathbf{v}}{i\omega}.$$

In matrix notation, this becomes

$$-T_{in} = \frac{n_{iK} c_{KL} k_{Lj}}{\omega} v_j,$$

where

$$n_{iK} \rightarrow \begin{bmatrix} n_x & 0 & 0 & 0 & n_z & n_y \\ 0 & n_y & 0 & n_z & 0 & n_x \\ 0 & 0 & n_z & n_y & n_x & 0 \end{bmatrix} \tag{7.114}$$

and k_{Lj} is defined by (6.9). The acoustic impedance matrix elements for the direction $\hat{\mathbf{n}}$ are therefore

$$(Z_a^n)_{ij} = \frac{n_{iK}c_{KL}k_{Lj}}{\omega}$$

$$i, j = x, y, z .\qquad (7.115)$$

These impedance matrix elements are easily obtained by a simple modification of the Christoffel matrix in (6.11).

EXAMPLE 12.   The acoustic impedance matrix (7.115) is easily evaluated by substituting appropriate stiffness constants into the expressions given in Problem 23 at the end of the chapter. For a cubic crystal

$$[Z_a^n] =$$

$$\frac{k}{\omega}\begin{bmatrix} c_{11}n_xl_x + c_{44}(n_zl_z + n_yl_y) & c_{12}n_xl_y + c_{44}n_yl_x & c_{12}n_xl_z + c_{44}n_zl_x \\ c_{12}n_yl_x + c_{44}n_xl_y & c_{11}n_yl_y + c_{44}(n_zl_z + n_xl_x) & c_{12}n_yl_z + c_{44}n_zl_y \\ c_{12}n_zl_x + c_{44}n_xl_z & c_{12}n_zl_y + c_{44}n_yl_z & c_{11}n_zl_z + c_{44}(n_yl_y + n_xl_x) \end{bmatrix}$$

L.   FARADAY ROTATION AND ROTARY ACTIVITY

In previous chapters a number of instances have been given of analogous physical effects in electromagnetic and acoustic wave propagation—combining of degenerate transverse waves to produce circular and elliptical polarizations, birefringence, polarization conversion by means of quarter-wave and half-wave plates†, and so on. Two other effects should also be mentioned. These are *Faraday rotation* and *rotary activity*. Phenomenologically they both cause the plane of polarization of a linearly polarized transverse wave to rotate as the wave propagates, but they are fundamentally quite different. This rotation phenomenon is distinguished from the polarization transformations occurring in a birefringent medium (Fig. 3.8) by the fact that the polarization remains linear and merely changes direction, rather than converting from linear to elliptical to circular and back again.

These polarization rotation effects occur when positive- and negative-circularly polarized waves propagate with different phase velocities. To illustrate this point, let $A_x e^{i\omega t}$ and $A_y e^{i\omega t}$ represent the x and y components of a vector field (**H** in the electromagnetic case and **v** in the acoustic case).

---

† A half-wave plate is a birefringent plate cut with the correct thickness for reversing the rotation direction of a circularly polarized wave. See Problem 10 in Chapter 3, and also F. A. Jenkins and H. E. White, *Fundamentals of Optics*, pp 529-530, McGraw-Hill, New York, 1950.

If the field is circularly polarized in the *positive* (or *clockwise*) sense with respect to the $+z$ *direction*

$$A_y = -iA_x ; \tag{7.116}$$

for circular polarization in the *negative* sense

$$A_y = +iA_x . \tag{7.117}$$

Circularly polarized field amplitudes can thus be defined as

$$A_\circlearrowright = A_x + iA_y \tag{7.118}$$

and

$$A_\circlearrowleft = A_x - iA_y , \tag{7.119}$$

since $A_\circlearrowleft$ is zero for (7.116) and $A_\circlearrowright$ is zero for (7.117). In a medium that exhibits Faraday rotation or rotary activity, traveling wave solutions propagating along z have

$$\begin{aligned} A_\circlearrowright &= e^{-ik_\circlearrowright z} \\ A_\circlearrowleft &= e^{-ik_\circlearrowleft z} \end{aligned} \tag{7.120}$$

with $k_\circlearrowright^2 \neq k_\circlearrowleft^2$. Because these solutions are nondegenerate they cannot be superimposed to produce a wave with *constant* linear polarization, as was done in Example 4 of Chapter 3. When (7.118) and (7.119) are solved for A_x, A_y and the wave functions (7.120) are substituted for $A_\circlearrowright$, $A_\circlearrowleft$, the result

$$A_x = \frac{e^{-ik_\circlearrowright z} + e^{-ik_\circlearrowleft z}}{2} = e^{-i(k_\circlearrowright + k_\circlearrowleft)z/2} \cos \frac{(k_\circlearrowright - k_\circlearrowleft)z}{2}$$

$$A_y = \frac{e^{-ik_\circlearrowright z} - e^{-ik_\circlearrowleft z}}{2i} = -e^{-i(k_\circlearrowright + k_\circlearrowleft)z/2} \sin \frac{(k_\circlearrowright - k_\circlearrowleft)z}{2} \tag{7.121}$$

is obtained. This shows that the polarization angle θ relative to the y axis,

$$\theta(z) = \tan^{-1} \frac{A_y}{A_x} = -\frac{(k_\circlearrowright - k_\circlearrowleft)z}{2} \tag{7.122}$$

changes continuously as a function of z. When $k_\circlearrowright < k_\circlearrowleft$ the polarization rotation with increasing z is clockwise with respect to the $+z$ direction (Fig. 7.17a); when $k_\circlearrowright > k_\circlearrowleft$ the rotation is counterclockwise (Fig. 7.17b). This effect should be contrasted with the behavior of a *birefringent* medium (Fig. 3.8).

Faraday rotation and rotary activity arise from frequency and wavelength dependence (*temporal* and *spatial dispersion*) of the constitutive parameters. Frequency dispersion is required for Faraday rotation, and spatial dispersion for rotary activity. However, the effects do not necessarily occur when these dispersions are present. Certain other conditions must also be satisfied.

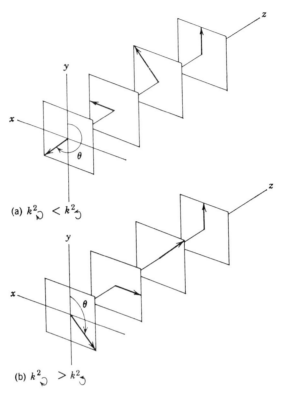

(a) $k^2_{\circlearrowleft} < k^2_{\circlearrowright}$

(b) $k^2_{\circlearrowleft} > k^2_{\circlearrowright}$

FIGURE 7.17. **Polarization rotation produced by combining nondegenerate circularly polarized shear waves.**

Physically, temporal and spatial dispersion of the constitutive parameters is due to interaction of the fields with "internal" degrees of freedom of the medium. In the permittivity matrix ϵ_{ij}, for example, *temporal dispersion* may be produced by interaction of the electric field with free charge carriers. Resonances of the carrier motion produce a frequency-dependent response to the field, and this appears as temporal dispersion in the permittivity. *Spatial dispersion* of the permittivity occurs when **D** at a given point depends on **E** not only at that point, but also at neighboring points. That is, the dielectric response is *nonlocal*. This phenomenon arises when the medium has "internal" degrees of freedom in the form of traveling waves. An example might be internal motions of molecules, coupled to each other by intermolecular forces. Dispersion of the permittivity matrix is expressed mathematically by writing the matrix elements as $\epsilon_{ij}(\omega, \mathbf{k})$. When the spatial dispersion is small, $\epsilon_{ij}(\omega, \mathbf{k})$ may be expanded in a power series in $\mathbf{k}$,

$$\epsilon_{ij}(\omega, \mathbf{k}) = \epsilon_{ij}(\omega) + ig_{ijl}k_l + \cdots. \tag{7.123}$$

In a similar way, temporal and spatial dispersion may appear in the compliance matrix s_{IJ}. The physical origins are, again, interactions of the field with "internal" degrees of freedom of the medium, and the effect is expressed formally by writing the compliance matrix elements in a form analogous to (7.123). That is,

$$s_{IJ}(\omega, \mathbf{k}) = s_{IJ}(\omega) + if_{IJl}(\omega)k_l + \cdots. \tag{7.124}$$

Chapter 5 presented arguments showing that the permittivity and compliance matrices are symmetric ($\epsilon_{ij} = \epsilon_{ji}$ and $s_{IJ} = s_{JI}$). An essential element in these arguments was the double requirement that the electric stored energy density be a function *only* of the electric field and that the strain stored energy density be a function *only* of the strain field. When the constitutive parameters are dispersive, owing to the above-mentioned interactions with "internal" degrees of freedom, energy is stored partially in these "internal" motions, and the previous proofs break down. For dispersive, *lossless* media, the conditions that $\boldsymbol{\epsilon}$ and $\mathbf{s}$ must be symmetric are then replaced by†

$$\epsilon_{ij}^*(\omega, \mathbf{k}) = \epsilon_{ji}(\omega, \mathbf{k}) \tag{7.125}$$

and

$$s_{IJ}^*(\omega, \mathbf{k}) = s_{JI}(\omega, \mathbf{k}). \tag{7.126}$$

That is, the permittivity and compliance matrices may now be *unsymmetric* ($\epsilon_{ij}(\omega, \mathbf{k}) \neq \epsilon_{ji}(\omega, \mathbf{k})$ and $s_{IJ}(\omega, \mathbf{k}) \neq s_{JI}(\omega, \mathbf{k})$). This asymmetry is a *necessary* condition for the existence of Faraday rotation and rotary activity. The quantities $g_{ijl}(i \neq j)$ and $f_{IJl}(I \neq J)$ in (7.123) and (7.124) must therefore be pure real and satisfy the conditions

$$g_{ijl}(\omega) = -g_{jil}(\omega) \tag{7.127}$$

and

$$f_{IJl}(\omega) = -f_{JIl}(\omega). \tag{7.128}$$

Crystal symmetry restrictions on the constitutive parameters in (7.123) and (7.124) can be deduced by using the method that was applied to the compliance matrix in Section 7.B. That is, the parameters must remain unchanged when transformed by the group generator operations listed in Table 7.2. Symmetry characteristics for ϵ_{ij} are given in Part C.1 of Appendix 2. For g_{ijl} and f_{IJl}, which are elements of third rank and fifth rank tensors respectively, the calculations are more difficult. It can, however, be shown

† Reference 19 at the end of the chapter.

rather easily that g_{ijl} and f_{IJl} are always zero for crystals that have inversion symmetry.†

L.1 Faraday Rotation

Electromagnetic Faraday rotation occurs when the permittivity (or permeability) matrix is temporally dispersive and contains unsymmetric elements. Suppose that matrix (4.22) is modified by two terms of this kind; that is ‡

$$
[\epsilon] = \begin{bmatrix} \epsilon_{xx} & -i\bar\epsilon_{xy} & 0 \\ +i\bar\epsilon_{xy} & \epsilon_{xx} & 0 \\ 0 & 0 & \epsilon_{zz} \end{bmatrix} \tag{7.129}
$$

For an electromagnetic wave represented by the complex exponential wave functions

$$
\begin{aligned} (\hat{\mathbf{x}}E_x + \hat{\mathbf{y}}E_y)e^{i(\omega t - kz)} \\ (\hat{\mathbf{x}}H_x + \hat{\mathbf{y}}H_y)e^{i(\omega t - kz)} \end{aligned} \tag{7.130}
$$

Maxwell's equations reduce to

$$
\begin{aligned} -kE_y &= \omega\mu_0 H_x & kH_y &= \omega(\epsilon_{xx}E_x - i\bar\epsilon_{xy}E_y) \\ kE_x &= \omega\mu_0 H_y & -kH_x &= \omega(+i\bar\epsilon_{xy}E_x + \epsilon_{xx}E_y). \end{aligned} \tag{7.131}
$$

If the components E_x and E_y are eliminated, these equations become

$$
(k^2 - \omega^2\mu_0\epsilon_{xx})H_x + i\omega^2\mu_0\bar\epsilon_{xy}H_y = 0 \tag{7.132}
$$

$$
(k^2 - \omega^2\mu_0\epsilon_{xx})H_y - i\omega^2\mu_0\bar\epsilon_{xy}H_x = 0. \tag{7.133}
$$

Multiplication of (7.133) by $\pm i$ and addition of the results to (7.132) then gives

$$
\begin{aligned} \{k_\circlearrowleft^2 - \omega^2\mu_0(\epsilon_{xx} - \bar\epsilon_{xy})\}\, H_\circlearrowleft &= 0, \\ \{k_\circlearrowright^2 - \omega^2\mu_0(\epsilon_{xx} + \bar\epsilon_{xy})\}\, H_\circlearrowright &= 0, \end{aligned} \tag{7.134}
$$

where $H_\circlearrowleft$, $H_\circlearrowright$ are defined as in (7.118) and (7.119). This shows that the wave solutions are a positive-circularly polarized wave

$$
\begin{aligned} &H_\circlearrowleft e^{i(\omega t - k_\circlearrowleft z)} \\ k_\circlearrowleft^2 &= \omega^2\mu_0(\epsilon_{xx} - \bar\epsilon_{xy}) \end{aligned} \tag{7.135}
$$

† Reference 18 at the end of the chapter.
‡ The form of the off-diagonal matrix elements is dictated by the condition (7.125).

and a negative-circularly polarized wave

$$H_\circlearrowright e^{i(\omega t - k_\circlearrowright z)}$$
$$k_\circlearrowright^2 = \omega^2 \mu_0(\epsilon_{xx} + \bar\epsilon_{xy}). \tag{7.136}$$

According to (7.121) and (7.122), superposition of these solutions with equal amplitudes produces a wave with a wave number

$$k_{AV} = \frac{k_\circlearrowleft + k_\circlearrowright}{2} \tag{7.137}$$

and a polarization angle

$$\theta(z) = -\frac{k_\circlearrowleft - k_\circlearrowright}{2} z = -\frac{k_\circlearrowleft^2 - k_\circlearrowright^2}{4 k_{AV}} z = \frac{\omega^2 \mu_0 \bar\epsilon_{xy}}{2 k_{AV}} z. \tag{7.138}$$

For propagation in the $+z$ direction ($k_{AV} > 0$) the polarization rotates as in Fig. (7.17a); for propagation in the $-z$ direction ($k_{AV} < 0$) the polarization rotates as in Fig. (7.17b). A wave propagating in the $+z$ *direction* experiences a *clockwise* polarization rotation relative to its *propagation direction* (Fig. 7.17a); and a wave propagating in the $-z$ *direction* experiences a *counterclockwise* polarization rotation relative to its *propagation direction* (Fig. 7.17b). Waves propagating in opposite directions therefore undergo opposite senses of rotation relative to their propagation directions. This is an instance of *nonreciprocity*, a phenomenon with many practical consequences and applications.

Faraday rotation of electromagnetic waves is produced by interactions with either current carriers or magnetic spins *in the presence of a dc magnetic field*. The same kinds of interactions are known to produce Faraday rotation of transverse acoustic waves.[†] This may be illustrated by introducing the unsymmetric matrix elements

$$s_{45} = -i\bar{s}_{45}$$
$$s_{54} = +i\bar{s}_{45}$$

into the compliance matrix (4.23). If transverse fields of the form

$$T_4 e^{i(\omega t - kz)}$$
$$T_5 e^{i(\omega t - kz)}$$
$$(\hat{x} v_x + \hat{y} v_y) e^{i(\omega t - kz)}$$

are assumed, the acoustic field equations can then be reduced to

$$v_y(k^2 - \omega^2 \rho s_{44}) + i\omega^2 \rho \bar{s}_{45} v_x = 0 \tag{7.139}$$
$$v_x(k^2 - \omega^2 \rho s_{44}) - i\omega^2 \rho \bar{s}_{45} v_y = 0. \tag{7.140}$$

† Reference 20 at the end of the chapter.

These are similar to (7.132) and (7.133), and solutions can be written down by direct analogy. The solution with positive-circular polarization is

$$v_\circlearrowright e^{i(\omega t - k_\circlearrowright z)} = (v_x + iv_y)e^{i(\omega t - k_\circlearrowright z)}$$
$$k_\circlearrowright^2 = \omega^2 \rho(s_{44} + \bar{s}_{45}) \tag{7.141}$$

and the solution with negative-circular polarization is

$$v_\circlearrowright e^{i(\omega t - k_\circlearrowright z)} = (v_x - iv_y)e^{i(\omega t - k_\circlearrowright z)}$$
$$k_\circlearrowright^2 = \omega^2 \rho(s_{44} - \bar{s}_{45}). \tag{7.142}$$

Faraday rotation follows from these results in the manner described above.

L.2   Rotary Activity

The equations derived in Part (1) also serve to illustrate rotary activity. In the electromagnetic (or optical) case the linear spatially dispersive terms in (7.123) are the significant ones. For a z-propagating wave, the substitution

$$\bar{\epsilon}_{xy} \mapsto g_{xyz}k$$

is made in (7.131). The calculation proceeds as before, giving

$$k_\circlearrowright^2 = \omega^2 \mu_0(\epsilon_{xx} - g_{xyz}k_\circlearrowright) \tag{7.143}$$

for the positive-circularly polarized solution and

$$k_\circlearrowright^2 = \omega^2 \mu_0(\epsilon_{xx} + g_{xyz}k_\circlearrowright) \tag{7.144}$$

for the negative-circularly polarized solution. From (7.137) and (7.138) a linear superposition of these solutions with equal amplitudes propagates with a wave number k_{AV} and a polarization angle

$$\theta(z) = \frac{(\omega^2 \mu_0 g_{xyz})}{2} z . \tag{7.145}$$

Since the polarization formula does not contain k_{AV}, the polarization rotates as in Fig. 7.17a for waves traveling along *both* $+z$ and $-z$ directions. Waves traveling in opposite directions therefore undergo the *same* sense of polarization rotation relative to their propagation directions. This is quite different from the nonreciprocal behavior of Faraday rotation. In the case of rotary activity the polarization rotation of a wave traveling toward $+z$ is completely removed when the wave is reflected and propagates back to its starting point again. For Faraday rotation, on the other hand, the polarization rotation *relative to the* $+z$ *axis* continues to increase in the reflected wave, and a net rotation remains when the wave has returned to its starting point. Acoustic rotary activity may be illustrated in the same way by simply substituting

$f_{45z}k$ for $\bar{s}_{45}$ in (7.141) and (7.142). That is,

$$k_{\circlearrowright}^2 = \omega^2 \rho(s_{44} + f_{45z}k_{\circlearrowright})$$
$$k_{\circlearrowleft}^2 = \omega^2 \rho(s_{44} - f_{45z}k_{\circlearrowleft}).$$

(7.146)

Although optical activity has been known since 1811, direct experimental observation of acoustical activity was not achieved until 1970.[†] These measurements were made for propagation along the Z axis of trigonal quartz, which belongs to one of the crystal classes (*32*) that is acoustically active for propagation along the principal symmetry axis. In the absence of spatial dispersion, shear waves propagating along the Z axis of a trigonal crystal are degenerate (Fig. 3.7 in Part B.4 of Appendix 2). For off-axis propagation the waves are nondegenerate and linearly polarized at right angles to each other. It has been seen from the examples above that spatial dispersion couples these two waves, leading to nondegenerate circularly polarized waves propagating along the Z axis. For off-axis propagation the coupled waves are elliptically polarized, becoming more and more linear as the propagation angle increases (Fig. 7.18). According to (7.146) the phase velocity difference for the circularly polarized waves is

$$(V_s^2)_{\circlearrowright} - (V_s^2)_{\circlearrowleft} = 2(V_s)_{AV}\,\delta V_s \approx -\frac{2f_{45z}k_{AV}}{\rho s_{44}^2}.$$

(7.147)

That is, the velocity difference δV_s is proportional to ω. The same conclusion applies to Z-propagating waves in quartz, where

$$\frac{\delta V_s}{(V_s)_{AV}} = 3.3 \times 10^{-4}\omega\,(GHz)$$

(7.148)

has been obtained from Brillouin scattering measurements. The dashed curves in Fig. 7.18 are computed from the elastic constants for quartz (as in Part A.2 of Appendix 2) and have the *linear* polarizations shown. The solid curves are for the frequency (28.9 GHz) at which the Brillouin measurements were performed.

PROBLEMS

1. Construct transformation matrices $[M]$ and $[N]$ for the two generator transformations *4* and $\bar{m}$ of the tetragonal crystal class $4/m$, and use (7.5) to show that the stiffness matrix has the form given in Part A.2 of Appendix 2.

2. Use (7.11) and Part A.2 of Appendix 2 to find the Christoffel matrix for a monoclinic crystal.

[†] References 21 and 22 at the end of the chapter.

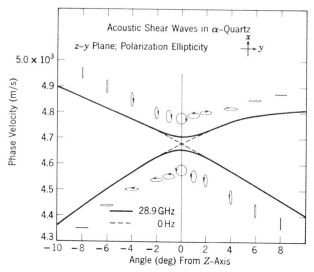

FIGURE 7.18. **Measured rotary activity in quartz (after A. S. Pine).**

3. Show that the plane wave solutions with **k** directed along the crystal axis Z of a monoclinic crystal are:

Y-polarized pure shear, $c_{44}k^2 = \rho\omega^2$

XZ-polarized quasishear, $\left(c_{33} + c_{55} - \sqrt{(c_{33} - c_{55})^2 + 4c_{35}^2}\right)k^2 = 2\rho\omega^2$

XZ-polarized quasilongitudinal, $\left(c_{33} + c_{55} + \sqrt{(c_{33} - c_{55})^2 + 4c_{35}^2}\right)k^2 = 2\rho\omega^2$.

4. Starting from the transformed stiffness matrix (3.44), derive the slowness curves (7.34) to (7.36) for propagation in a face diagonal plane of a cubic crystal.

5. Use the stiffness matrix (3.46) to derive slowness curves for propagation in the plane normal to [111] in Fig. 3.7b.

6. Taking material constants from Appendix 2, calculate the anisotropy factor A for yttrium aluminum garnet, yttrium iron garnet, aluminum, gold, and silver. Qualitatively sketch the slowness curves in Fig. 7.2, Fig. 7.3, and Problem 5 for all of these materials.

7. Transform the stiffness matrix for a hexagonal crystal to a coordinate system rotated through an angle ξ about the Z axis. Prove from this that the slowness surface is rotationally symmetric about Z.

8. Starting with the analysis of Problem 4, find the polarization directions for the quasilongitudinal and quasishear solutions as a function of the wave vector angle θ in Fig. 7.3. Verify that these waves are polarized at right angles to each other for all values of θ.

9.   Using the formulas derived in Problem 8, calculate and graph the quasishear wave polarization angle as a function of θ for yttrium aluminum garnet, yttrium iron garnet, aluminum, gold, and silver. What is the maximum angular deviation of $\mathbf{v}$ from $\mathbf{k}$ for each material?

10.   The particle velocity fields for quasilongitudinal and quasishear waves propagating in a cube diagonal plane were found in Problem 8. Calculate the complex Poynting vector $\mathbf{P} = -\mathbf{v}^* \cdot \mathbf{T}/2$ and the energy velocity $\mathbf{V}_e$ as a function of θ.

11.   Apply the method described in Example 8(b) to find the group velocity $\mathbf{V}_g$ for Problem 10 as a function of θ. Show that the expressions for $\mathbf{V}_g$ and $\mathbf{V}_e$ are equivalent.

12.   Use the group velocity formula obtained in Problem 11 and material constants given in Appendix 2 to graph the group velocity directions in gallium arsenide as a function of θ. (Neglect the piezoelectric effect.) Compare with group velocity directions obtained by measuring normal directions to the slowness curve in Fig. 7.3.

13.   Derive (7.76) for the two-dimensional case by superposing the four plane waves

$$\cos\left[(\omega - \delta\omega)t - (k - \delta k_z)z\right]$$
$$\cos\left[(\omega + \delta\omega)t - (k + \delta k_z)z\right]$$
$$\cos\left[(\omega - \delta\omega)t - (kz - \delta k_y y)\right]$$
$$\cos\left[(\omega + \delta\omega)t - (kz + \delta k_y y)\right]$$

and finding the motion of the modulation envelope.

14.   The *complex reciprocity relation* (Section 10.J in Volume II) states that

$$\nabla \cdot (-\mathbf{v}_2^* \cdot \mathbf{T}_1 - \mathbf{v}_1 \cdot \mathbf{T}_2^*) = -\frac{\partial}{\partial t}(\rho\mathbf{v}_1 \cdot \mathbf{v}_2^* + \mathbf{T}_1 : \mathbf{s} : \mathbf{T}_2^*)$$

for any two field solutions in a source-free nonpiezoelectric medium. Assume that the two solutions are plane waves of the same type, but with different frequencies and wave vector directions; that is,

$$\mathbf{v}_1 \rightarrow \mathbf{v}e^{i(\omega t - \mathbf{k}\cdot\mathbf{r})}$$
$$\mathbf{T}_1 \rightarrow \mathbf{T}e^{i(\omega t - \mathbf{k}\cdot\mathbf{r})}$$
$$\mathbf{v}_2 \rightarrow (\mathbf{v} + \delta\mathbf{v})e^{i[(\omega + \delta\omega)t - (\mathbf{k} + \delta\mathbf{k})\cdot\mathbf{r}]}$$
$$\mathbf{T}_2 \rightarrow (\mathbf{T} + \delta\mathbf{T})e^{i[(\omega + \delta\omega)t - (\mathbf{k} + \delta\mathbf{k})\cdot\mathbf{r}]},$$

where $\mathbf{v}$, $\mathbf{T}$, $\delta\mathbf{v}$, $\delta\mathbf{T}$ are complex constants. Show that

$$\delta\mathbf{k} \cdot (-\mathbf{v}^* \cdot \mathbf{T} - \mathbf{v} \cdot \mathbf{T}^*) = \delta\omega(\rho\mathbf{v} \cdot \mathbf{v}^* + \mathbf{T} : \mathbf{s} : \mathbf{T}^*)$$

to first order in the quantities $\delta\mathbf{k}$, $\delta\omega$, $\delta\mathbf{v}$, $\delta\mathbf{T}$, and use this result to prove that

$$\mathbf{V}_g = \hat{\mathbf{x}}\frac{\partial\omega}{\partial k_x} + \hat{\mathbf{y}}\frac{\partial\omega}{\partial k_y} + \hat{\mathbf{z}}\frac{\partial\omega}{\partial k_z} = \mathbf{V}_e\,.$$

15. Problem 10 showed that the quasilongitudinal and quasishear waves propagating in one of the symmetry planes of a cubic crystal have Poynting vectors that lie in the same plane. Use symmetry arguments, as in Section I, to show that this is always true for waves propagating in a crystal symmetry plane. Prove that $\mathbf{P}$ is always parallel to $\mathbf{k}$ when $\mathbf{k}$ lies along a 2-fold, 4-fold, or 6-fold symmetry axis and that $\mathbf{P}$ is parallel to $\mathbf{k}$ for a *longitudinal* wave when $\mathbf{k}$ lies along a 3-fold symmetry axis.

16. Derive the expression given in Fig. 7.3 for the crossing angle of the outer slowness curves at the [111] axis. (Assume $\mathbf{k} = \mathbf{k}_{[111]} + \delta\mathbf{k}$ and retain only first order terms in $\delta\mathbf{k}$.)

17. In Problem 11 of Chapter 3 it was seen that viscous damping can be taken into account by introducing the complex elastic stiffness constants

$$c_{IJ} + i\omega\eta_{IJ}\,,$$

where the viscosity matrix η_{IJ} has the same form as c_{IJ}. For cubic crystals there are, therefore, three independent viscosity constants η_{11}, η_{12}, and η_{44}. Starting with the solutions (7.24) and assuming that the damping is small, show that the attenuation of plane waves propagating in the XZ plane of a cubic crystal is given in the low-loss approximation by

$$\alpha = \frac{\omega^2}{2}\left(\frac{\rho}{c_{\text{eff}}}\right)^{1/2}\frac{\eta_{\text{eff}}}{c_{\text{eff}}}$$

where

$$c_{\text{eff}} = c_{44}$$

$$\eta_{\text{eff}} = \eta_{44}$$

for the Y-polarized pure shear wave and

$$c_{\text{eff}} = \frac{c_{11} + c_{44} \pm \sqrt{(c_{11} - c_{44})^2 \cos^2 2\phi + (c_{12} + c_{44})^2 \sin^2 2\phi}}{2}$$

$$\eta_{\text{eff}} = \frac{\eta_{11} + \eta_{44}}{2}$$
$$\pm\frac{((c_{11} - c_{44})(\eta_{11} - \eta_{44})\cos^2 2\phi + (c_{12} + c_{44})(\eta_{12} + \eta_{44})\sin^2 2\phi)}{2\sqrt{(c_{11} - c_{44})^2 \cos^2 2\phi + (c_{12} + c_{44})^2 \sin^2 2\phi}}$$

for the quasilongitudinal ($+$ sign) and quasishear ($-$ sign) solutions.

18. The elastic properties of crystals are usually determined by the pulse echo technique. An elastic plane wave pulse is launched at one end of a sample which has flat and parallel end faces, and the time delay and amplitude

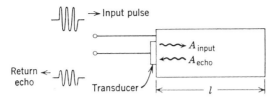

of the return echo from the other end are measured. (See, for example, Reference 9 in Chapter 3.) Show that the pulse delay is

$$\tau = \frac{2l}{V_p},$$

and the ratio of the echo amplitude to the input amplitude is

$$\frac{A_{echo}}{A_{input}} = e^{-2\alpha l}$$

where V_p and α are the phase velocity and attenuation of the plane wave involved.

In order to determine the various material constants needed to specify the elastic and viscous properties of a crystal, pulse echo measurements must be performed with several different wave types and crystal orientations. In these experiments it is desirable to have the Poynting vector parallel to $\mathbf{k}$, so that the situation illustrated in Fig. 7.8 may be avoided. Choose wave types and propagation directions to be used for measuring the stiffness and viscosity constants of crystals belonging to the cubic, hexagonal, and trigonal (32, 3m, 3̄m) systems.

19. Using Part C of Appendix 3 and material constants given in Appendix 2, find the pure mode directions for cadmium sulfide and locate these directions on Fig. 3.5 in Appendix 3. (Neglect the piezoelectric effect; that is, use the stiffness constants given in Part A.5 of Appendix 2.)

20. A set of rectangular Cartesian coordinate axes $(\hat{\mathbf{x}}, \hat{\mathbf{y}}, \hat{\mathbf{z}})$ is chosen so that $\hat{\mathbf{x}}$ lies along some pure mode direction. What conditions must be imposed on the stiffness matrix so that a wave propagating along $\hat{\mathbf{x}}$ has its Poynting vector $\mathbf{P}$ deflected from the wave vector direction? Are these conditions satisfied for the symmetric pure mode directions in Problem 19? Are they satisfied for the nonsymmetric pure mode directions in Problem 19?

21. Given

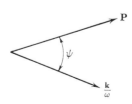

prove that

$$P_{\mathrm{AV}} = \frac{\rho V_p}{\cos \psi} \frac{\mathbf{v} \cdot \mathbf{v}^*}{2} .$$

Show how this formula and a set of slowness curves can be used to calculate the particle velocity amplitude of a wave from its power density.

22. For an acoustic wave with propagation direction $\hat{\mathbf{k}}$, the acoustic impedance measured in the direction $\hat{\mathbf{k}}$ is given by the scalar formula (7.109). When $\hat{\mathbf{n}} = \hat{\mathbf{k}}$, the more general expression (7.115) reduces to

$$(Z_a^k)_{ij} = \frac{\Gamma_{ij}}{\omega} k ,$$

where Γ_{ij} is the Christoffel matrix. Show that these results are consistent. Prove that

$$-T_{in} = \left((Z_a^n)_{ij} - \frac{\Gamma_{ij}}{\omega} + \rho V_p \, \delta_{ij} \right) v_j ,$$

where

$$\delta_{ij} = \begin{cases} 1, & i = j \\ 0, & i \neq j, \end{cases}$$

is an equivalent formulation of the impedance relation, which reduces to (7.108) when $\hat{\mathbf{n}} = \hat{\mathbf{k}}$.

23. Show that $[Z_a^n]$ in (7.115) has coefficients given by

$$\begin{aligned}
\omega(Z_a^n)_{xx} &= c_{11}n_x k_x + c_{66}n_y k_y + c_{55}n_z k_z + c_{56}(n_y k_z + k_y n_z) \\
&\quad + c_{15}(n_z k_x + k_z n_x) + c_{16}(n_x k_y + k_x n_y)
\end{aligned}$$

$$\begin{aligned}
\omega(Z_a^n)_{yy} &= c_{66}n_x k_x + c_{22}n_y k_y + c_{44}n_z k_z + c_{24}(n_y k_z + k_y n_z) \\
&\quad + c_{46}(n_z k_x + k_z n_x) + c_{26}(n_x k_y + k_x n_y)
\end{aligned}$$

$$\begin{aligned}
\omega(Z_a^n)_{zz} &= c_{55}n_x k_x + c_{44}n_y k_y + c_{33}n_z k_z + c_{34}(n_y k_z + k_y n_z) \\
&\quad + c_{35}(n_z k_x + k_z n_x) + c_{45}(n_x k_y + k_x n_y)
\end{aligned}$$

$$\begin{aligned}
\omega(Z_a^n)_{xy} &= c_{16}n_x k_x + c_{26}n_y k_y + c_{45}n_z k_z + c_{46}n_y k_z + c_{25}k_y n_z \\
&\quad + c_{14}k_z n_x + c_{56}n_z k_x + c_{12}n_x k_y + c_{66}k_x n_y
\end{aligned}$$

$$\begin{aligned}
\omega(Z_a^n)_{xz} &= c_{15}n_x k_x + c_{46}n_y k_y + c_{35}n_z k_z + c_{45}k_y n_z + c_{36}n_y k_z \\
&\quad + c_{13}k_z n_x + c_{55}n_z k_x + c_{14}n_x k_y + c_{56}k_x n_y
\end{aligned}$$

$$\begin{aligned}
\omega(Z_a^n)_{yz} &= c_{56}n_x k_x + c_{24}n_y k_y + c_{34}n_z k_z + c_{44}k_y n_z + c_{23}n_y k_z \\
&\quad + c_{36}k_z n_x + c_{45}n_z k_x + c_{25}k_x n_y + c_{46}n_z k_y
\end{aligned}$$

$$\begin{aligned}
\omega(Z_a^n)_{yx} &= c_{16}n_x k_x + c_{26}n_y k_y + c_{45}n_z k_z + c_{46}k_y n_z + c_{25}n_y k_z \\
&\quad + c_{14}n_z k_x + c_{56}k_z n_x + c_{12}k_x n_y + c_{66}n_x k_y , \quad \text{etc.}
\end{aligned}$$

Compare with (7.12).

24. In Example 12 assume that propagation is along $\hat{x}$ and that $\hat{n}$ lies in the xz plane, at an angle ϕ with $\hat{x}$. Evaluate the impedance matrix defined in Problem 22,

$$\left((Z_a^n)_{ij} - \frac{\Gamma_{ij}}{\omega} + \rho V_p\, \delta_{ij} \right),$$

for all three plane wave solutions.

25. Prove that the impedance transformation law derived for isotropic media in Problem 10 of Chapter 6 applies also to the impedances (7.109) for anisotropic media.

REFERENCES

Crystal Symmetries

1. S. Bhagavantam, *Crystal Symmetries and Physical Properties*, Ch. 5, Academic Press, New York, 1966.
2. W. Bond, "The Mathematics of the Physical Properties of Crystals," *BSTJ* **22**, 1–72, (1943).
3. M. J. Buerger, *Elementary Crystallography*, John Wiley & Sons, Inc., New York, 1956.
4. F. C. Phillips, An Introduction to Crystallography, Longmans, London, 1971.

Symmetry Properties of the Compliance and Stiffness Matrices

5. Reference 1, Ch. 6 and 11.
6. Reference 2, pp. 22–28.
7. J. F. Nye, *Physical Properties of Crystals*, pp. 137–142, Oxford, 1964.

Plane Wave Propagation

8. G. W. Farnell, "Elastic Waves in Trigonal Crystals," *Can. J. Phys.* **39**, 65–80 (1961).
9. F. I. Fedorov, *Theory of Elastic Waves in Crystals*, Ch. 3–7, Plenum, New York, 1968.
10. W. P. Mason, *Physical Acoustics and the Properties of Solids*, pp. 368–373, van Nostrand, New York, 1958.
11. M. J. P. Musgrave, *Crystal Acoustics*, Ch. 6–10, Holden-Day, San Francisco, 1970.
12. L. D. Landau and E. M. Lifshitz, *Theory of Elasticity*, pp. 103–105, Pergamon, New York, 1970.

13. A. Levelut, "Propagation de l'Energie Acoustique dans les Cristaux (Systemes Cubique, Hexagonal et Quadratique)," *Acta. Cryst.* **A25,** 553–563 (1969).

14. P. Waterman, "Orientation Dependence of Elastic Waves in Single Crystals," *Phys. Rev.* **113,** 1240–1253 (1959).

15. J. H. Staudt and B. D. Cook, "Visualization of Quasilongitudinal and Quasi-transverse Elastic Waves," *J. Acoust. Soc. Am.* **41,** 1547–1548 (1967).

Pure Mode Directions

16. F. E. Borgnis, "Specific Directions of Longitudinal Wave Propagation in Anisotropic Media," *Phys. Rev.* **98,** 1000–1005 (1955).

17. K. Brugger, "Pure Modes for Elastic Waves in Crystals," *J. Appl. Phys.* **36,** 759–768 (1965).

Faraday Rotation and Rotary Activity

18. D. L. Portigal and E. Burstein, "Acoustical Activity and other First-Order Spatial Dispersion Effects in Crystals," *Phys. Rev.* **170,** 673–678 (1968).

19. L. D. Landau and E. M. Lifshitz, *Statistical Physics* (translated by E. Peierls and R. F. Peierls), pp. 104 ff, Pergamon, New York, 1958.

20. R. C. LeCraw and R. L. Comstock, "Magnetoelastic Interactions in Ferro-magnetic Insulators," Chapter 4 in *Physical Acoustics,* Volume IIIB, W. P. Mason, Ed., Academic Press, New York, 1965.

21. A. S. Pine, "Direct Observation of Acoustical Activity in α Quartz," *Phys. Rev.* **B2,** 2049–2054 (1970).

22. J. Joffrin and A. Levelut, "Mise en Evidence et Mesure du Pouvoir Rotatoire Acoustique Naturel de Quartz-α," *Solid State Comm.* **8,** 1573–1575 (1970).

Chapter 8

PIEZOELECTRICITY

A. ONE-DIMENSIONAL MODEL OF THE PIEZOELECTRIC EFFECT

Section A of Chapter 4 introduced the concept of piezoelectric coupling between acoustic and electromagnetic fields, and then discussed in a qualitative fashion the physical reasons for the effect. This chapter will describe piezoelectric behavior in formal mathematical terms and then consider the properties of uniform plane wave propagation in piezoelectric solids. Before proceeding to the case of a continuous medium, however, it will be helpful to illustrate the basic piezoelectric mechanism by means of a simple model.

The model chosen (Figs. 8.1a and 8.2a) is an electrically-neutral system of charged particles connected by rigid and elastic bonds. All charges, which are supposed to represent the positive and negative ions in an actual solid, are constrained to move along the x axis, and the center point of the system is rigidly fixed at $x = 0$. It will be shown that this system,

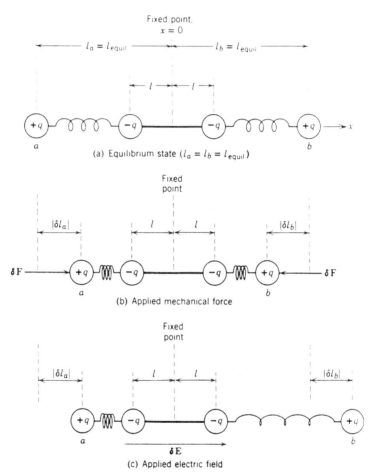

FIGURE 8.1 Model of a *nonpiezoelectric* solid

under the proper conditions, exhibits the kind of behavior characteristic of a piezoelectric solid, namely an electric response to applied mechanical forces and a mechanical response to applied electrical forces. By analogy with the stress applied to a solid, equal-and-opposite mechanical forces are applied to the ends of the model (Figs. 8.1b and 8.2b). Electrical forces are applied by means of an electric field (Figs. 8.1c and 8.2c). The mechanical response is defined as the change in length of the system,

$$\delta L = \delta l_a + \delta l_b \,, \tag{8.1}$$

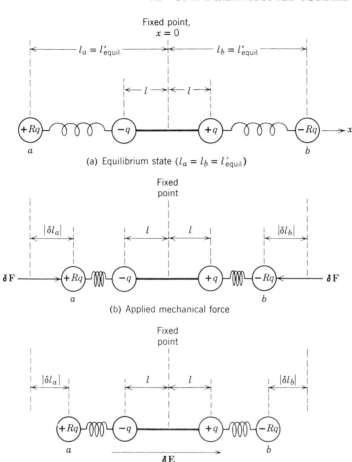

FIGURE 8.2. **Model of a *piezoelectric* solid.**

by analogy with the strain in a solid; and the electrical response is defined as the change in the total electric dipole moment P_x of the system,

$$P_x = q_n x_n \tag{8.2}$$

with n summed over all the charged particles.

The model is analyzed by first finding the equilibrium configuration and then calculating the configurational changes produced by the impressed forces. Equilibrium is determined by evaluating the combined spring and electrostatic forces acting on the movable charges a, b and setting these forces equal to zero. The springs are assumed to have *identical* force constants

K and unstrained lengths l_0. That is,

$$f_a = K(l_a - l - l_0) \tag{8.3}$$

for the left-hand spring, and

$$f_b = -K(l_b - l - l_0) \tag{8.4}$$

for the right-hand spring. The combined spring and electrostatic forces acting on charges a and b in Fig. 8.1a, for example, are therefore

$$(F_a)_x = K(l_a - l - l_0) + q^2\left(\frac{1}{(l_a - l)^2} + \frac{1}{(l_a + l)^2} - \frac{1}{(l_a + l_b)^2}\right)$$
$$(F_b)_x = -K(l_b - l - l_0) - q^2\left(\frac{1}{(l_b - l)^2} + \frac{1}{(l_b + l)^2} - \frac{1}{(l_a + l_b)^2}\right). \tag{8.5}$$

Consider first the system in Fig. 8.1. It is apparent from the symmetrical disposition of the components that the electrostatic forces will compress the left- and right-hand springs by equal amounts. The equilibrium spacing l_{equil} can thus be found by taking $l_a = l_b$ in (8.5) and equating either $(F_a)_x$ or $(F_b)_x$ to zero. An additional condition that must be imposed in order to ensure stability is

$$\left(\frac{\partial[(F_a)_x]_{l_a=l_b}}{\partial l_a}\right)_{l_a=l_{\text{equil}}} > 0.$$

This guarantees that any departure from equilibrium will produce a restoring force. For this discussion it is unnecessary to find an explicit solution of the equilibrium problem. All that is required is to note that a stable equilibrium state can always be obtained if the spring constant K is suitably chosen. Since the equilibrium configuration in Fig. 8.1a is symmetric, it follows that the equilibrium dipole moment (8.2) is zero,

$$(P_x)_{\text{equil}} = -ql_{\text{equil}} + ql - ql + ql_{\text{equil}} = 0. \tag{8.6}$$

This system, therefore, does not have a *spontaneous electrical polarization*.

The effect of applying mechanical forces will now be considered. Because the applied forces are symmetric (Fig. 8.1b), it is perfectly clear without calculation that the movable charges a and b are displaced inward by equal amounts

$$|\delta l_b| = |\delta l_a|. \tag{8.7}$$

The mechanical response, given by (8.1), is therefore

$$\delta L = -2|\delta l_a|. \tag{8.8}$$

From (8.2) and (8.7) the electrical response to mechanical forces is zero,

$$\delta P_x = q|\delta l_a| - q|\delta l_b| = 0. \tag{8.9}$$

The electrical forces applied to this system by means of an electric field are antisymmetric (Fig. 8.1c). In this case the calculation becomes slightly more difficult. When the field is applied, charges a and b are displaced until the total forces on each charge are again balanced. That is,

$$(F_a)_x + q\, \delta E_x = 0$$
$$(F_b)_x + q\, \delta E_x = 0. \tag{8.10}$$

For small departures from equilibrium one may use the approximations

$$(F_a)_x = (F_a)_{x_{\text{equil}}} + \left(\frac{\partial(F_a)_x}{\partial l_a}\right)_{\text{equil}} \delta l_a + \left(\frac{\partial(F_a)_x}{\partial l_b}\right)_{\text{equil}} \delta l_b$$
$$(F_b)_x = (F_b)_{x_{\text{equil}}} + \left(\frac{\partial(F_b)_x}{\partial l_a}\right)_{\text{equil}} \delta l_a + \left(\frac{\partial(F_b)_x}{\partial l_b}\right)_{\text{equil}} \delta l_b , \tag{8.11}$$

where the equilibrium forces equal zero. Since $l_a = l_b$ at equilibrium,

$$\left(\frac{\partial(F_a)_x}{\partial l_a}\right)_{\text{equil}} = -\left(\frac{\partial(F_b)_x}{\partial l_b}\right)_{\text{equil}} = A$$
$$\left(\frac{\partial(F_a)_x}{\partial l_b}\right)_{\text{equil}} = -\left(\frac{\partial(F_b)_x}{\partial l_a}\right)_{\text{equil}} = B, \tag{8.12}$$

from (8.5). Substitution of (8.11) and (8.12) into (8.10) then gives

$$A\, \delta l_a + B\, \delta l_b = -q\, \delta E_x$$
$$-B\, \delta l_a - A\, \delta l_b = -q\, \delta E_x , \tag{8.13}$$

which has solutions

$$\delta l_b = -\delta l_a = \frac{q\, \delta E_x}{A - B} \tag{8.14}$$

as illustrated in Part (c) of Fig. 8.1. The electrical response of the system is therefore

$$\delta P_x = -q\, \delta l_a + q\, \delta l_b = \frac{2q^2\, \delta E_x}{A - B}, \tag{8.15}$$

and the mechanical response is

$$\delta L = \delta l_a + \delta l_b = 0. \tag{8.16}$$

In Fig. 8.1 there is no mechanical response to applied electrical forces and no electrical response to applied mechanical forces. The situation is quite different in Fig. 8.2, where the mechanical components are still disposed symmetrically but the charge distribution is now antisymmetric. This system may also be analyzed in detail, but the main features of its behavior are readily apparent from inspection. At equilibrium it is clear that the two springs will again be symmetrically compressed by the electrostatic forces ($l_a = l_b = l'_{\text{equil}}$). From (8.2) the total electrical polarization at equilibrium is then

$$P_{x_{\text{equil}}} = -Rql'_{\text{equil}} + ql + ql - Rql'_{\text{equil}}$$
$$= 2q(l - Rl'_{\text{equil}}). \qquad (8.17)$$

This may be either zero or nonzero depending on the choice of R, which determines the charges on a and b. That is, there *may* be a spontaneous polarization in this case.

In Parts (b) and (c) of Fig. 8.2 it is seen that the mechanical and electrical applied forces are now *both* symmetric. Consequently,

$$|\delta l_a| = |\delta l_b| \qquad (8.18)$$

in both (c) and (b). The mechanical and electrical responses of the system,

$$\delta L = 2\delta l_a$$
$$\delta P_x = -Rq\,\delta l_a - qR\,\delta l_b = -2Rq\,\delta l_a\,, \qquad (8.19)$$

are now nonzero for both mechanical and electrical applied forces.†

To summarize, the behavior of the piezoelectric model (Fig. 8.2) may be described by relationships of the form

$$\delta P_x = \chi\,\delta E_x + d\,\delta F_x$$
$$\delta L = \underline{d}\,\delta E_x + s\,\delta F_x\,, \qquad (8.20)$$

where χ, d, etc. are parameters of the system. For the more symmetric system shown in Fig. 8.1 the electromechanical coupling terms in (8.20) go to zero. It will be seen in Section 8.E that this is a necessary consequence of the fact that Fig. 8.1a has inversion symmetry, while Fig. 8.2a does not. If the model is extended to three dimensions, the corresponding relations are

$$\delta P_i = \chi_{ij}\,\delta E_j + d_{ij}F_j$$
$$\delta L_i = d_{ij}\,\delta E_j + s_{ij}\,\delta F_j \qquad (8.21)$$
$$i, j = x, y, z.$$

The model might also be made more realistic in some other respects. For example, the electrical polarization changes calculated from (8.2) are due to changes in position of the charged particles, which are supposed to represent ions in an actual solid. In a real material this is called *ionic polarization*. There is another important effect (*electronic polarization*) whereby the ions themselves become polarized in an applied electric field. This could be

---

† This is true even when P_{equil} is zero in (8.17).

included in the model by allowing the charge elements in Figs. 8.1 and 8.2 to be polarizable, but this would not change the basic piezoelectric relation (8.21).

B. PIEZOELECTRIC CONSTITUTIVE RELATIONS

Within a solid medium, mechanical forces are described by the stress field components T_{ij}, and mechanical deformations by the strain field components S_{ij}. If the equilibrium state of all field variables is defined to be zero, one can write the field analogue of (8.21) as

$$P_i = \chi_{ij}E_j + d_{ijk}T_{jk}$$
$$S_{ij} = \underline{d}_{ijk}E_k + s_{ijkl}T_{kl} \, . \tag{8.22}$$

In modern engineering usage the dependent electrical variable is usually taken to be the electrical displacement,

$$D_i = \epsilon_0 E_i + P_i$$

in MKS units, and the piezoelectric equations are therefore written as

$$D_i = \epsilon_{ij}^T E_j + d_{ijk}T_{jk} \tag{8.23}$$

$$S_{ij} = \underline{d}_{ijk}E_k + s_{ijkl}^E T_{kl} \, , \tag{8.24}$$

where d_{ijk}, $\underline{d}_{ijk}$ are defined as the *piezoelectric strain constants*. Superscripts T and E have been added to ϵ_{ij} and s_{ijkl} to show that these constants describe dielectric and elastic properties measured under conditions of constant stress and constant electric field, respectively. Because of the coupling between electric and acoustic fields in a piezoelectric solid, measurements of the electrical properties depend upon the mechanical constraints imposed on the medium, and vice versa. Suppose, for example, that an electric field is applied to a medium. The first terms in (8.23) and (8.24) give, respectively, the electrical displacement and strain produced by the applied field. If the medium is confined mechanically, stress will then develop in response to the strain. This stress, which is related to the strain by the nature of the mechanical constraint, contributes to the second term in (8.23) and thereby modifies the relationship between E_j and D_i. A similar situation arises when a stress is applied to the medium.

Transformation laws for the piezoelectric strain coefficients can be deduced from (8.23) by the method used in Section B of Chapter 3 for the compliances and stiffnesses. In this case E_i, D_j transform according to (1.33) and S_{ij}, T_{jk} according to (1.39) and (2.13). The resulting transformation law,

$$d'_{mno} = a_{mi}a_{nj}a_{ok}d_{ijk} \, , \tag{8.25}$$

defines the d_{ijk}'s, and similarly the $\underline{d}_{ijk}$'s, as components of a *third rank tensor*. In symbolic notation these are denoted by bold face letters, and the *piezoelectric strain equations* take the form

$$\mathbf{D} = \boldsymbol{\epsilon}^T \cdot \mathbf{E} + \boldsymbol{d} : \mathbf{T} \tag{8.26}$$

$$\mathbf{S} = \underline{\boldsymbol{d}} \cdot \mathbf{E} + \mathbf{s}^E : \mathbf{T}. \tag{8.27}$$

Dot and double dot products represent, respectively, summations over single subscripts and pairs of subscripts in (8.23) and (8.24).

Since $S_{ij} = S_{ji}$, the arguments used in Section 3.C to show that $s_{ijkl} = s_{jikl}$ and $c_{ijkl} = c_{ijlk}$ can also be applied here. That is, the d_{ijk}'s can always be defined so that $\underline{d}_{ijk} = \underline{d}_{jik}$. When the stress is symmetric, $T_{ij} = T_{ji}$, the definition $d_{ijk} = d_{ikj}$ can also be made. This means that abbreviated subscripts can be introduced; and the piezoelectric strain equations then take the matrix form†

$$D_i = \epsilon_{ij}^T E_j + d_{iJ} T_J \tag{8.28}$$

$$S_I = \underline{d}_{Ij} E_j + s_{IJ}^E T_J , \tag{8.29}$$

where

$$[d_{iJ}] = \begin{bmatrix} d_{x1} & d_{x2} & d_{x3} & d_{x4} & d_{x5} & d_{x6} \\ d_{y1} & d_{y2} & d_{y3} & d_{y4} & d_{y5} & d_{y6} \\ d_{z1} & d_{z2} & d_{z3} & d_{z4} & d_{z5} & d_{z6} \end{bmatrix} \tag{8.30}$$

and

$$[\underline{d}_{Ij}] = \begin{bmatrix} \underline{d}_{1x} & \underline{d}_{1y} & \underline{d}_{1z} \\ \underline{d}_{2x} & \underline{d}_{2y} & \underline{d}_{2z} \\ \underline{d}_{3x} & \underline{d}_{3y} & \underline{d}_{3z} \\ \underline{d}_{4x} & \underline{d}_{4y} & \underline{d}_{4z} \\ \underline{d}_{5x} & \underline{d}_{5y} & \underline{d}_{5z} \\ \underline{d}_{6x} & \underline{d}_{6y} & \underline{d}_{6z} \end{bmatrix} \tag{8.31}$$

Transformation properties of the piezoelectric strain constants in abbreviated subscript notation are derived by using the Bond transformation matrices of Section 3.D. If the electric field is zero in (8.28)

$$[D] = [d][T]. \tag{8.32}$$

Multiplication of (8.32) by the coordinate transformation matrix $[a]$ gives

$$[D'] = [a][D] = [a][d][T]. \tag{8.33}$$

---

† In converting from the d_{ijk}'s to the d_{iJ}'s and from the $\underline{d}_{ijk}$'s to the $\underline{d}_{Ij}$'s care must be taken to keep track of factors 2 and 4, which appear here just as they did in (3.19).

From (3.31)

$$[T] = [M]^{-1}[T'];$$

and (8.33) is therefore equivalent to

$$[D'] = [a][d][M]^{-1}[T']. \tag{8.34}$$

The piezoelectric strain matrix in the new coordinate system is then

$$[d'] = [a][d][M]^{-1} = [a][d][\widetilde{N}], \tag{8.35}$$

where use has been made of the relation $[M]^{-1} = [\widetilde{N}]$ from Section D of Chapter 3.

EXAMPLE 1. It will be shown in Section D that (8.31) is always the transpose of (8.30). From Part B.1 of Appendix 2, cubic crystals of classes 23 and $\bar{4}3m$ have a piezoelectric strain matrix of the form

$$[d] = \begin{bmatrix} 0 & 0 & 0 & d_{x4} & 0 & 0 \\ 0 & 0 & 0 & 0 & d_{x4} & 0 \\ 0 & 0 & 0 & 0 & 0 & d_{x4} \end{bmatrix} = \widetilde{[d]} \tag{8.36}$$

referred to coordinate axes along the cube edge directions. Transformation to coordinate axes that are rotated through $45°$ about the Z axis is accomplished by multiplying (8.36) on the left with the coordinate transformation matrix $[a]$ in Example 6 of Chapter 3 (with $\xi = \pi/4$) and multiplying on the right by the transpose of the corresponding Bond matrix $[N]$. The latter matrix is found by substituting $\xi = \pi/4$ into (3.42) and then shifting factors of 2 in accord with (3.32) and (3.34). This gives

$$[\widetilde{N}] = \begin{bmatrix} \dfrac{1}{2} & \dfrac{1}{2} & 0 & 0 & 0 & -1 \\[2mm] \dfrac{1}{2} & \dfrac{1}{2} & 0 & 0 & 0 & 1 \\[2mm] 0 & 0 & 1 & 0 & 0 & 0 \\[2mm] 0 & 0 & 0 & \dfrac{1}{\sqrt{2}} & \dfrac{1}{\sqrt{2}} & 0 \\[2mm] 0 & 0 & 0 & -\dfrac{1}{\sqrt{2}} & \dfrac{1}{\sqrt{2}} & 0 \\[2mm] \dfrac{1}{2} & -\dfrac{1}{2} & 0 & 0 & 0 & 0 \end{bmatrix}$$

The transformed piezoelectric strain matrix is then found to be

$$[d'] = \begin{bmatrix} 0 & 0 & 0 & 0 & d_{x4} & 0 \\ 0 & 0 & 0 & -d_{x4} & 0 & 0 \\ \dfrac{d_{x4}}{2} & -\dfrac{d_{x4}}{2} & 0 & 0 & 0 & 0 \end{bmatrix} \tag{8.37}$$

In some problems it is necessary to use strain rather than stress as an independent variable. This is easily arranged by multiplying (8.29) on the left with the stiffness matrix, giving

$$c^E_{JI}S_I = c^E_{JI}\underline{d}_{Ij}E_j + T_J$$

or

$$T_J = -\underline{e}_{Jj}E_j + c^E_{JI}S_I, \tag{8.38}$$

where

$$\underline{e}_{Jj} = c^E_{JI}\underline{d}_{Ij}$$

are defined as the *piezoelectric stress constants*. Substitution of (8.38) into (8.28) then gives the companion equation,

$$D_i = (\epsilon^T_{ij} - d_{iJ}c^E_{JI}\underline{d}_{Ij})E_j + d_{iJ}c^E_{JI}S_I$$

or

$$D_i = \epsilon^S_{ij}E_j + e_{iI}S_I, \tag{8.39}$$

where

$$e_{iI} = d_{iJ}c^E_{JI} \tag{8.40}$$

and

$$\epsilon^S_{ij} = \epsilon^T_{ij} - d_{iJ}c^E_{JI}\underline{d}_{Ij} \tag{8.41}$$

is the permittivity at zero or constant strain. In symbolic notation the *piezoelectric stress equations* (8.38) and (8.39) appear as

$$\mathbf{T} = -\underline{\mathbf{e}} \cdot \mathbf{E} + \mathbf{c}^E : \mathbf{S} \tag{8.42}$$

$$\mathbf{D} = \boldsymbol{\epsilon}^S \cdot \mathbf{E} + \mathbf{e} : \mathbf{S}. \tag{8.43}$$

Other choices of independent variables may be made by rearranging the piezoelectric constitutive equations in this manner. (See Problem 1 at the end of the chapter).

EXAMPLE 2. The piezoelectric strain matrix for cubic piezoelectric crystals is given in (8.36) in Example 1, and the stiffness and permittivity matrices are found in Parts A.2 and C.1 of Appendix 2. Performance of the matrix multiplication (8.40) shows that $[e]$ is of the same form as $[d]$ in (8.36), with

$$e_{x4} = d_{x4}c^E_{44}. \tag{8.44}$$

From (8.41)

$$[\epsilon^S] = (\epsilon^T_{xx} - d^2_{x4}c^E_{44})[I], \tag{8.45}$$

where $[I]$ is the identity matrix.

The piezoelectric stress matrix $[e]$ does not always have exactly the same form as the piezoelectric strain matrix $[d]$. When the coordinates are rotated by $45°$ about the crystal Z axis, $[d']$ is given by (8.37) in Example 1 and the transformed stiffness matrix in this case is given by (3.44) in Example 7 of Chapter 3. Performance of the multiplications in (8.40) now gives

$$[e'] = \begin{bmatrix} 0 & 0 & 0 & 0 & e_{x4} & 0 \\ 0 & 0 & 0 & -e_{x4} & 0 & 0 \\ e_{x4} & -e_{x4} & 0 & 0 & 0 & 0 \end{bmatrix} \tag{8.46}$$

where e_{x4} is given by (8.44). Comparison with (8.37) shows that factors $\frac{1}{2}$ are absent from the matrix elements in the lower left hand corner. The same result could have been obtained by applying the transformation law

$$[e'] = [a][e][N]^{-1} = [a][e][\widetilde{M}] \tag{8.47}$$

which is derived in the same way as (8.35).

The physical significance of various piezoelectric strain constants d_{Ij} is illustrated in Fig. 8.3. From (8.28) one can see that the strain coefficients d_{iJ} are measured in coulomb/newton. The units of d_{Ij} in (8.29) are the same; namely

$$\frac{1}{\text{volt}} = \frac{\text{coulomb/m}^2}{\text{volt-coulomb/m}^2} = \frac{\text{coulomb/m}^2}{(\text{newton-meter})/\text{m}^3}.$$

Measured values may be as large as 10^{-10}. The piezoelectric stress constants e_{iJ}, e_{Ij} are measured in coulomb/m^2, and values as large as 10 are found in practice. At an elastic limit strain in the order of 10^{-3}, the piezoelectrically induced electrical displacement may be as large as 10^{-2} coulombs/m^2. Conversely, at breakdown electric field strengths on the order of 10^7 volts per meter, the piezoelectrically induced stress may be as large as 10^8 newtons per meter.

C. THERMODYNAMICS OF SOLIDS

Consider a unit mass of crystal. If an amount of work δW is done on the crystal by applied mechanical, electrical, and magnetic forces and if a quantity of heat δQ is applied, the total energy of the atoms in the crystal increases by

$$\delta U = \delta Q + \delta W, \tag{8.48}$$

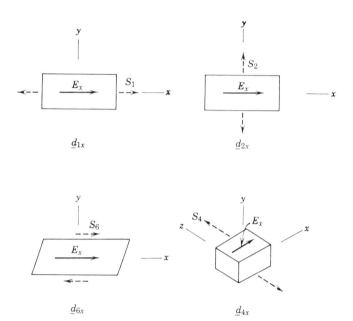

FIGURE 8.3.   Examples of the four basic types of piezoelectric coupling terms.

where U is the *internal energy* per unit mass of the crystal. In thermodynamics, internal energy is described in terms of averages over the atoms, expressed in terms of macroscopic field variables such as strain, electric displacement, etc. If a complete set of macroscopic *state variables* $X_1, \ldots, X_n$ is chosen, the internal energy is a function

$$U(X_1, \ldots, X_n)$$

and

$$dU = \frac{\partial U}{\partial X_1} dX_1 + \frac{\partial U}{\partial X_2} dX_2 + \cdots \frac{\partial U}{\partial X_n} dX_n. \tag{8.49}$$

In general, any set of variables which completely specifies the state of the crystal may be used. However, there is one particular set which gives simple physical interpretations to the partial derivatives in (8.49). In Chapter 5 it was seen that the change in stored energy produced by applying increments of strain, electric, and magnetic fields to a medium is[†]

$$dW = T_I \, dS_I + E_i \, dD_i + H_i \, dB_i.$$

† See, for example, (5.6).

From the second law of thermodynamics

$$dQ = \mathcal{T}(d\mathcal{S}), \tag{8.50}$$

where $\mathcal{T}$ is the temperature and $\mathcal{S}$ the entropy. Consequently,

$$dU = \mathcal{T} \, d\mathcal{S} + T_I \, dS_I + E_i \, dD_i + H_i \, dB_i , \tag{8.51}$$

from (8.48), and comparison with (8.49) shows that

$$\mathcal{T} = \frac{\partial U}{\partial \mathcal{S}}, \quad T_I = \frac{\partial U}{\partial S_I}, \quad E_i = \frac{\partial U}{\partial D_i}, \quad H_i = \frac{\partial U}{\partial B_i},$$

if $\mathcal{S}$, S_I, D_i, B_i are chosen as the state variables.

Suppose a different set of state variables, $\mathcal{T}$, S_I, D_i, B_i, is chosen. Then

$$dU = \frac{\partial U}{\partial \mathcal{T}} \, d\mathcal{T} + \frac{\partial U}{\partial S_I} \, dS_I + \frac{\partial U}{\partial D_i} \, dD_i + \frac{\partial U}{\partial B_i} \, dB_i.$$

In this case, the partial derivative with respect to S_I,

$$\left(\frac{\partial U}{\partial S_I} \right)_{\mathcal{T}, D_i, B_i}$$

is taken with temperature $\mathcal{T}$ held constant and is therefore not equal to T_I, which is the partial derivative

$$T_I = \left(\frac{\partial U}{\partial S_I} \right)_{\mathcal{S}, D_i, B_i}$$

with entropy held constant, according to (8.51). To use this new set of physical variables and still have an energy function whose partial derivatives have simple physical significance, it is necessary to define a new function such that the differentials of the desired variables appear appropriately in the differential of the energy function. For variables $\mathcal{T}$, S_I, D_i, B_i the appropriate function is the Helmholtz free energy,

$$F = U - \mathcal{T}\mathcal{S}.$$

In this case,

$$dF = \mathcal{T} \, d\mathcal{S} + T_I \, dS_I + E_i \, dD_i + H_i \, dB_i - d(\mathcal{T}\mathcal{S})$$
$$= \mathcal{T}\,d\mathcal{S} + T_I \, dS_I + E_i \, dD_i + H_i \, dB_i - \mathcal{T}\,d\mathcal{S} - \mathcal{S} \, d\mathcal{T}$$

and therefore

$$\mathcal{S} = -\frac{\partial F}{\partial \mathcal{T}}, \quad T_I = \frac{\partial F}{\partial S_I}, \quad E_i = \frac{\partial F}{\partial D_i}, \quad H_i = \frac{\partial F}{\partial B_i}.$$

The recipe for changing energy functions is clear from the illustration given. If a term $Y_1 \, dX_1$ appears in (8.49) and variable X_1 is to be replaced by

Y_1 in the new energy function, the desired function is

$$U - Y_1 X_1.$$

The reason for introducing these energy functions in acoustics problems is to demonstrate relations between the constitutive parameters describing the properties of the medium. Examples of the kind of argument used were given in Chapter 5, where symmetry of the permeability, permittivity, compliance, and stiffness matrices was demonstrated. There it was not necessary to consider in detail the properties of the thermodynamic energy functions. In the case of the piezoelectric strain and stress matrices both electrical and acoustic variables are involved and care must be taken to use the energy function suited to the variables. For the piezoelectric strain equations (8.28) and (8.29) the independent variables are T_I, E_i and a suitable energy function is the Gibbs function

$$G = U - \mathscr{T}\mathscr{S} - T_I S_I - E_i D_i - H_i B_i .$$

This gives

$$dG = -\mathscr{S}\, d\mathscr{T} - S_I\, dT_I - D_i\, dE_i - B_i\, dH_i$$

and

$$\mathscr{S} = -\frac{\partial G}{\partial \mathscr{T}}, \quad S_I = -\frac{\partial G}{\partial T_I}, \quad D_i = -\frac{\partial G}{\partial E_i}, \quad B_i = -\frac{\partial G}{\partial H_i}. \quad (8.52)$$

There are a great many possible energy functions, and several alternatives exist which have the required independent variables T_I, E_j. The subject is treated thoroughly in Reference 10 at the end of the chapter.

D. TRANSPOSE SYMMETRY OF THE PIEZOELECTRIC MATRICES

From (8.28) D_i is a linear function of E_j and T_J

$$D_i(E_j, T_J),$$

and therefore

$$d_{iJ} = \frac{\partial D_i}{\partial T_J}.$$

Similarly, $S_I(E_j, T_J)$ is a linear function of the same variables, and

$$\underline{d}_{Ij} = \frac{\partial S_I}{\partial E_j}.$$

By using (8.52), these constants may be defined in terms of the Gibbs function.

That is

$$d_{iJ} = -\frac{\partial^2 G}{\partial T_J \, \partial E_i}$$

$$\underline{d}_{Ij} = -\frac{\partial^2 G}{\partial E_j \, \partial T_I}.$$

Since the order of differentiation is not significant and I, j can be replaced by J, i in the second relation, it follows that the piezoelectric strain coefficients satisfy the transpose symmetry relationship

$$\underline{d}_{Ji} = d_{iJ}.$$

Thus the piezoelectric strain matrices in (8.30) and (8.31) are transposes of each other,

$$[\underline{d}] = \widetilde{[d]}, \tag{8.53}$$

and only one matrix $[d]$ is needed to describe the interaction. An analogous condition,

$$[\underline{e}] = \widetilde{[e]}, \tag{8.54}$$

applies to the piezoelectric stress constants. The piezoelectric strain equations (8.26) and (8.27) can therefore be written as

$$\mathbf{D} = \boldsymbol{\epsilon}^T \cdot \mathbf{E} + \mathbf{d} : \mathbf{T} \tag{8.55}$$

$$\mathbf{S} = \mathbf{d} \cdot \mathbf{E} + \mathbf{s}^E : \mathbf{T}, \tag{8.56}$$

and the piezoelectric stress equations (8.43) and (8.42) as

$$\mathbf{D} = \boldsymbol{\epsilon}^S \cdot \mathbf{E} + \mathbf{e} : \mathbf{S} \tag{8.57}$$

$$\mathbf{T} = -\mathbf{e} \cdot \mathbf{E} + \mathbf{c}^E : \mathbf{S}, \tag{8.58}$$

where the coupling terms are unambiguously defined by the dot product symbols.

From (8.29) it is also seen that

$$s_{IJ}^E = \frac{\partial S_I}{\partial T_J}$$

and, using (8.52) to express S_I in terms of the Gibbs function,

$$s_{IJ}^{E, \mathcal{T}} = -\left(\frac{\partial^2 G}{\partial T_J \, \partial T_I}\right)_{\mathcal{T}}.$$

Since the derivatives are taken at constant temperature this is the isothermal compliance. Because a compressed or stretched material undergoes a change in temperature, time must be allowed for it to return to ambient temperature if this value of compliance is to apply. In vibration problems this does not occur, except at very low frequencies, and the temperature therefore varies. Experimentally, it is found that the effective elastic constants in high frequency vibration are those calculated at constant entropy rather than constant temperature. Following the arguments of Section 8.C, it is found that the effective elastic *stiffness* at constant E and constant entropy will be obtained from the energy function $U - E_i D_i$, that is

$$c_{IJ}^{E, \mathscr{S}} = \left(\frac{\partial^2 (U - E_i D_i)}{\partial S_J \, \partial S_I} \right)_{\mathscr{S}}.$$

In what follows, it will be assumed that constant entropy parameters are used, and the superscript $\mathscr{S}$ will not be carried along.

E. CRYSTAL SYMMETRIES OF THE PIEZOELECTRIC MATRICES

If the transformation in (8.35) is a symmetry operation of the crystal, then

$$[d'] = [d]$$

and the symmetry condition

$$[d][M] = [a][d] \tag{8.59}$$

is obtained by multiplying (8.35) on the right with $[M]$. In the same way,

$$[e][N] = [a][e] \tag{8.60}$$

is derived from (8.47). As in Section 7.B, (8.59) and (8.60) need to be satisfied only for the symmetry group generators listed in Table 7.2. Part B.1 of Appendix 2 lists the piezoelectric matrices found in this manner for all of the piezoelectric crystal classes, and Parts B.2 and B.3 give numerical values of the matrix elements for various materials.

EXAMPLE 3.   An examination of the symmetry groups for piezoelectric crystal classes shows that none of them contains the inversion symmetry operator

$$[a] = \begin{bmatrix} -1 & 0 & 0 \\ 0 & -1 & 0 \\ 0 & 0 & -1 \end{bmatrix} \tag{8.61}$$

The fact that this symmetry forbids the piezoelectric effect was suggested by Section 8.A, and it is easily proven from the symmetry conditions (8.59) or (8.60). In Example

7 of Chapter 1 the strain matrix was seen to be unchanged by the inversion operation. This means that the matrix $[N]$ corresponding to (8.61) is

$$[N] = \begin{bmatrix} 1 & 0 & 0 & 0 & 0 & 0 \\ 0 & 1 & 0 & 0 & 0 & 0 \\ 0 & 0 & 1 & 0 & 0 & 0 \\ 0 & 0 & 0 & 1 & 0 & 0 \\ 0 & 0 & 0 & 0 & 1 & 0 \\ 0 & 0 & 0 & 0 & 0 & 1 \end{bmatrix} \tag{8.62}$$

which can also be calculated directly from (3.34). Consequently, (8.60) reads

$$[e] = -[e],$$

and this is satisfied only if all e_{jI} are zero. This shows that piezoelectricity cannot
exist in an *isotropic* solid.

F. UNIFORM PLANE WAVES IN PIEZOELECTRIC SOLIDS

F.1 General Characteristics

In Chapter 4 the electromagnetic field equations

$$-\nabla \times \mathbf{E} = \frac{\partial \mathbf{B}}{\partial t} \tag{8.63}$$

$$\nabla \times \mathbf{H} = \frac{\partial \mathbf{D}}{\partial t} + \mathbf{J}_c + \mathbf{J}_s \tag{8.64}$$

were shown to have a strong analogy with the acoustic field equations

$$\nabla \cdot \mathbf{T} = \frac{\partial \mathbf{p}}{\partial t} - \mathbf{F} \tag{8.65}$$

$$\nabla_s \mathbf{v} = \frac{\partial \mathbf{S}}{\partial t} ; \tag{8.66}$$

and, for nonpiezoelectric media, plane wave solutions to these two sets of
equations were found to have many characteristics in common. The most
striking difference is that the electromagnetic equations have two plane wave
solutions while the acoustic equations have three. In a nonpiezoelectric

medium the electromagnetic and acoustic solutions are completely independent of each other; but in the piezoelectric case, they are coupled together through the piezoelectric strain equations

$$\mathbf{D} = \boldsymbol{\epsilon}^T \cdot \mathbf{E} + \mathbf{d} : \mathbf{T} \tag{8.67}$$

$$\mathbf{S} = \mathbf{d} \cdot \mathbf{E} + \mathbf{s}^E : \mathbf{T}, \tag{8.68}$$

or the piezoelectric stress equations

$$\mathbf{D} = \boldsymbol{\epsilon}^S \cdot \mathbf{E} + \mathbf{e} : \mathbf{S} \tag{8.69}$$

$$\mathbf{T} = -\mathbf{e} \cdot \mathbf{E} + \mathbf{c}^E : \mathbf{S} \tag{8.70}$$

discussed in Section 8.B. Plane wave solutions in a piezoelectric solid are therefore coupled electromagnetic-acoustic waves. Since there were five plane wave solutions (two electromagnetic and three acoustic) to the uncoupled equations, there must also be five coupled-wave solutions. Before performing a detailed analysis of these coupled wave solutions it will be helpful to look at their characteristics in some specific cases. In all of the cases to be considered here and in other sections, the medium is assumed to be nonmagnetic; that is,

$$\mathbf{B} = \mu_0 \mathbf{H}. \tag{8.71}$$

The fourth constitutive relation is, of course,

$$\mathbf{p} = \rho \mathbf{v}. \tag{8.72}$$

EXAMPLE 4.   Propagation Along a Cube Edge Direction in a Cubic Piezoelectric Medium with No Losses.   Acoustic wave solutions for the nonpiezoelectric case were considered in Examples 4 and 5 of Chapter 3, and the same approach will be used here in order to bring out clearly the physical effects of adding piezoelectricity.

Assume that an x-polarized y-propagating particle displacement wave

$$\mathbf{u} = \hat{\mathbf{x}} \frac{k}{\rho \omega^2} \cos (\omega t - ky)$$

exists in the medium.[†] The corresponding particle velocity is

$$\mathbf{v} = -\hat{\mathbf{x}} \frac{k}{\rho \omega} \sin (\omega t - ky). \tag{8.73}$$

---

† Following Fig. 3.5, the coordinate axes x, y, z are chosen to coincide with the crystal axes X, Y, Z.

As in Example 5 of Chapter 2, the shear strain field associated with this displacement is

$$S_6 = \frac{\partial u_x}{\partial y} = \frac{k^2}{\rho\omega^2} \sin(\omega t - ky), \qquad (8.74)$$

from the strain-displacement relation (8.66). In the nonpiezoelectric case (Example 4, Chapter 3), T_6 and S_6 were related by Hooke's Law,

$$T_6 = c_{44}S_6, \qquad (8.75)$$

and this condition was used to find the dispersion relation (3.24). For a piezoelectric medium, one must use the piezoelectric strain or stress equations instead, and these introduce electrical quantities into the problem.

The electrical and mechanical responses to the strain field (8.74) are given by the piezoelectric stress equations (8.69) and (8.70). That is,

$$D_i = \epsilon^S_{ij}E_j + e_{iJ}S_J \qquad (8.76)$$

$$T_I = -e_{Ij}E_j + c^E_{IJ}S_J. \qquad (8.77)$$

The stiffness matrix for a cubic crystal was given in Example 4, Chapter 3. It is necessary only to add the superscript E for the piezoelectric case. The piezoelectric strain matrix is given by (8.44),

$$[e] = \begin{bmatrix} 0 & 0 & 0 & e_{x4} & 0 & 0 \\ 0 & 0 & 0 & 0 & e_{x4} & 0 \\ 0 & 0 & 0 & 0 & 0 & e_{x4} \end{bmatrix} \qquad (8.78)$$

and the permittivity matrix is

$$[\epsilon^S] = \begin{bmatrix} \epsilon^S_{xx} & 0 & 0 \\ 0 & \epsilon^S_{xx} & 0 \\ 0 & 0 & \epsilon^S_{xx} \end{bmatrix}, \qquad (8.79)$$

from (8.45). According to (8.76) and (8.77), then, the strain field S_6 generates a piezoelectric contribution to the electric displacement

$$(D_z)_{PE} = e_{x4}S_6 \qquad (8.80)$$

and an elastic contribution $(c^E_{44}S_6)$ to the stress field T_6. The electrical displacement $(D_z)_{PE}$ is produced in exactly the same way that deformation of the mechanical model in Fig. 8.2 produced a change in electrical polarization. When $(D_z)_{PE}$ is substituted into the electromagnetic equations it generates magnetic and electric fields, and the latter produces, in turn, a piezoelectric contribution to the stress field T_6. It is the sum of the elastic and piezoelectric contributions to T_6 that must be used for evaluating the plane wave dispersion relation. To do this, the electromagnetic field equations must be solved.

For a uniform plane wave propagating along the y axis $\partial/\partial x = \partial/\partial z = 0$ in (8.63) and (8.64), which then take the form

$$-\nabla \times \mathbf{E} = \mu_0 \frac{\partial \mathbf{H}}{\partial t} \qquad\qquad \nabla \times \mathbf{H} = \frac{\partial \mathbf{D}}{\partial t}$$

$$-\frac{\partial E_z}{\partial y} = \mu_0 \frac{\partial H_x}{\partial t} \quad \text{(a)} \qquad\qquad \frac{\partial H_z}{\partial y} = \frac{\partial D_x}{\partial t} \quad \text{(d)}$$

$$0 = \mu_0 \frac{\partial H_y}{\partial t} \quad \text{(b)} \qquad\qquad 0 = \frac{\partial D_y}{\partial t} \quad \text{(e)} \qquad (8.81)$$

$$\frac{\partial E_x}{\partial y} = \mu_0 \frac{\partial H_z}{\partial t} \quad \text{(c)} \qquad\qquad -\frac{\partial H_x}{\partial y} = \frac{\partial D_z}{\partial t} \quad \text{(f)}$$

when there are no conduction currents or source currents present. From (b) and (e) it follows that H_y and D_y are both zero, except for time-independent fields that are not of interest here. Following Example 1(a) of Chapter 4, the remaining equations are grouped in pairs containing E_z, H_x, D_z and E_x, H_z, D_x respectively. It is the first of these pairs,

$$-\frac{\partial E_z}{\partial y} = \mu_0 \frac{\partial H_x}{\partial t} \tag{8.82}$$

$$-\frac{\partial H_x}{\partial y} = \frac{\partial D_z}{\partial t}, \tag{8.83}$$

that contains the piezoelectric contribution to the electrical displacement due to S_6. Substitution of (8.76) into (8.83) then gives

$$-\frac{\partial H_x}{\partial y} = \epsilon_{xx}^S \frac{\partial}{\partial t} E_z + e_{x4} \frac{\partial}{\partial t} S_6, \tag{8.84}$$

from (8.80); and an equation for the electric field is obtained by eliminating H_x from (8.82) and (8.84),

$$\frac{\partial^2 E_z}{\partial y^2} - \mu_0 \epsilon_{xx}^S \frac{\partial^2 E_z}{\partial t^2} = \mu_0 e_{x4} \frac{\partial^2}{\partial t^2} S_6. \tag{8.85}$$

This is a wave equation for E_z, with a "source" term

$$\mu_0 e_{x4} \frac{\partial^2}{\partial t^2} S_6 = -\frac{\mu_0 e_{x4} k^2}{\rho} \sin(\omega t - ky).$$

The particular integral of (8.85) must have the same wave behavior (frequency ω and wavenumber k) as S_6, and the electric field is therefore

$$E_z = -\frac{\mu_0 e_{x4} k^2}{\rho(\mu_0 \epsilon_{xx}^S \omega^2 - k^2)} \sin(\omega t - ky) = -\frac{\mu_0 e_{x4} \omega^2}{(\mu_0 \epsilon_{xx}^S \omega^2 - k^2)} S_6. \tag{8.86}$$

The relationship between T_6 and S_6 can now be written down by substituting (8.86) into the piezoelectric stress equation (8.77), which gives

$$T_6 = \left(c_{44}^E + \frac{\mu_0 e_{x4}^2 \omega^2}{(\mu_0 \epsilon_{xx}^S \omega^2 - k^2)} \right) S_6. \tag{8.87}$$

The equation of motion (8.65), with $F = 0$, is satisfied by (8.87) and (8.73) only if the condition

$$-\frac{\rho\omega}{k} = \left(c_{44}^E + \frac{\mu_0 e_{x4}^2 \omega^2}{(\mu_0\epsilon_{xx}^S\omega^2 - k^2)}\right)\left(-\frac{k}{\omega}\right)$$

is applied. After a simple rearrangement this becomes the dispersion relation

$$(\rho\omega^2 - c_{44}^E k^2)(\mu_0\epsilon_{xx}^S\omega^2 - k^2) = \mu_0 e_{x4}^2\omega^2 k^2. \qquad (8.88)$$

In the absence of piezoelectric coupling, $e_{x4} = 0$, (8.88) reduces to

$$\rho\omega^2 = c_{44}^E k^2$$

for the acoustic wave and

$$\mu_0\epsilon_{xx}^S\omega^2 = k^2$$

for the electromagnetic wave. For the piezoelectric case, there are still two solutions but they are now coupled (or hybrid) waves, each having both electromagnetic and acoustic fields. The hybrid wave that reduces to a purely acoustic wave when the coupling goes to zero is called *quasiacoustic*, and the other is called *quasielectromagnetic*. In both cases the relationship between the electric and acoustic fields is given by (8.86). The effect of wave coupling on the phase velocity of the wave solutions is easily pictured by dividing (8.88) by k^4 and graphing both sides of the equation as a function of $(\omega/k)^2 = V_p^2$. This is shown in Fig. 8.4, where solutions to the dispersion relation are given by intersections of the solid and dashed curves. When the coupling goes to zero these intersections occur at the acoustic velocity $(c_{44}/\rho)^{1/2}$ and the electromagnetic velocity $(\mu_0\epsilon_{xx})^{-1/2}$. In the presence of piezoelectric coupling, the quasiacoustic phase velocity shifts to a lower value than the acoustic velocity and the quasielectromagnetic phase velocity shifts to a higher value than the electromagnetic velocity. It was noted in Example 4, Chapter 3 that the acoustic and electromagnetic velocities differ by approximately five orders of magnitude. One would therefore expect that the phase velocity shifts are very small, and this is easily

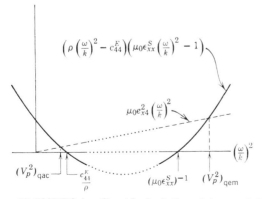

FIGURE 8.4.   Graphical solution of the coupled-wave dispersion relation (8.88).

checked by substituting typical numbers into (8.88). For a quasiacoustic wave the elastic energy density is much greater than the electric energy density, and the reverse applies for a quasielectromagnetic wave.

The calculation for a z-polarized y-propagating particle displacement field follows exactly the same pattern and the same result is obtained, with appropriate changes in subscripts. As in the nonpiezoelectric case, the x- and z-polarized quasiacoustic and quasielectromagnetic waves are degenerate and can be combined to form arbitrary elliptical and circular polarizations (Fig. 3.4).

The solution for a y-polarized y-propagating wave, with a particle displacement field

$$\mathbf{u} = \hat{\mathbf{y}} \cos(\omega t - ky),$$

is quite different. From (3.26) in Example 5 of Chapter 3, the associated compressional strain field is

$$S_2 = k \sin(\omega t - ky).$$

Substitution of S_2 into (8.76), and use of (8.78) for the piezoelectric matrix, shows that $(\mathbf{D})_{\mathrm{PE}} = 0$. There is therefore no coupling of this displacement field with the electromagnetic field, and the solution is a purely acoustic compressional wave—even in a piezoelectric medium. As in Examples 4 and 5 of Chapter 3, these conclusions apply for propagation along *any* crystal axis of a cubic medium.

To summarize, uniform plane waves propagating along any cube edge direction of a cubic piezoelectric solid consist of *two quasielectromagnetic waves, two quasiacoustic waves,* and *one purely acoustic wave.* The purely acoustic wave has pure longitudinal polarization and the other four waves have purely transverse polarizations. Illustrations of the fields and their relationships are given in Fig. 8.5.

EXAMPLE 5. **Propagation Along a Face Diagonal Direction in a Cubic Piezoelectric Medium with No Losses.** Another interesting feature of piezoelectric plane waves may be illustrated by examining a different direction of propagation. Consider the rotated coordinate system in Example 6 of Chapter 3, with $\xi = 45°$, and assume wave propagation along the x' axis in Fig. 3.7(a). For the nonpiezoelectric case, this problem was treated in Examples 8 and 9 of Chapter 3. The relevant stiffness matrix for the piezoelectric problem is given by (3.44), with superscript E appended to each constant. There is no change in the permittivity matrix (8.79), and the piezoelectric stress matrix is given by (8.46) in Example 2.

For a y' (or $[\bar{1}10]$)-polarized x'-propagating particle displacement

$$\mathbf{u} = \hat{\mathbf{y}}' \cos(\omega t - kx'),$$

the strain field is

$$S_{6'} = k \sin(\omega t - kx')$$

from Example 9 of Chapter 3; and reference to (8.46) shows that

$$\mathbf{e}:\mathbf{S} = e_{iJ}S_J = 0 \tag{8.89}$$

in (8.69). That is, there is no piezoelectric contribution to the electric displacement, and therefore no coupling of the strain to the electric field. The y'-polarized shear wave is thus purely acoustic.

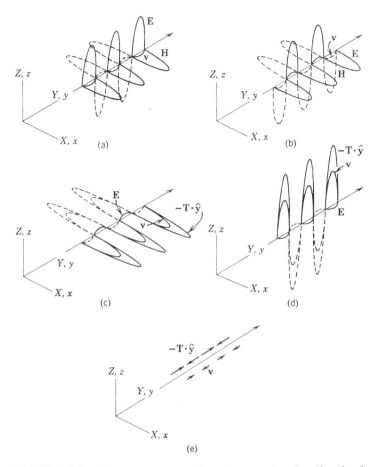

FIGURE 8.5. Plane wave propagation along a cube edge direction in a cubic piezoelectric medium. (a), (b) quasielectromagnetic waves; (c), (d) quasiacoustic waves; (e) acoustic wave. For clarity, the stress field has been omitted in (a), (b) and the magnetic field in (c), (d).

In the case of a z' (or [001])-polarized x'-propagating particle displacement

$$\mathbf{u} = \hat{\mathbf{z}}' \cos{(\omega t - kx')}, \tag{8.90}$$

and the strain field is

$$S_{5'} = \frac{\partial u_z}{\partial x'} = k \sin{(\omega t - kx')} \tag{8.91}$$

from (8.66). The electric displacement field associated with (8.91) is calculated from (8.69), where the term $\mathbf{e}:\mathbf{S}$ is now

$$(D_{x'})_{\mathrm{PE}} = e_{x\mathbf{4}}S_{5'}. \tag{8.92}$$

For this problem $\partial/\partial y' = \partial/\partial z' = 0$, and the electromagnetic field equations take the form

$$-\nabla \times \mathbf{E} = \mu_0 \frac{\partial \mathbf{H}}{\partial t} \qquad\qquad \nabla \times \mathbf{H} = \frac{\partial \mathbf{D}}{\partial t}$$

$$0 = \mu_0 \frac{\partial H_{x'}}{\partial t} \quad \text{(a)} \qquad\qquad 0 = \frac{\partial D_{x'}}{\partial t} \quad \text{(d)}$$

$$\frac{\partial E_{z'}}{\partial x'} = \mu_0 \frac{\partial H_{y'}}{\partial t} \quad \text{(b)} \qquad\qquad -\frac{\partial H_{z'}}{\partial x'} = \frac{\partial D_{y'}}{\partial t} \quad \text{(e)}$$

$$-\frac{\partial E_{y'}}{\partial x'} = \mu_0 \frac{\partial H_{z'}}{\partial t} \quad \text{(c)} \qquad\qquad \frac{\partial H_{y'}}{\partial x'} = \frac{\partial D_{z'}}{\partial t} \quad \text{(f)}$$

$$(8.93)$$

From (a) and (d) $H_{x'}$ and $D_{x'}$ are both zero, except for time-independent fields that are not of interest. According to (8.92) the piezoelectric effect makes a contribution to $D_{x'}$. One must therefore write

$$D_{x'} = \epsilon_{xx}^S E_{x'} + e_{x4}S_{5'} = 0 \tag{8.94}$$

from (8.69) and (8.92). The electric field is then

$$E_{x'} = -\frac{e_{x4}}{\epsilon_{xx}^S} S_{5'}; \tag{8.95}$$

and substitution into (8.70) gives

$$T_{5'} = -e_{x4}E_{x'} + c_{44}^E S_{5'}$$
$$= \left(\frac{e_{x4}^2}{\epsilon_{xx}^S} + c_{44}^E\right)S_{5'}, \tag{8.96}$$

with the use of (3.44), (8.46), and (8.95). The equation of motion (8.65), with $\mathbf{F} = 0$, is consistent with (8.96) and (8.90) only if the dispersion relation

$$\rho\omega^2 = \left(c_{44}^E + \frac{e_{x4}^2}{\epsilon_{xx}^S}\right)k^2 \tag{8.97}$$

is satisfied.

It should be noted that the dispersion relation (8.97) shows no evidence of the coupled-wave behavior described by (8.88). There is only one wave solution, which behaves as if the elastic stiffness were increased from c_{44}^E to

$$\bar{c}_{44} = c_{44}^E + \frac{e_{x4}^2}{\epsilon_{xx}^S} \tag{8.98}$$

by the piezoelectric coupling. This phenomenon is called *piezoelectric stiffening* and (8.98) is called a *stiffened elastic constant*. The effect is caused by the generation of an electric field from the strain in (8.95) and subsequent generation of a piezoelectric stress by the first term in (8.96). It is also noteworthy that the electric field $E_{x'}$ appears only in a component of the electromagnetic field equations that plays no part in the propagation of electromagnetic waves. This is why there is no coupled wave behavior. The field

$$\mathbf{E} = \hat{\mathbf{x}}'E_{x'}(x', t)$$

has, in fact, zero curl and can therefore be represented as the gradient of a scalar potential—even though it is time-varying, not static. For this reason it is called a *quasistatic field*. Thus,

$$E(x', t) = -\nabla\Phi(x', t) \tag{8.99}$$

and

$$\Phi = -\frac{e_{x4}}{\epsilon_{xx}^S}\cos(\omega t - kx'), \tag{8.100}$$

from (8.91) and (8.95). Piezoelectric stiffening is always a weak effect. Even for the largest observed piezoelectric constants (e_{iJ} of order 10 coulomb/m^2), the stiffening term in (8.98) is only a fraction of c_{44}^E.

For the x'-polarized x'-propagating compressional wave of Example 8 in Chapter 3,

$$\mathbf{u} = \hat{\mathbf{x}}'\cos(\omega t - kx')$$

and the strain field is

$$S_{1'} = \frac{\partial u_{x'}}{\partial x'} = k\sin(\omega t - kx'). \tag{8.101}$$

Accordingly, the piezoelectric contribution to the electrical displacement in (8.69) is now

$$(D_{z'})_{PE} = e_{x4}S_{1'}$$

from (8.46) and (8.101). This appears on the right-hand side in part (f) of (8.93); and elimination of $H_{y'}$ from parts (b) and (f) gives

$$\frac{\partial^2 E_{z'}}{\partial x'^2} - \mu_0\epsilon_{xx}^S\frac{\partial^2 E_{z'}}{\partial t^2} = \mu_0 e_{x4}\frac{\partial^2 S_{1'}}{\partial t^2},$$

corresponding to (8.85) in Example 4. The analysis proceeds as before and leads to quasielectromagnetic and quasiacoustic wave solutions with z'-polarized electric fields and x'-polarized acoustic fields. Again, the coupled waves are very slightly perturbed versions of the electromagnetic and acoustic waves for the nonpiezoelectric case.

There remain two electromagnetic equations ((c) and (e) in (8.93)) that have not yet been used. These equations do not interact in any way with the acoustic field, because the piezoelectric term in (8.69) has no y' component; and they therefore lead to a purely electromagnetic y'-polarized wave. As in Examples 8 and 9 of Chapter 3, this and the other results obtained in this example apply for propagation along *any* cube face-diagonal direction in a piezoelectric cubic crystal.

Summarizing, uniform plane waves propagating along a face-diagonal direction in a cubic piezoelectric crystal consist of *one purely acoustic wave, one purely electromagnetic wave, one quasiacoustic wave, one quasielectromagnetic wave*, and *one stiffened acoustic wave*. The quasiacoustic and quasielectromagnetic waves have pure longitudinal acoustic polarizations and pure transverse electric polarizations. All other waves have pure transverse polarizations. Illustrations of the fields and their relationships are given in Fig. 8.6. It is interesting to note the physical difference between the quasiacoustic wave and the stiffened acoustic wave. In the first case (Fig. 8.7a) the piezoelectric contribution to **D** excites a magnetic field **H** by means of

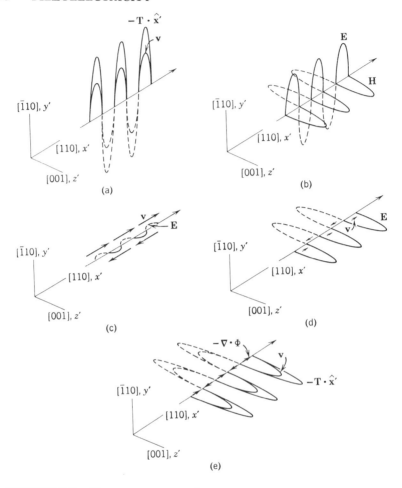

FIGURE 8.6.   Plane wave propagation along a face diagonal direction in
a cubic piezoelectric medium. (a) acoustic wave; (b) electromagnetic wave;
(c) quasiacoustic wave; (d) quasielectromagnetic wave; (e) stiffened acoustic
wave. For clarity, the stress and magnetic fields have been omitted in (c) and
(d). *There is no magnetic field in* (e).

the curl relation (8.64). For the stiffened acoustic wave (Fig. 8.7b) D_{PE} does not
have the correct orientation to excite **H**, but it does generate a distribution of bound
electric charges. These are arranged in layers, as shown in the figure, and produce an
electric field as in an ordinary electrostatic problem.

F.2   Formal Coupled-Wave Theory

Examples 4 and 5 have demonstrated the characteristics of piezoelectric
plane waves, using a method of analysis that brings out the physical

significance of the solutions. This is, however, not the most efficient way of approaching the problem. It is better to follow the treatment of non-piezoelectric plane waves in Section 6.A and eliminate unnecessary variables by manipulating the field equations in symbolic form. This procedure leads eventually to the piezoelectric counterpart of the Christoffel equation (6.10).

The first step is to eliminate $\mathbf{T}$ from the acoustic field equations, as in Section 6.A, and then to eliminate $\mathbf{H}$ from the electromagnetic field equations. Multiplication of (8.66) by $\mathbf{c}^E$ leads to

$$\mathbf{c}^E : \nabla_s \mathbf{v} = \mathbf{c}^E : \frac{\partial \mathbf{S}}{\partial t} .$$

The time derivative of the piezoelectric stress equation (8.70) can therefore

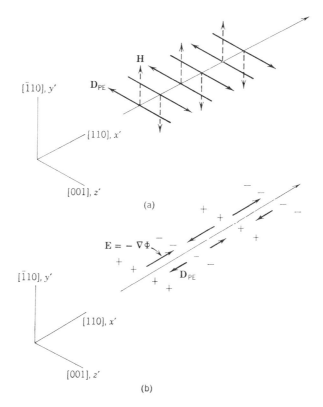

FIGURE 8.7. (a) Generation of a magnetic field by $\mathbf{D}_{PE}$ in the *quasiacoustic* wave of Fig. 8.6(c). (b) Generation of bound electric charges by $\mathbf{D}_{PE}$ in the stiffened acoustic wave of Fig. 8.6(e).

be converted to

$$\frac{\partial \mathbf{T}}{\partial t} = \mathbf{c}^E : \nabla_s \mathbf{v} - \mathbf{e} \cdot \frac{\partial \mathbf{E}}{\partial t} , \tag{8.102}$$

and substitution into the time derivative of (8.65) gives

$$\nabla \cdot \mathbf{c}^E : \nabla_s \mathbf{v} = \rho \frac{\partial^2 \mathbf{v}}{\partial t^2} + \nabla \cdot \left(\mathbf{e} \cdot \frac{\partial \mathbf{E}}{\partial t} \right) - \frac{\partial \mathbf{F}}{\partial t} , \tag{8.103}$$

after use of (8.72).

The magnetic field is eliminated by taking the curl of (8.63), with $\mathbf{B} = \mu_0 \mathbf{H}$ from (8.71),

$$-\nabla \times \nabla \times \mathbf{E} = \mu_0 \frac{\partial}{\partial t} \nabla \times \mathbf{H} ,$$

and substituting the time derivative of (8.64). This gives

$$-\nabla \times \nabla \times \mathbf{E} = \mu_0 \left(\frac{\partial^2 \mathbf{D}}{\partial t^2} + \frac{\partial \mathbf{J}_c}{\partial t} + \frac{\partial \mathbf{J}_s}{\partial t} \right) .$$

The time derivative of $\mathbf{D}$ can be expressed in terms of $\mathbf{E}$ and $\mathbf{v}$ by means of (8.69) and (8.66). Therefore

$$-\nabla \times \nabla \times \mathbf{E} = \mu_0 \boldsymbol{\epsilon}^S \cdot \frac{\partial^2 \mathbf{E}}{\partial t^2} + \mu_0 \mathbf{e} : \nabla_s \frac{\partial \mathbf{v}}{\partial t} + \mu_0 \frac{\partial \mathbf{J}_c}{\partial t} + \mu_0 \frac{\partial \mathbf{J}_s}{\partial t} . \tag{8.104}$$

For a lossless medium with no sources, the coupled field equations (8.103) and (8.104) become

$$\nabla \cdot \mathbf{c}^E : \nabla_s \mathbf{v} = \rho \frac{\partial^2 \mathbf{v}}{\partial t^2} + \nabla \cdot \left(\mathbf{e} \cdot \frac{\partial \mathbf{E}}{\partial t} \right) \tag{8.105}$$

$$-\nabla \times \nabla \times \mathbf{E} = \mu_0 \boldsymbol{\epsilon}^S \cdot \frac{\partial^2 \mathbf{E}}{\partial t^2} + \mu_0 \mathbf{e} : \nabla_s \frac{\partial \mathbf{v}}{\partial t} . \tag{8.106}$$

These are coupled wave equations for $\mathbf{v}$ and $\mathbf{E}$. When the piezoelectric coupling goes to zero, (8.105) reduces to (6.5) with $\mathbf{F} = 0$, and (8.106) reduces to the usual wave equation for $\mathbf{E}$.

To solve a specific problem, (8.105) and (8.106) must be expressed in terms of field components and are most conveniently written as matrix equations. The left-hand side of (8.105) is treated as in Section 6.A,

$$\nabla \cdot \mathbf{c}^E : \nabla_s \mathbf{v} \rightarrow \nabla_{iK} c^E_{KL} \nabla_{Lj} v_j ,$$

where the matrix-differential operators ∇_{iK} and ∇_{Lj} are defined in (2.36) and (1.53); and the left-hand side of (8.106) is converted to matrix form by using the matrix-differential operator defined in (4.11). There is no difficulty in evaluating the coupling terms, so long as the physical meanings of the symbols are kept clearly in mind. The coupling term in (8.105) has a *single* dot between **e** and $\partial \mathbf{E}/\partial t$. This means that

$$\mathbf{e} \cdot \frac{\partial \mathbf{E}}{\partial t} \rightarrow e_{Jk} \frac{\partial E_k}{\partial t}$$

is the time derivative of a stress. Consequently, the divergence operation must be represented by the matrix-differential operator ∇_{iJ} defined in (2.36); that is,

$$\nabla \cdot \left(\mathbf{e} \cdot \frac{\partial \mathbf{E}}{\partial t} \right) \rightarrow \nabla_{iJ} e_{Jk} \frac{\partial E_k}{\partial t}. \tag{8.107}$$

For the coupling term in (8.106) the operator ∇_s is represented in matrix-differential form by ∇_{Ij} in (1.53). The *double* dot between **e** and ∇_s signifies summation over the abbreviated subscript I. Therefore

$$\mu_0 \mathbf{e} : \nabla_s \frac{\partial \mathbf{v}}{\partial t} \rightarrow \mu_0 e_{iI} \nabla_{Ij} \frac{\partial v_j}{\partial t}. \tag{8.108}$$

EXAMPLE 6. To illustrate the technique for evaluating piezoelectric coupling terms, consider the problem treated in Example 4. In this case the fields vary only with y, and

$$[\nabla_{iJ}] = \begin{bmatrix} 0 & 0 & 0 & 0 & 0 & \dfrac{\partial}{\partial y} \\ 0 & \dfrac{\partial}{\partial y} & 0 & 0 & 0 & 0 \\ 0 & 0 & 0 & \dfrac{\partial}{\partial y} & 0 & 0 \end{bmatrix} \tag{8.109}$$

From (8.78)

$$[e_{Jk}] = \begin{bmatrix} 0 & 0 & 0 \\ 0 & 0 & 0 \\ 0 & 0 & 0 \\ e_{x4} & 0 & 0 \\ 0 & e_{x4} & 0 \\ 0 & 0 & e_{x4} \end{bmatrix} \tag{8.110}$$

and $\partial \mathbf{E}/\partial t$ is given in matrix form as

$$\frac{\partial}{\partial t} \begin{bmatrix} E_x \\ E_y \\ E_z \end{bmatrix} \tag{8.111}$$

The coupling term (8.107) is then written down by multiplying matrices (8.109), (8.110), and (8.111) in that order. This gives

$$
\nabla \cdot \mathbf{e} \cdot \frac{\partial \mathbf{E}}{\partial t} \rightarrow
\begin{bmatrix}
0 & 0 & e_{x4}\dfrac{\partial}{\partial y} \\[2mm]
0 & 0 & 0 \\[2mm]
e_{x4}\dfrac{\partial}{\partial y} & 0 & 0
\end{bmatrix}
\frac{\partial}{\partial t}
\begin{bmatrix} E_x \\[2mm] E_y \\[2mm] E_z \end{bmatrix}
= e_{x4}
\begin{bmatrix} \dfrac{\partial^2 E_z}{\partial y\,\partial t} \\[2mm] 0 \\[2mm] \dfrac{\partial^2 E_x}{\partial y\,\partial t} \end{bmatrix}
\tag{8.112}
$$

In (8.108) ∇_{Ij} is represented by the transpose of (8.109) and e_{iI} by the transpose of (8.110). The matrix product $e_{iI}\nabla_{Ij}$ is then simply the transpose of the 3×3 matrix in (8.112), and

$$
\mu_0 \mathbf{e} : \nabla_s \frac{\partial \mathbf{v}}{\partial t} \rightarrow \mu_0 e_{x4}
\begin{bmatrix}
\dfrac{\partial^2 v_z}{\partial y\,\partial t} \\[2mm]
0 \\[2mm]
\dfrac{\partial^2 v_x}{\partial y\,\partial t}
\end{bmatrix}
\tag{8.113}
$$

These results show that v_x, E_z are coupled together and v_z, E_x are coupled together by the equations (8.105) and (8.106), in agreement with Example 4.

The coupled field equations (8.105) and (8.106) give a general mathematical description of the coupled-wave behavior illustrated in Examples 4 and 5. These examples also demonstrated that two very different kinds of electric field distributions can occur in piezoelectric wave problems. In some instances, such as the waves with x- and z-polarized particle displacements in Example 4, the curl of $\mathbf{E}$ was not zero. This is called a *rotational field* $\mathbf{E}^{(r)}$, and is characteristic of electromagnetic waves. Other solutions, such as the wave with z'-polarized particle displacement in Example 5, had an electric field with zero curl. This is called a *quasistatic (or irrotational) field*, and can be represented as the gradient of a potential,

$$
\mathbf{E}^{(i)} = -\nabla\Phi ,
\tag{8.114}
$$

even though it is a time-varying field. For more general problems it is reasonable to expect that both of these kinds of fields may occur in the same wave solution. The electric field in (8.105) and (8.106) is therefore separated into rotational and irrotational parts

$$
\mathbf{E} = \mathbf{E}^{(r)} - \nabla\Phi .
\tag{8.115}
$$

When the piezoelectric coupling is reduced to zero the rotational part of $\mathbf{E}$ in (8.115) becomes a purely electromagnetic solution. The condition

$$
\nabla \cdot \boldsymbol{\epsilon}^S \cdot \mathbf{E}^{(r)} = 0
\tag{8.116}
$$

that is characteristic of such solutions will therefore be imposed on this part of the field.† With the substitution (8.115), and use of the identity $\nabla \times \nabla\Phi = 0$, the coupled field equations (8.105) and (8.106) reduce to

$$\nabla \cdot \mathbf{c}^E : \nabla_s \mathbf{v} - \rho \frac{\partial^2 \mathbf{v}}{\partial t^2} = \nabla \cdot \left(\mathbf{e} \cdot \frac{\partial \mathbf{E}^{(r)}}{\partial t} \right) - \nabla \cdot \left(\mathbf{e} \cdot \frac{\partial}{\partial t} \nabla\Phi \right) \qquad \text{(a)}$$

$$-\nabla \times \nabla \times \mathbf{E}^{(r)} - \mu_0 \boldsymbol{\epsilon}^S \cdot \frac{\partial^2 \mathbf{E}^{(r)}}{\partial t^2} = -\mu_0 \boldsymbol{\epsilon}^S \cdot \nabla \frac{\partial^2 \Phi}{\partial t^2} + \mu_0 \mathbf{e} : \nabla_s \frac{\partial \mathbf{v}}{\partial t} \qquad \text{(b)}$$

$$0 = -\mu_0 \nabla \cdot \left(\boldsymbol{\epsilon}^S \cdot \frac{\partial^2 \nabla\Phi}{\partial t^2} \right) + \mu_0 \nabla \cdot \left(\mathbf{e} : \nabla_s \frac{\partial \mathbf{v}}{\partial t} \right) \qquad \text{(c)}$$

$$\nabla \cdot \boldsymbol{\epsilon}^S \cdot \mathbf{E}^{(r)} = 0, \qquad \text{(d)}$$

$$(8.117)$$

where part (d) of (8.117) is just the condition (8.116). Part (c) is obtained by taking the divergence of (8.106) and then applying (8.116) to the term containing $\mathbf{E}^{(r)}$ on the right-hand side.

The equations (8.117) have a rather forbidding appearance but they are not, in fact, difficult to convert to matrix form. In parts (a) and (b) the piezo-electric terms have the same forms as (8.107) and (8.108), and are treated exactly as in Example 6. Since the gradient operation on a scalar quantity has the representation

$$\nabla = \hat{\mathbf{x}} \frac{\partial}{\partial x} + \hat{\mathbf{y}} \frac{\partial}{\partial y} + \hat{\mathbf{z}} \frac{\partial}{\partial z} = \hat{\mathbf{r}}_i \frac{\partial}{\partial r_i}, \qquad (8.118)$$

the first term on the right-hand side of part (b) is

$$-\mu_0 \boldsymbol{\epsilon}^S \cdot \nabla \frac{\partial^2 \Phi}{\partial t^2} \rightarrow -\mu_0 \epsilon_{ij}^S \frac{\partial^3 \Phi}{\partial t^2 \partial r_j}. \qquad (8.119)$$

The divergence operation on a vector quantity,

$$\nabla \cdot \mathbf{A} = \frac{\partial}{\partial r_i} A_i, \qquad (8.120)$$

is represented by the row matrix-differential operator

$$[\nabla_i] = \begin{bmatrix} \dfrac{\partial}{\partial x} & \dfrac{\partial}{\partial y} & \dfrac{\partial}{\partial z} \end{bmatrix}. \qquad (8.121)$$

† This ensures that the separation of $\mathbf{E}$ into rotational and irrotational parts is unique.

Accordingly, the electrical terms in parts (c) and (d) of (8.117) are

$$\nabla \cdot \boldsymbol{\epsilon}^S \cdot \nabla \frac{\partial^2 \Phi}{\partial t^2} \to \epsilon_{ij} \frac{\partial^4 \Phi}{\partial r_i \, \partial r_j \, \partial t^2} \tag{8.122}$$

and

$$\nabla \cdot \boldsymbol{\epsilon}^S \cdot \mathbf{E}^{(r)} \to \epsilon_{ij}^S \frac{\partial E_j^{(r)}}{\partial r_i}, \tag{8.123}$$

where the constants ϵ_{ij}^S have been shifted from under the derivatives. The *double* dot between $\mathbf{e}$ and ∇_s on the right-hand side of part (c) indicates that the quantity

$$\mathbf{e} : \nabla_s \frac{\partial}{\partial t} \mathbf{v} \to e_{iJ} \nabla_{Jk} \frac{\partial}{\partial t} v_k$$

is a vector. This means that the divergence must be calculated according to (8.120). That is,

$$\nabla \cdot \left(\mathbf{e} : \frac{\partial}{\partial t} \nabla_s \mathbf{v} \right) \to \frac{\partial}{\partial r_i} e_{iJ} \nabla_{Jk} \frac{\partial}{\partial t} v_k. \tag{8.124}$$

EXAMPLE 7. *Piezoelectric Wave Propagation Along the X Axis of a Hexagonal (6mm) Crystal.* In Section E of this chapter it was noted that isotropic solids cannot be piezoelectric. This makes it difficult to find simple illustrative problems. Hexagonal crystals are, however, elastically and electrically isotropic in the XY plane† and may also be piezoelectric. The piezoelectric matrices given for the hexagonal classes in Part B.1 of Appendix 2 are not obviously isotropic with respect to coordinate rotations about the Z axis, but application of the transformation laws (8.35) or (8.47) shows that they are. These crystal classes therefore provide some relatively simple problems. Although *6mm* is not the simplest hexagonal class in terms of its piezoelectric behavior, it has been chosen as an example because many common piezoelectric materials have this symmetry.

From Appendix 2, the constitutive matrices for a hexagonal (*6mm*) medium are

$$[c_{IJ}^E] = \begin{bmatrix} c_{11}^E & c_{12}^E & c_{13}^E & 0 & 0 & 0 \\ c_{12}^E & c_{11}^E & c_{13}^E & 0 & 0 & 0 \\ c_{13}^E & c_{13}^E & c_{33}^E & 0 & 0 & 0 \\ 0 & 0 & 0 & c_{44}^E & 0 & 0 \\ 0 & 0 & 0 & 0 & c_{44}^E & 0 \\ 0 & 0 & 0 & 0 & 0 & c_{66}^E \end{bmatrix}, \quad c_{66}^E = \frac{c_{11}^E - c_{12}^E}{2} \tag{8.125}$$

† See footnote in Example 2 of Chapter 4.

$$[\epsilon_{ij}^S] = \begin{bmatrix} \epsilon_{xx}^S & 0 & 0 \\ 0 & \epsilon_{xx}^S & 0 \\ 0 & 0 & \epsilon_{zz}^S \end{bmatrix} \tag{8.126}$$

$$[e_{iJ}] = \begin{bmatrix} 0 & 0 & 0 & 0 & e_{x5} & 0 \\ 0 & 0 & 0 & e_{x5} & 0 & 0 \\ e_{z1} & e_{z1} & e_{z3} & 0 & 0 & 0 \end{bmatrix} \tag{8.127}$$

Because the medium is isotropic in the XY plane, there is no loss of generality in assuming propagation along X. Therefore†

$$[\nabla_{iJ}] = \begin{bmatrix} \dfrac{\partial}{\partial x} & 0 & 0 & 0 & 0 & 0 \\ 0 & 0 & 0 & 0 & 0 & \dfrac{\partial}{\partial x} \\ 0 & 0 & 0 & 0 & \dfrac{\partial}{\partial x} & 0 \end{bmatrix} \tag{8.128}$$

If the field solutions are assumed proportional to

$$e^{i(\omega t - kx)}$$

the substitutions

$$\frac{\partial}{\partial x} = -ik$$

$$\frac{\partial}{\partial t} = i\omega$$

can be made in all equations. By following the steps outlined in Example 6 and in (8.119) to (8.124), the equations in (8.117) are then reduced to the matrix form

$$\begin{bmatrix} \omega^2\rho - k^2c_{11}^E & 0 & 0 \\ 0 & \omega^2\rho - k^2c_{66}^E & 0 \\ 0 & 0 & \omega^2\rho - k^2c_{44}^E \end{bmatrix} \begin{bmatrix} v_x \\ v_y \\ v_z \end{bmatrix} = \omega k \begin{bmatrix} e_{z1}E_z^{(r)} \\ 0 \\ e_{x5}E_x^{(r)} \end{bmatrix} + i\omega k^2 \begin{bmatrix} 0 \\ 0 \\ e_{x5}\Phi \end{bmatrix} \tag{a}$$

$$\begin{bmatrix} \omega^2\mu_0\epsilon_{xx}^S & 0 & 0 \\ 0 & \omega^2\mu_0\epsilon_{xx}^S - k^2 & 0 \\ 0 & 0 & \omega^2\mu_0\epsilon_{zz}^S - k^2 \end{bmatrix} \begin{bmatrix} E_x^{(r)} \\ E_y^{(r)} \\ E_z^{(r)} \end{bmatrix} = -i\omega^2 k\mu_0 \begin{bmatrix} \epsilon_{xx}^S\Phi \\ 0 \\ 0 \end{bmatrix} + \omega k\mu_0 \begin{bmatrix} e_{x5}v_z \\ 0 \\ e_{z1}v_x \end{bmatrix} \tag{b}$$

$$i\omega k^2 \epsilon_{xx}^S \Phi = k^2 e_{x5} v_z \tag{c}$$

$$-ik\epsilon_{xx}^S E_x^{(r)} = 0. \tag{d} \tag{8.129}$$

† The coordinate axes x, y, z are chosen to coincide with the crystal axes X, Y, Z, as in Fig. 4.1.

It is evident from part (d) of (8.129) that

$$E_x^{(r)} = 0;$$ (8.130)

and substitution of this result into part (b) gives

$$i\omega^2 k \mu_0 \epsilon_{xx}^S \Phi = \omega k \mu_0 e_{x5} v_z ,$$

which is simply equal to $\mu_0\omega/k$ times part (c). There are therefore only six independent equations, in addition to (8.130). From the first line of part (a) and the third line of part (b),

$$(\omega^2 \rho - k^2 c_{11}^E)v_x - \omega k e_{z1} E_z^{(r)} = 0$$

$$-\omega k \mu_0 e_{z1} v_x + (\omega^2 \mu_0 \epsilon_{xx}^S - k^2)E_z^{(r)} = 0.$$ (8.131)

Solution of these coupled-wave equations gives one *quasiacoustic* wave and one *quasielectromagnetic* wave, with **v** polarized along x and $\mathbf{E}^{(r)}$ polarized along z. The dispersion relation is obtained by setting the determinant of the equations equal to zero

$$(\omega^2 \rho - k^2 c_{11}^E)(\omega^2 \mu_0 \epsilon_{xx}^S - k^2) = \mu_0 e_{z1}^2 \omega^2 k^2.$$ (8.132)

The second lines of parts (a) and (b) in (8.129) give *purely acoustic* and *purely electromagnetic* equations

$$(\omega^2 \rho - k^2 c_{66}^E)v_y = 0$$ (8.133)

and

$$(\omega^2 \mu_0 \epsilon_{xx}^S - k^2)E_y = 0.$$ (8.134)

In the third line of part (a), $E_x^{(r)} = 0$ (from (8.130)) and

$$(\omega^2 \rho - k^2 c_{44}^E)v_z - i\omega k^2 e_{x5} \Phi = 0.$$ (8.135)

This equation is paired with the remaining equation in part (c),

$$-k^2 e_{x5} v_z + i\omega k^2 \epsilon_{xx}^S \Phi = 0.$$ (8.136)

The solution to these equations is a *stiffened acoustic* wave, with **v** polarized along z and a quasistatic electric field $-\nabla\Phi$ polarized along x. This has a dispersion relation

$$\omega^2 \rho - k^2\left(c_{44} + \frac{e_{x5}^2}{\epsilon_{xx}^S}\right) = 0.$$ (8.137)

F.3   The Quasistatic Approximation

It was noted in Example 4 that the effects of piezoelectric coupling between electromagnetic and acoustic *uniform* plane waves in unbounded media are completely negligible by comparison with the influence of the quasistatic electric field.† Consequently, insignificant errors are introduced if the

---

† This is not always true in the case of problems involving boundary conditions but, even there, the quasistatic approximation may be used in all but a very few cases. Problems of this kind will be considered in Volume II.

rotational (or electromagnetic) part of **E** is neglected in (8.115). This is called the *quasistatic approximation*, and it leads to a very great simplification of the analysis. Removal of all terms in $\mathbf{E}^{(r)}$ from (8.117) reduces the set of equations to two,

$$\nabla \cdot \mathbf{c}^E : \nabla_s \mathbf{v} - \rho \frac{\partial^2 \mathbf{v}}{\partial t^2} = -\nabla \cdot \left(\mathbf{e} \cdot \frac{\partial \, \nabla \Phi}{\partial t} \right) \tag{8.138}$$

$$0 = -\mu_0 \nabla \cdot \left(\mathbf{\epsilon}^S \cdot \frac{\partial^2 \, \nabla \Phi}{\partial t^2} \right) + \mu_0 \nabla \cdot \left(\mathbf{e} : \nabla_s \frac{\partial \mathbf{v}}{\partial t} \right). \tag{8.139}$$

These equations govern plane wave solutions that travel at velocities comparable to acoustic velocities. In the quasistatic approximation quasielectromagnetic waves are regarded as purely electromagnetic.

Conversion of (8.138) and (8.139) to matrix form is carried out in the same way as for (8.117). This gives†

$$\nabla_{iK} c^E_{KL} \nabla_{Lj} v_j - \rho \frac{\partial^2 v_i}{\partial t^2} = -\nabla_{iK} e_{Kj} \nabla_j \frac{\partial \Phi}{\partial t} \tag{8.140}$$

$$\nabla_i \epsilon^S_{ij} \nabla_j \frac{\partial^2 \Phi}{\partial t^2} = \nabla_i e_{iL} \left(\nabla_{Lj} \frac{\partial v_j}{\partial t} \right), \tag{8.141}$$

where ∇_i signifies either (8.121) or its transpose. These equations are not restricted to uniform plane waves and will be applied to more general problems in Volume II. For plane wave solutions proportional to the complex wave function $e^{i(\omega t - k\hat{\mathbf{l}} \cdot \mathbf{r})}$, where $\hat{\mathbf{l}}$ is a unit vector in the propagation direction, they may be reduced still further. First, one may write

$$-k^2 (l_{iK} c^E_{KL} l_{Lj}) v_j + \rho \omega^2 v_i = i\omega k^2 (l_{iK} e_{Kj} l_j) \Phi \tag{8.142}$$

$$\omega^2 k^2 (l_i \epsilon^S_{ij} l_j) \Phi = -i\omega k^2 (l_i e_{iL} l_{Lj}) v_j , \tag{8.143}$$

where l_{iK} and l_{Lj} were defined in (6.8) and (6.9). The factor multiplying Φ on the left-hand side of (8.143) is a scalar and may be divided out, giving the potential in terms of the particle velocity. That is,

$$\Phi = \frac{1}{i\omega} \frac{(l_i e_{iL} l_{Lj})}{l_i \epsilon^S_{ij} l_j} v_j . \tag{8.146}$$

---

† The quasistatic equations (8.140) and (8.141) are frequently given in the literature in full subscript (or tensor) notation, using the acoustic variable **u** rather than **v**. That is,

$$c^E_{iklj} u_{j,kl} - \rho \ddot{u}_i = -e_{ikj} \Phi_{,kj} \tag{8.144}$$

$$\epsilon^S_{ij} \Phi_{,ij} = e_{ikj} u_{j,ik} , \tag{8.145}$$

where a dot over a symbol denotes $\partial/\partial t$ and a subscript i preceded by a comma denotes $\partial/\partial r_i$.

After substitution into (8.142) and some rearrangement of terms, one has

$$k^2 \left(l_{iK} \left\{ c^E_{KL} + \frac{[e_{Kj}l_j][l_i e_{iL}]}{l_i \epsilon^S_{ij} l_j} \right\} l_{Lj} \right) v_j = \rho \omega^2 v_i . \tag{8.147}$$

This has exactly the same form as the Christoffel equation (6.10) but with c_{KL} replaced by the expression in curly brackets,

$$\left\{ c^E_{KL} + \frac{[e_{Kj}l_j][l_i e_{iL}]}{l_i \epsilon^S_{ij} l_j} \right\},$$

which is called a *piezoelectrically stiffened elastic constant*. Plane wave solutions for **v** are found by exactly the same method used in Chapter 7 for the nonpiezoelectric case. There are three uniform plane wave solutions for each propagation direction $\hat{\mathbf{l}}$. They have particle velocity polarizations that are at right angles to each other and are pure transverse or pure longitudinal only for certain propagation directions. Once **v** has been found from the *stiffened Christoffel equation* (8.147), the electric potential is easily calculated from (8.146).

EXAMPLE 8.   The Quasistatic Approximation for Propagation Along the X Axis of a Hexagonal (6mm) Crystal. It will be instructive to apply the quasistatic approximation first to the problem that was solved exactly in Example 7. Since

$$\hat{\mathbf{l}} = \hat{\mathbf{x}}$$

the matrix l_{iK} is

$$l_{iK} = \begin{bmatrix} 1 & 0 & 0 & 0 & 0 & 0 \\ 0 & 0 & 0 & 0 & 0 & 1 \\ 0 & 0 & 0 & 0 & 1 & 0 \end{bmatrix} \tag{8.148}$$

and l_{Lj} is the transpose. Constitutive matrices were given in (8.125), (8.126), and (8.127). The second factor in the numerator of the stiffening term in (8.147) is therefore

$$[l_i e_{iL}] = \left\{ \begin{bmatrix} 1 & 0 & 0 \end{bmatrix} \begin{bmatrix} 0 & 0 & 0 & 0 & e_{x5} & 0 \\ 0 & 0 & 0 & e_{x5} & 0 & 0 \\ e_{z1} & e_{z1} & e_{z3} & 0 & 0 & 0 \end{bmatrix} \right\}$$

$$= \begin{bmatrix} 0 & 0 & 0 & 0 & e_{x5} & 0 \end{bmatrix}, \tag{8.149}$$

and the first factor, $e_{Kj}l_j$, is just the transpose of (8.149). That is,

$$[e_{Kj}l_j] = \begin{bmatrix} 0 \\ 0 \\ 0 \\ 0 \\ e_{x5} \\ 0 \end{bmatrix} \tag{8.150}$$

If the six-element row matrix (8.149) is multiplied on the left by the six-element column matrix (8.150), the 6×6 matrix

$$[e_{Kj}l_j][l_i e_{iL}] = \begin{bmatrix} 0 & 0 & 0 & 0 & 0 & 0 \\ 0 & 0 & 0 & 0 & 0 & 0 \\ 0 & 0 & 0 & 0 & 0 & 0 \\ 0 & 0 & 0 & 0 & 0 & 0 \\ 0 & 0 & 0 & 0 & e_{x5}^2 & 0 \\ 0 & 0 & 0 & 0 & 0 & 0 \end{bmatrix} \tag{8.151}$$

is obtained. This is the numerator of the stiffening term in (8.147). The denominator is a scalar

$$l_i \epsilon_{ij}^S l_j = \epsilon_{xx}^S , \tag{8.152}$$

from (8.126) and $\hat{\mathbf{l}} \cdot \hat{\mathbf{l}} = 1$. Combining of these results with (8.125) gives the matrix of stiffened elastic constants in (8.147). That is

$$[c]_{\text{stiffened}} = \begin{bmatrix} c_{11}^E & c_{12}^E & c_{13}^E & 0 & 0 & 0 \\ c_{12}^E & c_{11}^E & c_{13}^E & 0 & 0 & 0 \\ c_{13}^E & c_{13}^E & c_{33}^E & 0 & 0 & 0 \\ 0 & 0 & 0 & c_{44}^E & 0 & 0 \\ 0 & 0 & 0 & 0 & c_{44}^E + \dfrac{e_{x5}^2}{\epsilon_{xx}^S} & 0 \\ 0 & 0 & 0 & 0 & 0 & c_{66}^E \end{bmatrix} \tag{8.153}$$

The stiffened Christoffel matrix is obtained by multiplying (8.153) on the left and right respectively by l_{iK} and l_{Kj}; and the stiffened Christoffel equation is therefore

$$k^2 \begin{bmatrix} c_{11}^E & 0 & 0 \\ & & \\ 0 & c_{66}^E & 0 \\ & & \\ 0 & 0 & c_{44}^E + \dfrac{e_{x5}^2}{\epsilon_{xx}^S} \end{bmatrix} \begin{bmatrix} v_x \\ v_y \\ v_z \end{bmatrix} = \rho \omega^2 \begin{bmatrix} v_x \\ v_y \\ v_z \end{bmatrix} \tag{8.154}$$

The first two solutions are an *unstiffened* compressional wave (v_x) with dispersion relation

$$\omega^2 \rho = k^2 c_{11}^E \tag{8.155}$$

and an *unstiffened* shear wave (v_y) with dispersion relation

$$\omega^2 \rho = k^2 c_{66}^E . \tag{8.156}$$

These correspond to the quasiacoustic and purely acoustic solutions in Example 7. The quasistatic approximation does not distinguish between these two cases. Finally, the third solution is a *stiffened* shear wave (v_z) with dispersion relation

$$\omega^2 \rho = k^2 \left(c_{44}^E + \frac{e_{x5}^2}{\epsilon_{xx}^S} \right) \tag{8.157}$$

To completely characterize this solution it is necessary to find the electric potential from (8.146). The factor $l_i e_{iL}$ has already been calculated in (8.149), and the potential is found to be

$$\Phi = \frac{e_{x5}}{i\omega \epsilon_{xx}^S} v_z . \tag{8.158}$$

EXAMPLE 9.   The Quasistatic Approximation for Propagation in the XZ Plane of a Hexagonal (6mm) Crystal.   To illustrate the method with a somewhat more difficult case, consider propagation in the *XZ* plane of a *6mm* crystal. This problem may be approached in either of two ways: (1) assume a unit direction vector $\hat{\mathbf{l}} = \hat{\mathbf{x}} l_x + \hat{\mathbf{z}} l_z$ and proceed, using the constitutive matrices of the previous example, or (2) transform the constitutive matrices to the rotated coordinate system of Example 2 in Chapter 4 and assume $\hat{\mathbf{l}} = \hat{\mathbf{x}}'$.† If the coordinate transformation has not already been made, the first calculation is the simpler of the two. In this case, however, the permittivity and stiffness transformations are given in Example 2 of Chapter 4, and the second approach will be used. This has the advantage of presenting the solutions in a simpler form and also gives a ready comparison with the nonpiezoelectric case treated in Chapter 4.

---

† The transformed coordinate system is as shown in Fig. 4.4.

The piezoelectric stress matrix transforms according to (8.47),

$$[e'] = [a][e][\widetilde{M}];$$

and, using the transformation matrices $[a]$ and $[M]$ from Example 2 in Chapter 4,

$$[e'] = \begin{bmatrix} e'_{x1} & e'_{x2} & e'_{x3} & 0 & e'_{x5} & 0 \\ 0 & 0 & 0 & e'_{y4} & 0 & e'_{y6} \\ e'_{z1} & e'_{z2} & e'_{z3} & 0 & e'_{z5} & 0 \end{bmatrix} \qquad (8.159)$$

with

$$e'_{x1} = -e_{z1} \sin \eta \cos^2 \eta - e_{z3} \sin^3 \eta - e_{x5} \cos \eta \sin 2\eta$$

$$e'_{x2} = -e_{z1} \sin \eta$$

$$e'_{x3} = -e_{z1} \sin^2 \eta - e_{z3} \cos^2 \eta \sin \eta + e_{x5} \cos \eta \sin 2\eta$$

$$e'_{x5} = -e_{z1} \frac{\sin \eta \sin 2\eta}{2} + e_{z3} \frac{\sin \eta \sin 2\eta}{2} + e_{x5} \cos \eta \cos 2\eta$$

$$e'_{y4} = e_{x5} \cos \eta$$

$$e'_{y6} = -e_{x5} \sin \eta$$

$$e'_{z1} = e_{z1} \cos^3 \eta + e_{z3} \cos \eta \sin^2 \eta - e_{x5} \sin \eta \sin 2\eta$$

$$e'_{z2} = e_{z1} \cos \eta$$

$$e'_{z3} = e_{z1} \cos \eta \sin^2 \eta + e_{z3} \cos^3 \eta + e_{x5} \sin \eta \sin 2\eta.$$

$$e'_{z5} = e_{z1} \frac{\cos \eta \sin 2\eta}{2} - e_{z3} \frac{\cos \eta \sin 2\eta}{2} + e_{x5} \sin \eta \cos 2\eta$$

Since the wave propagates along the x' axis of the transformed coordinate system, l_{iK} is still given by (8.148). The second factor in the numerator of the stiffening term (8.147) is then

$$[l_i e'_{iL}] = \left\{ \begin{bmatrix} 1 & 0 & 0 \end{bmatrix} \begin{bmatrix} e'_{x1} & e'_{x2} & e'_{x3} & 0 & e'_{x5} & 0 \\ 0 & 0 & 0 & e'_{y4} & 0 & e'_{y6} \\ e'_{z1} & e'_{z2} & e'_{z3} & 0 & e'_{z5} & 0 \end{bmatrix} \right\}$$

$$= \begin{bmatrix} e'_{x1} & e'_{x2} & e'_{x3} & 0 & e'_{x5} & 0 \end{bmatrix} \qquad (8.160)$$

and the first factor in the numerator is the transpose,

$$[e_{Kj} l_j] = \begin{bmatrix} e'_{x1} \\ e'_{x2} \\ e'_{x3} \\ 0 \\ e'_{z5} \\ 0 \end{bmatrix} \qquad (8.161)$$

From (4.43) the denominator is

$$l_i \epsilon^{S'}_{ij} l_j = \epsilon^{S'}_{xx}. \qquad (8.162)$$

Multiplication of (8.160) on the left with (8.161) and division by (8.162) gives the piezoelectric stiffening matrix. This is then combined with (4.50) to give the matrix of stiffened elastic constants

$$[c]_{\text{stiffened}} = \begin{bmatrix} \bar{c}'_{11} & \bar{c}'_{12} & \bar{c}'_{13} & 0 & \bar{c}'_{15} & 0 \\ \bar{c}'_{12} & \bar{c}'_{22} & \bar{c}'_{23} & 0 & \bar{c}'_{25} & 0 \\ \bar{c}'_{13} & \bar{c}'_{23} & \bar{c}'_{33} & 0 & \bar{c}'_{35} & 0 \\ 0 & 0 & 0 & c'_{44} & 0 & c'_{46} \\ \bar{c}'_{15} & \bar{c}'_{25} & \bar{c}'_{35} & 0 & \bar{c}'_{55} & 0 \\ 0 & 0 & 0 & c'_{46} & 0 & c'_{66} \end{bmatrix} \tag{8.163}$$

with stiffened constants

$$\bar{c}'_{IJ} = c^{E'}_{IJ} + \frac{e'_{xI} e'_{xJ}}{\epsilon^{S'}_{xx}}. \tag{8.164}$$

This has the same form as (4.50), with stiffening terms added to some of the matrix elements.

The stiffened Christoffel equation is now written down by multiplying (8.163) on the left with (8.148) and on the right with its transpose. This gives

$$k^2 \begin{bmatrix} \bar{c}'_{11} & 0 & \bar{c}'_{15} \\ 0 & c'_{66} & 0 \\ \bar{c}'_{15} & 0 & \bar{c}'_{55} \end{bmatrix} \begin{bmatrix} v_{x'} \\ v_{y'} \\ v_{z'} \end{bmatrix} = \rho\omega^2 \begin{bmatrix} v_{x'} \\ v_{y'} \\ v_{z'} \end{bmatrix} \tag{8.165}$$

As in Example 2(b) of Chapter 4, one solution is a pure shear wave polarized along y', with dispersion relation

$$\omega^2 \rho = k^2 c'_{66}. \tag{8.166}$$

The other two solutions have quasishear and quasilongitudinal polarizations in the xz plane, with a dispersion relation

$$(\bar{c}'_{11} k^2 - \rho\omega^2)(\bar{c}'_{55} k^2 - \rho\omega^2) - (\bar{c}'_{15})^2 k^4 = 0. \tag{8.167}$$

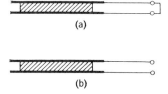

(a)

(b)

FIGURE 8.8. **(a) In a small thin piezoelectric disk with short-circuited electrodes the electric field E may be accurately assumed to be zero, and stiffness constants c^E_{KL} are applicable. (b) When the electrodes are open-circuited the electrical displacement D is very nearly zero, and stiffness constants c^D_{KL} are applicable.**

Only the second and third solutions are piezoelectrically stiffened. From (8.146), (8.160), and (8.162)

$$\Phi = \frac{e'_{x1}}{i\omega\epsilon^{S'}_{xx}} v_{x'} + \frac{e'_{x5}}{i\omega\epsilon^{S'}_{xx}} v_{z'}. \qquad (8.168)$$

F.4 Piezoelectric Stiffening and Electromechanical Coupling Constants

It has been seen that the Christoffel equation for a piezoelectric solid involves the stiffened elastic constants

$$c^E_{KL} + \frac{[e_{Kj}l_j][l_ie_{iL}]}{l_i\epsilon^S_{ij}l_j}, \qquad (8.169)$$

where the change in elastic stiffness is a function of the piezoelectric constants, the permittivity, *and the wave vector direction* $\hat{\mathbf{l}}$. Another instance of piezo-electrically induced change in the elastic constants can be derived by re-arranging the piezoelectric stress equations (8.69) and (8.70) so that $\mathbf{D}$ is an independent variable. Multiplication of (8.69) by the inverse of $\boldsymbol{\epsilon}^S$ and re-arrangement of terms gives

$$\mathbf{E} = (\boldsymbol{\epsilon}^S)^{-1} \cdot \mathbf{D} - (\boldsymbol{\epsilon}^S)^{-1} \cdot \mathbf{e} : \mathbf{S}.$$

When this is substituted into (8.70),

$$\mathbf{T} = -\mathbf{e} \cdot (\boldsymbol{\epsilon}^S)^{-1} \cdot \mathbf{D} + (\mathbf{c}^E + \mathbf{e} \cdot (\boldsymbol{\epsilon}^S)^{-1} \cdot \mathbf{e}) : \mathbf{S}. \qquad (8.170)$$

The factor multiplying $\mathbf{S}$ in (8.170) is the elastic stiffness tensor at *zero electric displacement*,

$$\mathbf{c}^D = \mathbf{c}^E + \mathbf{e} \cdot (\boldsymbol{\epsilon}^S)^{-1} \cdot \mathbf{e}.$$

In matrix notation

$$c^D_{KL} = c^E_{KL} + e_{Kj}(\epsilon^S)^{-1}_{ji} e_{iL}. \qquad (8.171)$$

The piezoelectric stiffening in (8.171) is relevant to static measurements or to the dynamic behavior of crystal samples that are very small compared to the acoustic wavelength (Fig. 8.8), but it is not applicable to wave propagation problems. It is therefore important to determine the precise electrical conditions applicable to (8.169), so that these piezoelectrically stiffened constants may be distinguished from the ones given by (8.171). The electric charge density ρ_e that appears in the divergence relation (4.14),

$$\nabla \cdot \mathbf{D} = \rho_e,$$

relates to *free* charge and not to the *bound* charge that is produced by strain in a piezoelectric (Fig. 8.7). In a source-free medium, therefore, one has

$$\nabla \cdot \mathbf{D} = 0. \qquad (8.172)$$

For fields represented by the complex exponential wave function $e^{i(\omega t - k\hat{\mathbf{i}} \cdot \mathbf{r})}$

$$\nabla \rightarrow -ik\hat{\mathbf{i}},$$

and (8.172) becomes

$$-ik\hat{\mathbf{i}} \cdot \mathbf{D} = 0. \tag{8.173}$$

This defines the electrical constraint relevant to the elastic constants in (8.169). That is

$$c_{KL}^{\hat{\mathbf{i}} \cdot \mathbf{D}} = c_{KL}^{E} + \frac{[e_{Kj}l_j][l_i e_{iL}]}{l_i \epsilon_{ij}^{S} l_j}. \tag{8.174}$$

EXAMPLE 10. The physical difference between (8.171) and (8.174) is best illustrated by considering the specific case treated in Example 9. In the case of (8.171) the *total* electric displacement **D** is zero. For the wave propagation problem, the electrical displacement is calculated from (8.69). That is

$$D_i = -\epsilon_{ij}^{S} \frac{\partial \Phi}{\partial r_j} + e_{iJ} S_J. \tag{8.175}$$

In Example 9 the stiffened quasishear and quasilongitudinal solutions have a potential function given by (8.168) and the strain field has two components

$$S_{1'} = \frac{1}{i\omega} \frac{\partial v_{x'}}{\partial x'} = -\frac{k}{\omega} v_{x'}$$

$$S_{5'} = \frac{1}{i\omega} \frac{\partial v_{z'}}{\partial x'} = -\frac{k}{\omega} v_{z'}. \tag{8.176}$$

Using the transformed permittivity matrix (4.43) and the transformed piezoelectric matrix (8.159), the electric displacement field is calculated to be

$$D_{x'} = -\epsilon_{xx}^{S'}(-ik\Phi) + e_{x1}' S_{1'} + e_{x5}' S_{5'}$$

$$D_{y'} = 0$$

$$D_{z'} = -\epsilon_{xz}^{S}(-ik\Phi) + e_{z1}' S_{1'} + e_{z5}' S_{5'}.$$

Substitution of (8.168) and (8.176) then shows that the electrical displacement component along **k** is zero,

$$D_{x'} = 0, \tag{8.177}$$

as required by (8.173); but

$$D_{z'} = \frac{k}{\omega \epsilon_{xx}^{S}} \{(\epsilon_{xz}^{S} e_{x1}' - \epsilon_{xx}^{S} e_{z1}')v_{x'} + (\epsilon_{xz}^{S} e_{x5}' - \epsilon_{xx}^{S} e_{z5}')v_{z'}\} \tag{8.178}$$

is not equal to zero. This result will be seen to have a very important consequence with regard to the power flow in this wave.

If (8.174) is rewritten as

$$c_{KL}^{\hat{i} \cdot D} = c_{KL}^{E}\left(1 + \frac{[e_{Kj}l_j][l_i e_{iL}]}{(c_{KL}^{E})(l_i \epsilon_{ij}^{S} l_j)}\right)$$

it is seen that the fractional change in the KL elastic constant is equal to

$$(K_{KL}(\hat{i}))^2 = \frac{[e_{Kj}l_j][l_i e_{iL}]}{(c_{KL}^{E})(l_i \epsilon_{ij}^{S} l_j)} . \tag{8.179}$$

The quantities $K_{KL}(\hat{i})$, often called *electromechanical coupling constants*, indicate the strength of the piezoelectric interaction and often have a simple and direct physical interpretation. In Example 8, for instance, (8.157) shows that piezoelectric stiffening changes the (phase velocity)^2 of the z-polarized x-propagating shear wave by a fractional amount given by the square of the electromechanical coupling constant

$$K_{55} = \frac{e_{x5}}{(c_{44}^{E}\epsilon_{xx}^{S})^{1/2}} . \tag{8.180}$$

In other cases, such as (8.167) in Example 9, the phase velocity is a rather complicated function of several coupling constants. Section 8.J will show that electromechanical coupling constants also enter in an important way into the theory of piezoelectric transducers.

Typical values of electromechanical coupling constants for some common materials are given in Table 8.1. In most materials K^2 is not more than a few percent and the stiffening effect may be neglected when calculating the phase velocity. This is the case for curves shown in Part B of Appendix 3. For lithium niobate, however, the stiffening is appreciable and must be taken into account. The magnitude of the stiffening effect in this material can be seen by comparing Figs. 8.9 and 8.10, in which piezoelectricity is ignored, with the exact curves given by Figs. 8.11 and 8.12.

G. POYNTING'S THEOREM   FOR PIEZOELECTRIC MEDIA

Poynting's Theorems were derived in Chapter V for electromagnetic and acoustic fields in *nonpiezoelectric* media. The derivations of the *real* forms of these theorems were based on two relations,

$$\nabla \cdot (\mathbf{E} \times \mathbf{H}) = -\mathbf{H} \cdot \frac{\partial \mathbf{B}}{\partial t} - \mathbf{E} \cdot \frac{\partial \mathbf{D}}{\partial t} - \mathbf{E} \cdot \mathbf{J}_c - \mathbf{E} \cdot \mathbf{J}_s \tag{8.181}$$

and

$$\nabla \cdot (-\mathbf{v} \cdot \mathbf{T}) = -\mathbf{v} \cdot \frac{\partial \mathbf{p}}{\partial t} - \mathbf{T} : \frac{\partial \mathbf{S}}{\partial t} + \mathbf{v} \cdot \mathbf{F}, \tag{8.182}$$

TABLE 8.1. Electromechanical Coupling Constants

Bismuth germanium oxide (Cubic *23*)

[110]-propagation, longitudinal: No coupling

[110]-propagation, [001]-shear: $\left(\dfrac{e_{4x}^2}{c_{44}^E \epsilon_{xx}^S}\right)^{1/2} = 0.338$

[110]-propagation, [$\bar{1}$10]-shear: No coupling

Cadmium sulfide (Hex. *6mm*)

Z-propagation, longitudinal: $\left(\dfrac{e_{z3}^2}{c_{33}^E \epsilon_{zz}^S}\right)^{1/2} = 0.156$

Z-propagation, shear: No coupling

X-propagation, Z-shear: $\left(\dfrac{e_{x5}^2}{c_{44}^E \epsilon_{xx}^S}\right)^{1/2} = 0.192$

X-propagation, Y-shear: No coupling

Lithium niobate (Trig. *3m*)

Z-propagation, longitudinal: $\left(\dfrac{e_{z3}^2}{c_{33}^E \epsilon_{zz}^S}\right)^{1/2} = 0.163$

Z-propagation, shear: No coupling

Quartz (Trig. *32*)

Z-propagation, longitudinal: No coupling
Z-propagation, shear: No coupling

Zinc oxide (Hex. *6mm*)

Z-propagation, longitudinal: $\left(\dfrac{e_{z3}^2}{c_{33}^E \epsilon_{zz}^S}\right)^{1/2} = 0.302$

Z-propagation, shear: No coupling

X-propagation, Z-shear: $\left(\dfrac{e_{x5}^2}{c_{44}^E \epsilon_{xx}^S}\right)^{1/2} = 0.268$

X-propagation, Y-shear: No coupling

---

obtained from the general field equations (8.63) to (8.66) without specifying any particular kind of medium. These relations apply therefore to piezoelectric, as well as nonpiezoelectric, materials. In the nonpiezoelectric case there is no coupling between the field variables in (8.181) and those in (8.182), and separate electromagnetic and acoustic Poynting's Theorems are obtained. In piezoelectric materials coupling is provided by the piezoelectric constitutive relations, and a single *piezoelectric* Poynting's Theorem results.

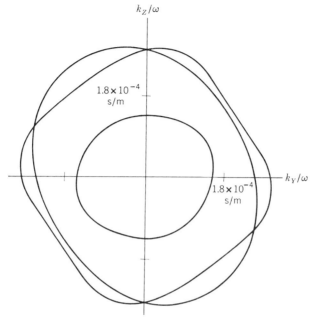

FIGURE 8.9. Inverse velocity (or slowness) curves for propagation in the YZ plane of lithium niobate, with the piezoelectric effect ignored.

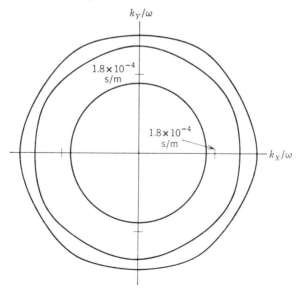

FIGURE 8.10. Inverse velocity (or slowness) curves for propagation in the XY plane of lithium niobate, with the piezoelectric effect ignored.

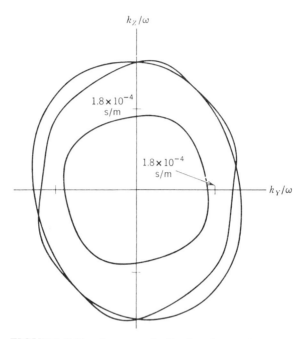

FIGURE 8.11. Inverse velocity (or slowness) curves for propagation in the YZ plane of lithium niobate, including piezoelectricity.

This is derived by first adding (8.181) to (8.182),

$$\nabla \cdot (-\mathbf{v} \cdot \mathbf{T} + \mathbf{E} \times \mathbf{H}) = -\mathbf{v} \cdot \frac{\partial \mathbf{p}}{\partial t} - \mathbf{T} : \frac{\partial \mathbf{S}}{\partial t} - \mathbf{H} \cdot \frac{\partial \mathbf{B}}{\partial t} - \mathbf{E} \cdot \frac{\partial \mathbf{D}}{\partial t}$$
$$+ \mathbf{v} \cdot \mathbf{F} - \mathbf{E} \cdot \mathbf{J}_c - \mathbf{E} \cdot \mathbf{J}_s.$$

Integration over a volume V and application of the divergence theorem then gives

$$\oint_S (-\mathbf{v} \cdot \mathbf{T} + \mathbf{E} \times \mathbf{H}) \cdot \hat{n} \, dS$$
$$= -\int_V \left(\mathbf{v} \cdot \frac{\partial \mathbf{p}}{\partial t} + \mathbf{T} : \frac{\partial \mathbf{S}}{\partial t} + \mathbf{H} \cdot \frac{\partial \mathbf{B}}{\partial t} + \mathbf{E} \cdot \frac{\partial \mathbf{D}}{\partial t} \right) dV$$
$$- \int_V \mathbf{E} \cdot \mathbf{J}_c \, dV + \int_V (\mathbf{v} \cdot \mathbf{F} - \mathbf{E} \cdot \mathbf{J}_s) \, dV, \qquad (8.183)$$

where S is the volume enclosing V. Terms on the right-hand side of (8.183) are identified by following the same arguments used in Chapter 5. The third term is the instantaneous power supplied by mechanical and electrical sources, and the second term is minus the instantaneous power dissipation in conductive losses. For lossless media the first term is minus the time rate of change of stored energy. When viscous damping is present, the second quantity under the integral also includes viscous damping loss. From energy conservation the left-hand side must represent mechanical and electrical power flow outward through the surface S. The complete expression is therefore a statement of the real Poynting's Theorem for piezoelectric media, and the *instantaneous* piezoelectric Poynting vector is

$$\mathbf{P}(t) = -\mathbf{v}(t) \cdot \mathbf{T}(t) + \mathbf{E}(t) \times \mathbf{H}(t). \tag{8.184}$$

The right-hand side of (8.183) can be expressed entirely in terms of the independent field variables $\mathbf{v}$, $\mathbf{T}$, $\mathbf{H}$, $\mathbf{E}$ by means of the constitutive equations (8.67) to (8.72).

A *complex* Poynting's Theorem for piezoelectric media is obtained in a similar way, starting from

$$\nabla \cdot \frac{(\mathbf{E} \times \mathbf{H}^*)}{2} = i\omega\left(\frac{\mathbf{E} \cdot \mathbf{D}^*}{2} - \frac{\mathbf{H}^* \cdot \mathbf{B}}{2}\right) - \frac{\mathbf{E} \cdot \mathbf{J}_c^*}{2} - \frac{\mathbf{E} \cdot \mathbf{J}_s^*}{2} \tag{8.185}$$

and

$$\nabla \cdot \left(-\frac{\mathbf{v}^* \cdot \mathbf{T}}{2}\right) = i\omega\left(\frac{\mathbf{T}:\mathbf{S}^*}{2} - \frac{\mathbf{v}^* \cdot \mathbf{p}}{2}\right) + \frac{\mathbf{v}^* \cdot \mathbf{F}}{2}. \tag{8.186}$$

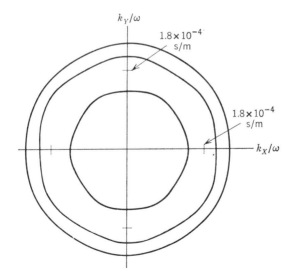

FIGURE. 8.12   Inverse velocity (or slowness) curves for propagation in the XY plane of lithium niobate, including piezoelectricity.

The result is,

$$\oint_S \left(-\frac{\mathbf{v}^* \cdot \mathbf{T}}{2} + \frac{\mathbf{E} \times \mathbf{H}^*}{2} \right) \cdot \hat{\mathbf{n}} \, dS$$

$$= i\omega \int_V \left(\frac{\mathbf{T}:\mathbf{S}^*}{2} - \frac{\mathbf{v}^* \cdot \mathbf{p}}{2} + \frac{\mathbf{E} \cdot \mathbf{D}^*}{2} - \frac{\mathbf{H}^* \cdot \mathbf{B}}{2} \right) dV$$

$$- \int_V \frac{\mathbf{E} \cdot \mathbf{J}_c^*}{2} \, dV + \int_V \left(\frac{\mathbf{v}^* \cdot \mathbf{F}}{2} - \frac{\mathbf{E} \cdot \mathbf{J}_s^*}{2} \right) dV, \qquad (8.187)$$

and the *complex* piezoelectric Poynting vector is therefore

$$\mathbf{P} = -\frac{\mathbf{v}^* \cdot \mathbf{T}}{2} + \frac{\mathbf{E} \times \mathbf{H}^*}{2}. \qquad (8.188)$$

It was seen in the preceding section that use of the quasistatic approximation, in which

$$\mathbf{E} = -\nabla\Phi, \qquad (8.189)$$

greatly simplifies analysis of the piezoelectric interaction. Since this approach can be applied to essentially all problems involving piezoelectric media, it is of the greatest importance to have a method for calculating power flow within the framework of the quasistatic approximation. This is easily obtained by making some simple modifications of (8.183) and (8.187).

In deriving (8.183), use was made of the identity

$$\nabla \cdot (\mathbf{E} \times \mathbf{H}) = \mathbf{H} \cdot \nabla \times \mathbf{E} - \mathbf{E} \cdot \nabla \times \mathbf{H}.$$

When the quasistatic approximation (8.189) is made, the first term on the right-hand side goes to zero, and

$$\nabla \cdot (\mathbf{E} \times \mathbf{H}) = \nabla\Phi \cdot (\nabla \times \mathbf{H}). \qquad (8.190)$$

The elementary vector identity

$$\nabla \cdot \Phi\mathbf{A} = \Phi\nabla \cdot \mathbf{A} + \nabla\Phi \cdot \mathbf{A} \qquad (8.191)$$

can then be used to convert (8.190) into

$$\nabla \cdot (\mathbf{E} \times \mathbf{H}) = \nabla \cdot \Phi(\nabla \times \mathbf{H}),$$

since $\nabla \cdot (\nabla \times \mathbf{H}) = 0$; and substitution of $\nabla \times \mathbf{H}$ from (8.64) gives

$$\nabla \cdot (\mathbf{E} \times \mathbf{H}) = \nabla \cdot \Phi \left(\frac{\partial \mathbf{D}}{\partial t} + \mathbf{J}_c + \mathbf{J}_s \right). \qquad (8.192)$$

It was the integration of the left-hand side of this equation over the volume V, followed by use of the divergence theorem, that led to the $\mathbf{E} \times \mathbf{H}$ term in the Poynting vector (8.184). One finds from (8.192), then, that the quasistatic approximation to the instantaneous Poynting vector (8.184) is

$$\mathbf{P}(t) = -\mathbf{v}(t) \cdot \mathbf{T}(t) + \Phi(t) \frac{\partial \mathbf{D}(t)}{\partial t}$$

when the currents $\mathbf{J}_c$ and $\mathbf{J}_s$ are zero on the boundary surface S. The same kind of argument applied to $\mathbf{E} \times \mathbf{H}^*/2$ in (8.187) shows that the quasistatic approximation for the *complex* Poynting vector is

$$\mathbf{P} = -\frac{\mathbf{v}^* \cdot \mathbf{T}}{2} + \frac{\Phi(i\omega \mathbf{D})^*}{2}$$

in current-free regions. Since $\partial \mathbf{B}/\partial t = -\nabla \times \mathbf{E}$ is zero in the quasistatic approximation, the real form of Poynting's Theorem becomes

$$\oint_S \mathbf{P} \cdot \hat{\mathbf{n}}\, dS = \oint_S \left(-\mathbf{v} \cdot \mathbf{T} + \Phi \frac{\partial \mathbf{D}}{\partial t} \right) \cdot \hat{\mathbf{n}}\, dS$$

$$= -\int_V \left(\mathbf{v} \cdot \frac{\partial \mathbf{p}}{\partial t} + \mathbf{T}:\frac{\partial \mathbf{S}}{\partial t} - \nabla \Phi \cdot \frac{\partial \mathbf{D}}{\partial t} \right) dV$$

$$+ \int_V \left(\mathbf{v} \cdot \mathbf{F} + \Phi \frac{\partial \rho_e}{\partial t} \right) dV \qquad (8.193)$$

and the complex form becomes

$$\oint_S \mathbf{P} \cdot \hat{\mathbf{n}}\, dS = \oint_S \left(-\frac{\mathbf{v}^* \cdot \mathbf{T}}{2} + \frac{\Phi(i\omega \mathbf{D})^*}{2} \right) \cdot \hat{\mathbf{n}}\, dS$$

$$= i\omega \int_V \left(\frac{\mathbf{T}:\mathbf{S}^*}{2} - \frac{\mathbf{v}^* \cdot \mathbf{p}}{2} - \frac{\nabla \Phi \cdot \mathbf{D}^*}{2} \right) dV$$

$$- \int_V -\frac{\nabla \Phi \cdot \mathbf{J}_c^*}{2}\, dV + \int_V \left(\frac{\mathbf{v}^* \cdot \mathbf{F}}{2} + \frac{\Phi(i\omega \rho_e)^*}{2} \right) dV. \quad (8.194)$$

Note that the current sources $\mathbf{J}_s$ in (8.183) and (8.187) have now been replaced with free charge sources ρ_e. This result follows from (8.191), the equation of charge continuity (4.12), and the assumption that $\mathbf{J}_s = 0$ on the boundary S.

EXAMPLE 11.   According to (8.146), the quasistatic potential cannot exist unless there is an acoustic wave present. Despite this parasitic relationship, the quasistatic field may still manage to contribute to the power flow and do this in a very interesting way. Consider wave propagation in the *XZ* plane of a hexagonal (*6mm*) crystal, Example 9. When the piezoelectric coupling is small, the power flow may be evaluated approximately by using the acoustic Poynting vector calculation of Example 4 in Chapter 5,

$$\mathbf{P}_{\text{acoust}}(t) = \hat{\mathbf{x}}' \frac{\omega}{k} \rho(v_{x'}^2 + v_{z'}^2)$$

$$+ \hat{\mathbf{z}}'\left\{\left(\frac{\omega\rho}{k} + \frac{k}{\omega} c_{13}'\right) v_{x'}v_{z'} + \frac{k}{\omega} c_{35}'(v_{z'})^2\right\}. \qquad (8.195)$$

Piezoelectric coupling modifies this result in two different ways. In the first place, the elastic constants in the Christoffel equation are stiffened, and this changes the values of ω/k, $v_{x'}$, and $v_{z'}$. Secondly, an electric contribution to the power flow may appear. This is the second term in the integrand on the left-hand side of (8.193). For the problem at hand, Φ and $\mathbf{D}$ are given by (8.168) and (8.178). Using instantaneous values of Φ and $\mathbf{D}$, one finds that

$$\mathbf{P}_{\text{electric}}(t) = \hat{\mathbf{z}}'\Phi(t)\frac{\partial}{\partial t} D_{z'}(t). \qquad (8.196)$$

Since the wave vector $\mathbf{k}$ is along $\hat{\mathbf{x}}'$ in this problem, *the electric power flow is seen to be orthogonal to* $\mathbf{k}$. This is characteristic of quasistatic fields. There can be no electric power flow along $\mathbf{k}$ because

$$\mathbf{k} \cdot \mathbf{D} = 0,$$

from (8.173). Addition of (8.195) and (8.196) gives the total instantaneous Poynting vector

$$\mathbf{P}(t) = \hat{\mathbf{x}}' \frac{\omega}{k} \rho(v_{x'}^2 + v_{z'}^2)$$

$$+ \hat{\mathbf{z}}'\left\{\left(\frac{\omega\rho}{k} + \frac{k}{\omega} c_{13}'\right) v_{x'}v_{z'} + \frac{k}{\omega} c_{35}'(v_{z'})^2 + \Phi \frac{\partial D_{z'}}{\partial t}\right\}, \qquad (8.197)$$

with Φ and $D_{z'}$ expressed in terms of $v_{x'}$ and $v_{z'}$ by means of (8.168) and (8.178).

It was seen in Example 4 of Chapter 5 that the acoustic Poynting vector deflects from $\mathbf{k}$ for all angles η except 0 and $\pi/2$. Depending on the relative signs of the electric and acoustic parts of the z' component in (8.197), the electric power flow may either increase or decrease this deflection. The electric power flow goes to zero at $\eta = 0$, $\pi/2$. Consequently, the total piezoelectric power flow is along $\mathbf{k}$ for these two limiting cases. For strongly piezoelectric materials the change in power flow deflection due to electric power flow is large enough to be experimentally significant. In the case of the trigonal crystal lithium niobate, for example, the power flow angle for the Y-propagating quasilongitudinal wave shifts from $8.8°$ to $9.2°$ when the electric contribution is included. This calculation has been found to agree with the normal direction to the slowness surface and also with experiment.†

---

† Reference 16 at the end of the chapter.

H.   TRANSMISSION LINE MODEL FOR PIEZOELECTRIC SOLIDS

In the quasistatic approximation, (8.102) takes the form

$$\mathbf{c}^E : \nabla_s \mathbf{v} = \frac{\partial}{\partial t} \mathbf{T} - \mathbf{e} \cdot \frac{\partial}{\partial t} \nabla \Phi, \tag{8.198}$$

and its companion equation

$$\nabla \cdot \mathbf{T} = \rho \frac{\partial \mathbf{v}}{\partial t} - \mathbf{F} \tag{8.199}$$

is unchanged. A transmission line model for these equations can be constructed by the same technique used in Sections 6.D and 7.J. That is, they are written out in component form and separated into sets of independent equations. For simplicity, only the special case of cubic piezoelectrics will be considered here and propagation will be restricted to the xz plane, with particle displacement along the y direction, as in Fig. 7.15. The constitutive matrices are the same as in Example 4, and the relevant equations are therefore

$$\frac{\partial}{\partial z} T_4 + \frac{\partial}{\partial x} T_6 = \rho \frac{\partial}{\partial t} v_y - F_y$$

$$c_{44}^E \frac{\partial v_y}{\partial z} = \frac{\partial}{\partial t} T_4 - e_{x4} \frac{\partial}{\partial t} (\nabla \Phi)_x$$

$$c_{44}^E \frac{\partial v_y}{\partial x} = \frac{\partial}{\partial t} T_6 - e_{x4} \frac{\partial}{\partial t} (\nabla \Phi)_z . \tag{8.200}$$

In (8.200) the electric field $-\nabla \Phi$ includes both an applied field $-\nabla \Phi_a$ and a strain-induced field $-\nabla \Phi_s$.[†] For fields proportional to $e^{i(\omega t - \hat{k}\mathbf{l} \cdot \mathbf{r})}$, $\nabla \Phi_s$ is calculated from (8.146); and

$$i\omega \nabla \Phi_s = ik\hat{\mathbf{l}} \frac{e_{x4}}{\epsilon_{xx}^S} \sin 2\theta \, v_y . \tag{8.201}$$

Time and space derivatives can be introduced into (8.201) by making the substitutions

$$i\omega \rightarrow \frac{\partial}{\partial t}$$

$$-ik\hat{\mathbf{l}} \rightarrow \nabla.$$

[†] The applied potential is defined as a solution to (8.141) with the piezoelectric constants equal to zero.

This means that

$$-\frac{\partial}{\partial t}\nabla\Phi = -\frac{\partial}{\partial t}\nabla\Phi_a - \frac{e_{x4}}{\epsilon_{xx}^S}\sin 2\theta\,\nabla v_y\,,$$

and (8.200) can be written as

$$\frac{\partial}{\partial z}T_4 + \frac{\partial}{\partial x}T_6 = \rho\frac{\partial}{\partial t}v_y - F_y$$

$$c_{44}^E\frac{\partial v_y}{\partial z} = \frac{\partial}{\partial t}T_4 - e_{x4}\frac{\partial}{\partial t}(\nabla\Phi_a)_x - \frac{e_{x4}^2}{\epsilon_{xx}^S}\sin 2\theta\frac{\partial v_y}{\partial x}$$

$$c_{44}^E\frac{\partial v_y}{\partial x} = \frac{\partial}{\partial t}T_6 - e_{x4}\frac{\partial}{\partial t}(\nabla\Phi_a)_z - \frac{e_{x4}^2}{\epsilon_{xx}^S}\sin 2\theta\frac{\partial v_y}{\partial z}. \qquad (8.202)$$

The applied potential Φ_a is assumed to be a function only of t and z', consistent with excitation of a uniform plane wave propagating along z' in Fig. 7.15.

As in Section 7.J, the equations (8.202) are now transformed to the rotated coordinate system x', y', z' in Fig. 7.15. Spatial derivatives in (8.202) are transformed according to (7.92), and the second and third lines are combined so as to introduce the variable

$$T_{4'} = T_4\cos\theta + T_6\sin\theta.$$

The resulting equations are then

$$\frac{\partial}{\partial z'}T_{4'} = \rho\frac{\partial v_{y'}}{\partial t} - F_{y'}, \qquad (8.203)$$

and

$$\left(c_{44}^E + \frac{e_{x4}^2}{\epsilon_{xx}^S}\sin^2 2\theta\right)\frac{\partial v_{y'}}{\partial z'} = \frac{\partial}{\partial t}T_{4'} - e_{x4}\frac{\partial}{\partial t}(\nabla\Phi_a)_{z'}\sin 2\theta, \qquad (8.204)$$

where the coefficient on the left-hand side is the stiffened elastic constant $\hat{c}_{44}^{\text{l·D}}$.

These equations differ from (7.93) and (7.94) only in a change of the stiffness constant and addition of a piezoelectric source term. The previous analogy with a transmission line therefore remains valid, and only the stiffening term and the piezoelectric source term need to be identified in the model. The shunt capacitance per unit length is now the inverse of the stiffened elastic constant,

$$C = \frac{1}{c_{44}^E + (e_{x4}^2/\epsilon_{xx}^S)\sin^2 2\theta}.$$

This is equivalent to the series combination of a capacitor $(c_{44}^E)^{-1}$ and a capacitor $((e_{x4}^2/\epsilon_{xx}^S) \sin^2 2\theta)^{-1}$. That is,

$$\frac{1}{C} = \frac{1}{(c_{44}^E)^{-1}} + \frac{1}{[(e_{x4}^2/\epsilon_{xx}^S) \sin^2 2\theta]^{-1}},$$

as in Fig. 8.13. Comparison with Fig. 7.16a shows that the effect of the piezoelectric stiffening is to add a capacitance

$$\left(\frac{e_{x4}^2}{\epsilon_{xx}^S} \sin^2 2\theta\right)^{-1}$$

in series with the shunt arm. Since the shunt equivalent current per unit length is

$$-\frac{\partial v_y}{\partial z'} = (c_{44}^{\hat{1}\cdot D})^{-1} \frac{\partial}{\partial t} [-T_{4'} + e_{x4}(\nabla \Phi_a)_{z'} \sin 2\theta]$$

from (8.204), the equivalent piezoelectric generator appears in the shunt arm of the transmission line.

I. EXCITATION OF PLANE WAVES BY DISTRIBUTED PIEZO-ELECTRIC SOURCES

When the piezoelectric equations are to be solved for plane waves excited by a given distribution of applied electric field, it is most convenient to use the normal mode technique described in Section 6.E.

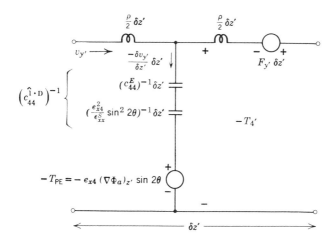

FIGURE 8.13. **Transmission line model for plane wave propagation in a piezoelectric medium.**

EXAMPLE 12. Consider the specific case treated in the previous section. Symbols are first introduced for the phase velocity and impedance of the stiffened shear wave governed by (8.203) and (8.204). That is,

$$\bar{V}_s = \left(\frac{\bar{c}_{44}}{\rho}\right)^{1/2} \tag{8.205}$$

and

$$\bar{Z}_s = (\rho \bar{c}_{44})^{1/2}. \tag{8.206}$$

In both equations

$$\bar{c}_{44} = c_{44}^E + \frac{e_{x4}^2}{\epsilon_{xx}^S} \sin^2 2\theta.$$

Normal mode amplitudes are then defined as

$$a_+ = -T_{4'} + v_{y'}\bar{Z}_s$$
$$a_- = -T_{4'} - v_{y'}\bar{Z}_s, \tag{8.207}$$

by analogy with (6.53). If (8.203) is subtracted from $1/\bar{V}_s$ times (8.204), the result can be rearranged in the normal mode form

$$\left(\frac{\partial}{\partial z'} + \frac{1}{\bar{V}_s}\frac{\partial}{\partial t}\right)a_+ = b_+, \tag{8.208}$$

where

$$b_+ = F_{y'} - \frac{e_{x4}}{\bar{V}_s}\frac{\partial}{\partial t}(\nabla\Phi_a)_{z'}\sin 2\theta \ .$$

Similarly the equation for the negative-traveling wave amplitude is

$$\left(\frac{\partial}{\partial z'} - \frac{1}{\bar{V}_s}\frac{\partial}{\partial t}\right)a_- = b_-, \tag{8.209}$$

with

$$b_- = F_{y'} + \frac{e_{x4}}{\bar{V}_s}\frac{\partial}{\partial t}(\nabla\Phi_a)_{z'}\sin 2\theta.$$

Except for the source terms, these equations are of the same form as (6.65); and solutions are obtained in exactly the same way. Suppose that the piezoelectric medium is unbounded, that $F_{y'} = 0$, and that there is a time-harmonic piezoelectric source distribution

$$\frac{i\omega T_{\mathrm{PE}}(z')}{\bar{V}_s} = \begin{cases} -\dfrac{i\omega}{\bar{V}_s}e_{x4}E_a\sin 2\theta & |z'| < |l| \\[2mm] 0 & |l| < |z'|. \end{cases} \tag{8.210}$$

where $E_a = -(\nabla\Phi_a)_{z'}$ is a constant. This has the same form as the body force source distribution in (6.73) of Example 3 in Chapter 4. Solutions to the present problem can thus be obtained by making the substitution

$$F \rightarrow \frac{i\omega}{\bar{V}_s}e_{x4}E_a\sin 2\theta$$

in the previously obtained expressions for $a_+(z)$ and

$$F \rightarrow -\frac{i\omega}{\bar{V}_s} e_{x4} E_a \sin 2\theta$$

in the expressions for $a_-(z)$. Also, the velocity V_l is replaced by $\bar{V}_s$. From (6.75), then,

$$a_+(+l) = 2i(e_{x4} E_a \sin 2\theta) \sin \frac{\omega l}{\bar{V}_s} e^{-i(\omega l/\bar{V}_s)}$$

$$a_-(-l) = 2i(e_{x4} E_a \sin 2\theta) \sin \frac{\omega l}{\bar{V}_s} e^{-i(\omega l/\bar{V}_s)}. \qquad (8.211)$$

EXAMPLE 13. Piezoelectric materials are of great practical importance because they provide the simplest and most efficient method for exciting high frequency acoustic waves. The technique usually used is to attach a plate of piezoelectric material to the surface of the nonpiezoelectric medium in which the waves are to propagate and then to apply a voltage V across electrodes attached to the surface of the plate (Fig. 8.14). In problems of this kind, the normal mode method can be used to calculate the amplitude of the acoustic wave excited by the applied electric field,

$$E_a = (-\nabla \Phi_a)_{z'} = \frac{V}{2l},$$

in the piezoelectric.

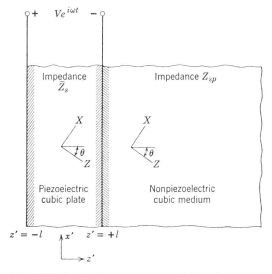

FIGURE 8.14. **Shear wave excitation by means of a piezoelectric plate bonded to an unbounded medium.**

The y'-polarized shear wave treated in Example 12 will be used as an illustration. In Fig. 8.14 both the piezoelectric plate and the underlying propagation medium are assumed to be unbounded in the x' and y' directions. Both media are cubic and have the crystallographic orientation shown in Fig. 8.14. The field quantities vary only with z' and are time-harmonic with frequency ω. Normal mode equations (8.208) and (8.209), with

$$\frac{\partial}{\partial t} \to i\omega,$$

are therefore applicable within the piezoelectric plate.

In the case of an unbounded medium (Example 3 of Chapter 6) it was seen that $a_+(z')$ must satisfy the boundary condition

$$a_+(-l) = 0$$

at the left-hand end of the excitation region and $a_-(z')$ must satisfy the boundary condition

$$a_-(+l) = 0$$

at the right-hand end of the excitation region. A different set of boundary conditions must be used in Fig. 8.14. If the electrodes at $z' = -l$ and $z' = +l$ are very thin, their influence on the mechanical boundary conditions may be ignored.[†] The boundary condition at the mechanically free left-hand side of the plate is therefore

$$T_{4'} = T_{y'z'} = 0. \tag{8.212}$$

If the plate is *rigidly* bonded to the underlying medium, $v_{y'}$ and $T_{4'}$ must be continuous at $z' = +l$. By analogy with (4.87) in Example 3(b) of Chapter 4, it follows that the reflection coefficient at this point is

$$R_v(+l) = \frac{v_{y-}(+l)}{v_{y+}(+l)} = -\frac{Z_{sp} - \bar{Z}_s}{Z_{sp} + \bar{Z}_s}, \tag{8.213}$$

where $\bar{Z}_s$ (defined in (8.206)) is the acoustic impedance in the piezoelectric plate and

$$Z_{sp} = (\rho c_{44})_p^{1/2}$$

is the acoustic impedance in the underlying medium. From (8.207), the particle velocity of a positive-traveling wave ($a_- = 0$) is

$$v_{y+} = \frac{a_+}{2\bar{Z}_s},$$

and the particle velocity of a negative-traveling wave is

$$v_{y-} = -\frac{a_-}{2\bar{Z}_s}.$$

[†] For plane waves propagating along z', the strain-induced electric field (8.201) is normal to the electrodes and the *electrical* boundary conditions are automatically satisfied.

The boundary condition (8.213) is therefore equivalent to

$$\frac{a_-(+l)}{a_+(+l)} = \frac{Z_{sp} - \bar{Z}_s}{Z_{sp} + \bar{Z}_s}. \tag{8.214}$$

Since

$$-T_{4'} = \frac{a_+ + a_-}{2}$$

from (8.207), the boundary condition (8.212) is equivalent to

$$\frac{a_+(-l)}{a_-(-l)} = -1. \tag{8.215}$$

The normal mode equations (8.208) and (8.209) must now be solved with a source distribution (8.210) and boundary conditions (8.214), (8.215). By analogy with (6.60) and (6.61), general solutions for $a_+(z)$ and $a_-(z)$ are

$$a_+(z) = -\frac{i\omega}{\bar{V}_s} \int_{-l}^{z} T_{PE}(\zeta) e^{-(i\omega/\Gamma_s)(z-\zeta)} \, d\zeta + a_+(-l) e^{-(i\omega/\Gamma_s)(z+l)}, \quad z > -l$$

$$a_-(z) = -\frac{i\omega}{\bar{V}_s} \int_{z}^{+l} T_{PE}(\zeta) e^{(i\omega/\Gamma_s)(z-\zeta)} \, d\zeta + a_-(+l) e^{(i\omega/\Gamma_s)(z-l)}, \quad z < +l$$

After substituting for T_{PE} and $a_+(-l)$ from (8.210) and (8.215), one finds that

$$a_+(+l) = 2i(e_{x4}E_a \sin 2\theta) \sin\frac{\omega l}{\bar{V}_s} e^{-i(\omega l/\Gamma_s)} - a_-(-l)e^{-i(2\omega l/\Gamma_s)}. \tag{8.216}$$

Similarly

$$a_-(-l) = 2i(e_{x4}E_a \sin 2\theta) \sin\frac{\omega l}{\bar{V}_s} e^{-i\omega l/\Gamma_s} + a_+(+l)\left(\frac{Z_{sp} - \bar{Z}_s}{Z_{sp} + \bar{Z}_s}\right) e^{-i2\omega l/\Gamma_s}. \tag{8.217}$$

These equations may be solved simultaneously for the unknown amplitudes $a_+(+l)$ and $a_-(-l)$. The solution for $a_+(+l)$ is

$$a_+(+l) = -\frac{Z_{sp} + \bar{Z}_s}{\bar{Z}_s}(e_{x4}E_a \sin 2\theta) \frac{\left(1 - \cos\frac{2\omega l}{\bar{V}_s}\right)}{\frac{Z_{sp}}{\bar{Z}_s}\cos\frac{2\omega l}{\bar{V}_s} + i\sin\frac{2\omega l}{\bar{V}_s}}. \tag{8.218}$$

The field problem within the transducer plate has now been solved, and the radiated wave in the underlying medium can be found by matching boundary conditions at $z' = +l$. At this plane, the transmission coefficient associated with $R_v(+l)$ in (8.213) is

$$T_v = \frac{(v_{y+})\rho}{v_{y+}} = \frac{2\bar{Z}_s}{Z_{sp} + \bar{Z}_s}. \tag{8.219}$$

This can be stated in terms of normal mode amplitudes by means of the relations

$$v_{y+} = \frac{a_+}{2\bar{Z}_s}$$

$$(v_{y+})_p = \frac{(a_+)_p}{2Z_{sp}}$$

derived from (8.207). The amplitude of the normal mode propagating into the underlying medium is therefore

$$(a_+)_p = -\frac{2Z_{sp}}{\bar{Z}_s}(e_{x4}E_a \sin 2\theta)\frac{\left(1 - \cos\dfrac{2\omega l}{V_s}\right)}{\dfrac{Z_{sp}}{\bar{Z}_s}\cos\dfrac{2\omega l}{\bar{V}_s} + i\sin\dfrac{2\omega l}{\bar{V}_s}}. \qquad (8.220)$$

Using the complex Poynting's Theorem, the average radiated power density is

$$P_{AV} = \frac{|v_{y+}|_p^2 Z_{sp}}{2} = \frac{|a_+|_p^2}{8Z_{sp}}.$$

After substitution of (8.220) and some algebraic manipulation, this gives

$$P_{AV} = \frac{Z_{sp}}{\bar{Z}_s}\frac{\omega_0 V^2}{2\pi}\left(\frac{\epsilon_{xx}^S}{2l}\right)\left(\frac{(e_{x4}\sin 2\theta)^2}{\epsilon_{xx}^S \bar{c}_{44}}\right)\left(\frac{\left(1 - \cos\dfrac{2\omega l}{\bar{V}_s}\right)^2}{\sin^2\dfrac{2\omega l}{\bar{V}_s} + \left(\dfrac{Z_{sp}}{\bar{Z}_s}\cos\dfrac{2\omega l}{\bar{V}_s}\right)^2}\right) \qquad (8.221)$$

where ω_0 is the frequency at which the electrode spacing is one-half wavelength,

$$\frac{2\omega_0 l}{\bar{V}_s} = \pi$$

and

$$V = E_a 2l$$

is the voltage applied across the electrodes. The first term in brackets is the *clamped* (or zero strain) electrode capacitance per unit area, and the second bracket is the square of an electromechanical coupling constant

$$K_t = \frac{e_{x4}\sin 2\theta}{(\bar{c}_{44}\epsilon_{xx}^S)^{1/2}}. \qquad (8.222)$$

Comparison with (8.180) shows that this coupling constant, which relates to piezoelectric excitation problems, is not exactly the same as the coupling constant discussed in Section F.4. The difference lies in the use of

$$\bar{c}_{KL} = \hat{c}_{KL}^{\mathbf{i\cdot D}}$$

in one case and

$$c_{KL}^E$$

in the other. For weakly piezoelectric materials the difference is small.

It should be noted that the radiated acoustic power density goes to zero when

$$\frac{2\omega l}{\overline{V}_s} = 2n\pi$$

$$n = 1, 2, 3, \ldots,$$

that is, when the thickness $2l$ of the piezoelectric plate is equal to an integral number of wavelengths. This result may be interpreted in terms of the interference phenomenon discussed in Example 3 of Chapter 6. Another physical picture is obtained by regarding the plate as a *piezoelectric standing wave resonator* that is excited by an applied electric field and coupled with a propagation medium to which it is bonded. The third bracketed term in (8.221) is then looked upon as the frequency response curve of the resonator. The shape of the frequency response curve near $\omega = \omega_0$ is shown in Fig. 8.15 for the range of impedance ratios typical of available materials. The behavior is quite complicated. For impedance ratios lower than

$$\frac{Z_{sp}}{\overline{Z}_s} = \sqrt{2}$$

there is a single maximum at $\omega = \omega_0$. Above this value the response becomes

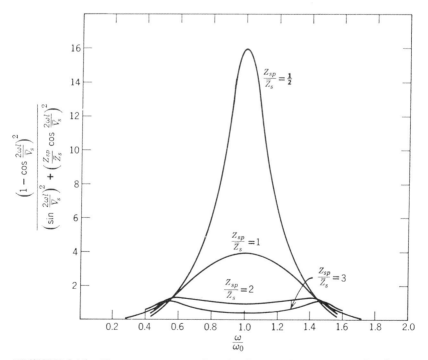

FIGURE 8.15. Frequency response function in (8.221). At $\omega = \omega_0$ the piezoelectric plate in Fig. 8.14 is one-half wavelength thick.

double-peaked and very broad. This characteristic repeats periodically with center (or resonant) frequencies determined by the condition

$$\frac{2\omega\ell}{V_s} = (2n + 1)\pi$$

$$n = 0, 1, 2, \ldots$$

Piezoelectric resonators of this and other kinds will be discussed further in Chapter 11.

J. PIEZOELECTRIC THIN DISK TRANSDUCERS

Electromechanical transduction is an essential element in technological applications of acoustic waves—pulse echo systems for ranging and acoustic imaging, delay lines for signal storage and signal processing, frequency filtering, etc. In designing devices for these applications it is of the utmost importance to have a quantitative characterization of the electromechanical transducer itself. Attention is focused here on piezoelectric transduction. But magnetostrictive, electrostrictive, and especially electromagnetic-acoustic (EMAT) transducers are also widely used. (The coupling mechanism in an EMAT is the Lorentz force acting on a current element in a magnetic field. Details are available in the Supplementary Reference List.)

Piezoelectric transducers are used in both bulk wave and surface wave devices. This section treats the basic principles of piezoelectric bulk wave transduction and its characterization by an electromechanical transducer impedance matrix. More detailed treatments of bulk wave transducer operation and design are presented in recent books on acoustics (E. Dieulesaint and D. Royer, "Elastic Waves in Solids," Wiley, New York 1980; V. M. Ristic, "Principles of Acoustic Devices," Wiley, New York 1983; G. S. Kino, "Acoustic Waves: Devices, Imaging and Analog Signal Processing," Prentice-Hall, Englewood Cliffs, N.J. 1987). Surface wave transducers will be considered in Section L of Chapter 10, and applications of bulk and surface wave transduction to acoustic resonators in Chapter 11.

The one-dimensional piezoelectric excitation problem treated in Example 13 is an idealization of the simplest bulk wave transducer geometry—a thin disk with lateral dimensions large relative to the thickness. Under these conditions the edges of the plate have a small effect on the behavior of the fundamental thickness resonance, and the transduction into this mode can be analyzed to a good approximation by the one-dimensional model of Fig. 8.16. The validity of this approximation depends on the width-to-thickness ratio of the actual transducer disk. As this ratio decreases, coupling to spurious modes increases and can seriously affect the frequency response of the transducer (Fig. 8.23 and Chapter 11).

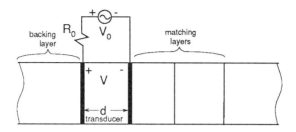

(a) Geometric structure

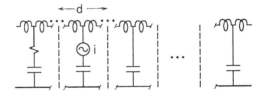

(b) Transmission line model

FIGURE 8.16.   **One-dimensional model of a bulk wave
transducer.**

In the model of Example 13 it is assumed that the voltage across the
electrodes is known. This fixes the electric field applied to the piezoelectric
plate. In a real transducer a voltage V_0 is applied from a generator
with finite internal impedance (Fig. 8.16(a)). If the transducer plate is
considered as a capacitor C_0, the voltage across the electrodes

$$V = \frac{1/(i\omega C_0)}{R_0 + (1/i\omega C_0)} V_0.$$

is not constant with frequency. In addition, the transducer plate cannot
be modeled electrically as just a capacitor. Since the transducer radiates
average acoustic power given by (8.221), power conservation requires that
the input impedance have a real part accounting for the electrical power
absorbed from the source, and an additional imaginary part accounting for
stored acoustic energy. This added term—the radiation impedance of the
transducer—must also be included in the above equation for V. Since the
radiated power itself depends on V, a complete self-consistent analysis must
be performed.

J.1 Electromechanical Impedance Matrix

As illustrated in Fig. 8.16, practical transducer structures may include a lossy backing layer and one or more impedance-matching front layers. These are added to adjust the frequency response of the transducer for optimum performance. Part (b) of the figure shows an equivalent electrical representation of the transducer, based on the acoustic transmission line models of Figs. 6.2, 7.16, and 8.13. Only the transducer plate is considered to be piezoelectric. Since the piezoelectric material itself is necessarily anisotropic, the piezoelectric model in Fig. 8.13 must sometimes be extended to include general acoustic polarizations, as in Fig. 7.16(b). Inclusion of these anisotropy effects may also be required in the propagation medium and the front and back layers. In general, the applied voltage couples to more than one acoustic polarization. When this occurs the inclusion of even more transmission lines is required in Fig. 8.16(b). However, since transducers are designed specifically to couple to only one wave polarization, these extra lines can usually be neglected in the model. Pure mode polarizations can also be assumed, because many of the popular crystal cuts used for transducer plates have this property.

Figure 8.17(a) illustrates an X-cut hexagonal (6mm) piezoelectric transducer satisfying the above assumptions. Among other important materials corresponding to this class are the lead titanate-zirconate ferroelectric ceramics (Appendix B), with the poling (or Z) axis in the plane of the transducer. From Example 8, the x component of electric field couples only to the z-polarized shear wave, and the acoustic field variables at the terminal planes are as shown in part (b) of the figure. (The terminal planes are placed *inside* the electrodes, so that mechanical loading by the electrodes can be included with the effects of the external layers in Fig. 8.16.) A general characterization of the transducer is given by the electromechanical transducer impedance matrix $\mathbf{Z}_T$ relating the electrical and mechanical terminal variables in Fig. 8.17(b). That is,

$$\begin{bmatrix} F_1 \\ F_2 \\ V_3 \end{bmatrix} = \begin{bmatrix} Z_{11} & Z_{12} & Z_{13} \\ Z_{21} & Z_{22} & Z_{23} \\ Z_{31} & Z_{32} & Z_{33} \end{bmatrix} \begin{bmatrix} v_1 \\ v_2 \\ I_3 \end{bmatrix} \qquad (8.223)$$

Elements of $\mathbf{Z}_T$ are evaluated by solving the full acoustic and electric field problem in the plate, and matching boundary conditions to the independent terminal variables in (8.223). The dependent variables are then calculated at the terminal planes from this complete field solution.

Taking the divergence of (4.5), with no source current ($\mathbf{J}_s = 0$), gives

$$\nabla \cdot (i\omega \mathbf{D} + \mathbf{J}_c) = 0 \,.$$

Integrating over the dashed surfaces under the electrodes in Fig. 8.17(a),

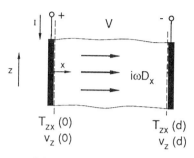

(a) Geometric structure

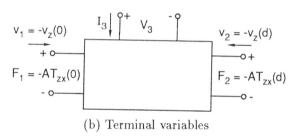

(b) Terminal variables

FIGURE 8.17. **Pure bulk shear wave transducer.**

and applying the divergence theorem, gives the current boundary condition

$$I_3 = (J_c)_x A = i\omega D_x A \qquad (8.224(a))$$

at $x = 0, d$. The voltage at the electrical terminals is

$$V_3 = \int_0^d E_x \, dx \qquad (8.224(b))$$

From Example 8, the piezoelectric equations for the z-polarized shear wave are

$$D_x = \epsilon_{xx}^S E_x + e_{x5} S_5$$
$$T_5 = -e_{x5} E_x + c_{44}^E S_5 \qquad (8.225(a))$$

Since the current boundary condition (8.224(a)) is stated in terms of D_x, it is convenient to rearrange the piezoelectric equations as (see Problem 1)

$$E_x = \left(\epsilon_{xx}^S\right)^{-1} D_x - h_{x5} S_5$$
$$T_5 = -h_{x5} D_x + c_{44}^D S_5 \qquad (8.225(b))$$

with

$$h_{x5} = \frac{e_{x5}}{\epsilon_{xx}^S}$$

The dependent variable E_x then matches with (8.224(b)).

The $\mathbf{v}$ and Φ fields of the z-polarized shear wave were obtained in Examples 7 and 8, and the x-component of $\mathbf{D}$ is zero from (8.136). That is,

$$v_z = e^{i\omega t} e^{\mp i \overline{k} x}$$

$$\Phi = \frac{e_{x5}}{i\omega \epsilon_{xx}^S} v_z \qquad (8.226(a))$$

$$D_x = 0$$

where

$$\overline{c}_{44} = c_{44}^{\hat{x} \cdot D} = c_{44}^E + \frac{E_{x5}^2}{\epsilon_{xx}^S}, \qquad \overline{k} = \omega \left(\frac{\rho}{\overline{c}_{44}} \right)^{1/2}$$

Since the fields (8.226(a)) have zero D_x they are not sufficient for satisfying the boundary condition in (8.224(a)) for the *total* D_x. The required quantity is contained in the particular solution

$$\mathbf{v} = 0$$
$$E_x, D_x = \text{constant in space} \qquad (8.226(b))$$

which can be shown by substitution to satisfy (8.140) and (8.141).

A superposition of (8.226(a)) and (8.226(b)) gives, after regrouping the exponentials, the particle velocity and displacement fields

$$v_z = a \sin \overline{k} x + b \cos \overline{k} x$$
$$D_x = c \qquad (8.227)$$

where the constants a, b, c can be obtained from (8.224(a)) and the boundary conditions on v_z, defined in Fig. 8.17. Then, from (8.224(b)) and (8.225(b)), the dependent terminal variables F_1, F_2, V_3 in the figure can be expressed as linear functions of the independent terminal variables v_1, v_2, I_3, with coefficients that are elements of $\mathbf{Z}_T$ in (8.223). The resulting matrix is

$$\mathbf{Z}_T = -i\overline{Z}_0 \begin{bmatrix} \cot \overline{k}d & \csc \overline{k}d & (h_{x5}/\omega Z_0) \\ \csc \overline{k}d & \cot \overline{k}d & (h_{x5}/\omega Z_0) \\ (h_{x5}/\omega Z_0) & (h_{x5}/\omega Z_0) & (1/\omega C_0 Z_0) \end{bmatrix} \qquad (8.228)$$

where

$$\overline{Z}_0 = A(\rho \overline{c}_{44})^{1/2}$$

is the stiffened mechanical impedance of the transducer disk, and

$$C_0 = \frac{\epsilon_{xx}^S A}{d}$$

is the clamped (or zero strain) geometric capacitance. Notice that $Z_{ij} = Z_{ji}$ in (8.223). This property will have important consequences in part 5 of this section.

A completely analogous analysis can be performed for pure longitudinal wave transducers—for example, a lead titanate-zirconate ceramic plate cut normal to the poling (or Z) axis. (See Appendix B.)

J.2 Equivalent Circuits

It is often convenient to interpret the basic transducer equation (8.223) in terms of an equivalent electrical circuit, by treating all the independent variables as "currents" and all the dependent variables as "voltages." Referring the transducer properties to the familiar context of an electrical circuit diagram often makes it easier to visualize interactions of the transducer plate with the external layers and the propagation medium in Fig. 8.16.

The oldest and most frequently used equivalent circuit is the Mason model shown in Fig. 8.18. Here, the upper left-hand (or mechanical) part of Z_T in (8.223) is modeled by a standard electrical T-network representation. The more subtle part is the introduction of the current coupling in the top two rows of the matrix. These appear as a current injection in series with the shunt arm of the T-network. This is consistent with the corresponding positioning of the piezoelectric drive term in the elemental piezoelectric transmission line section of Figs. 8.13 and 8.16(b).

Since the transducer plate thickness d in (8.228) is arbitrary, it can be reduced to a differential dx. In this limit it is easy to show that the mechanical T-network in Fig. 8.18 reduces to a differential transmission line T-element analogous to that in Fig. 8.13. By extension, the T-network in Fig. 8.18 represents a transmission line of length d. According to Sections 6.D and 7.J this is a correct physical model for the acoustic behavior of the plate. Two other transducer models, the Redwood model and the KLM model, build this distributed transmission line element directly into the equivalent circuit. Figure 8.19 illustrates the KLM model, which is especially useful for broadband transducer design. A detailed discussion and comparison with the Redwood model is given in the Kino book cited at the beginning of this section.

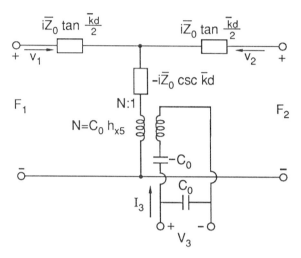

FIGURE 8.18. Mason equivalent circuit for a piezoelectric thin disk transducer.

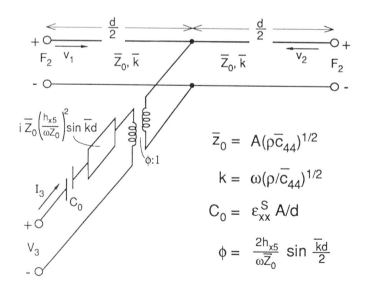

FIGURE 8.19. KLM distributed equivalent circuit for a piezoelectric thin disk transducer.

J.3 Transmitting Transducer

Figure 8.20(a) is a schematic representation, in terms of Z_T, of a transmitting transducer driven by a voltage source with internal impedance R_0 and radiating directly into an infinite propagating medium, as in Fig. 8.14. It was seen at the beginning of this section that the voltage at the input terminals of the transducer, and the acoustic power radiating into the load medium, can be evaluated only after finding the electrical input impedance of the loaded transducer.

Electrical Input Impedance

From Fig. 8.17(b) the electrical input impedance is defined as

$$Z_{\text{INt}} = V_3/I_3 \tag{8.229}$$

Assuming that the propagating medium is attached at terminals 2 in Fig. 8.17(b), the *mechanical load impedance* (i.e., the impedance looking *out* from these terminals) is

$$Z_{\text{Lt}} = -\frac{F_2}{v_2} = A(\rho c_{44})^{1/2} = Z_{0p} \tag{8.230}$$

This is A times the acoustic impedance of the radiating shear wave. As in Fig. 8.14, there is no backing plate. Terminals 1 in Fig. 8.17(b) are therefore unloaded, and F_1 is zero.

To evaluate (8.229),

$$F_1 = 0$$

for an air-backed transducer and F_2 from (8.230) are substituted, with (8.228), into (8.223). The particle velocities v_1 and v_2 are then eliminated from the resulting set of linear equations, and V_3 is found in terms of I_3. The resulting electrical input impedance (8.229) can then be written as

$$\begin{aligned}
Z_{\text{INt}} &= \frac{1}{i\omega C_0} + \frac{P(\omega)}{R(\omega C_0)^2} \\
&= \frac{1}{i\omega C_0} + Z_a \, .
\end{aligned} \tag{8.231}$$

Here

$$R = \frac{\pi}{\omega_0 C_0 k_t^2} \, , \tag{8.232}$$

where

$$k_t = \frac{e_{x5}}{(\bar{c}_{44}\epsilon_{xx}^S)^{1/2}} \tag{8.233}$$

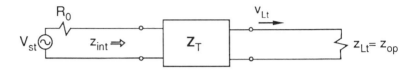

(a) Transmitting transducer.

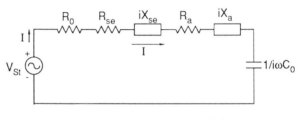

(b) Input circuit detail for (a).

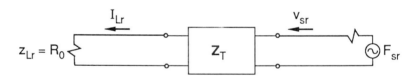

(c) Receiving transducer.

FIGURE 8.20. **Comparison of receiving and transmitting transducers.**

is the same kind of coupling constant as in (8.222), and ω_0 is the half-wavelength resonant frequency,

$$\overline{k}d = \frac{\omega_0 d}{\overline{V}_s} = \pi$$

In the second (acoustic radiation or motional impedance) term Z_a of (8.231) the frequency response function is

$$P(\omega) = \frac{2i(1 - \cos \overline{k}d) + \frac{Z_{0p}}{Z_0} \sin \overline{k}d}{\sin \overline{k}d - i\frac{Z_{0p}}{Z_0} \cos \overline{k}d} . \qquad (8.234)$$

The acoustic radiation impedance term can be separated into real and imaginary parts,

$$Z_a = R_a + iX_a = \hat{R}_a \left(\frac{\omega_0}{\omega}\right)^2 (H_r(\omega) + iH_i(\omega)), \qquad (8.235(a))$$

with

$$\hat{R}_a = \frac{4}{\pi} \frac{\overline{Z}_0}{Z_{0p}} \frac{k_t^2}{\omega_0 C_0} \qquad (8.235(b))$$

$$H_r(\omega) = \frac{1}{4} \left(\frac{Z_{0p}}{\overline{Z}_0}\right)^2 \frac{(1 - \cos \overline{k}d)^2}{(\sin \overline{k}d)^2 + \left(\frac{Z_{0p}}{Z_0} \cos \overline{k}d\right)^2} \qquad (8.235(c))$$

$$H_i(\omega) = \frac{1}{2} \frac{Z_{0p}}{\overline{Z}_0} \frac{\sin \overline{k}d \left[1 + \left(\frac{1}{2}\left(\frac{Z_{0p}}{Z_0}\right)^2 - 1\right) \cos \overline{k}d\right]}{(\sin \overline{k}d)^2 + \left(\frac{Z_{0p}}{Z_0} \cos \overline{k}d\right)^2} \qquad (8.235(d))$$

At the half-wavelength (or fundamental) resonance of the transducer ($\overline{k}d = \pi$)

$$H_r = 1 , \quad H_i = 0$$

and the radiation impedance is pure real. The above equations form the basis of the Reeder-Winslow design method for air-backed transducers. This method provides a very convenient characterization of conversion loss and frequency response for such transducers. It can also be extended to backed transducers (Reference 30).

Conversion Loss

Figure 8.20(b) shows in detail the input circuit corresponding to (a), incorporating the radiative part Z_a and the capacitive part of (8.231) and a lossy inductor element

$$Z_{se} = R_{se} + i\omega L_{se}$$

used to tune out C_0. The conversion loss L of the transducer is defined as the power from the source into a matched electrical load

$$P_0 = \frac{V_{st}^2}{8R_0} , \qquad (8.236)$$

divided by the radiated acoustic power

$$P_a = \frac{1}{2} I^2 R_a, \tag{8.237}$$

In the Reeder-Winslow design method (8.237) is evaluated by using the circuit diagram in Fig. 8.20(b) and the expressions (8.235). The ratio of (8.236) to (8.237) then gives the conversion loss as a function of ω. This can be broken down as

$$L(\omega) = M_e(\omega) M_a(\omega), \tag{8.238}$$

where

$$M_e(\omega) = \frac{(R_0 + R_{se} + R_a)^2 + \left(X_{se} + X_a - \frac{1}{\omega C_0}\right)^2}{4 \left(\frac{\omega_0}{\omega}\right)^2 R_0 \hat{R}_a} \tag{8.239(a)}$$

is the *circuit bandshape function* and

$$M_a(\omega) = \frac{\sin^2 \overline{k} d + \left(\frac{Z_{0p}}{Z_0} \cos \overline{k} d\right)^2}{\frac{1}{4} \left(\frac{Z_{0p}}{Z_0}\right)^2 (1 - \cos \overline{k} d)^2} \tag{8.239(b)}$$

is the *acoustic bandshape function*. At the fundamental resonance frequency ω_0, and with C_0 tuned out by L_{se},

$$L(\omega_0) = \frac{(R_0 + R_{se} + \hat{R}_a)^2}{4 R_0 \hat{R}_a} \tag{8.240}$$

from (8.238) and (8.239).

In (8.240) the conversion loss ratio is unity (0 dB) if the tuning inductor has no resistive loss ($R_{se} = 0$) and the center frequency radiation resistance $\hat{R}_a$ is impedance matched to the source resistance R_0. Since K_t^2 in (8.235(b)) is usually smaller than ≈ 0.1, this matching condition cannot be achieved without use of an electrical impedance transformation circuitry. Even under the assumption that all circuit elements are lossless, there remain electric and acoustic losses in the piezoelectric transducer medium itself. As a result,

$$\epsilon_{xx}^s, \overline{k}_d, \quad \text{and} \quad \overline{Z}_0$$

are all complex in (8.225(a)) to (8.239(b)); and $L(\omega_0)$ remains greater than unity, even when the electrical matching circuit is lossless.

Transducers can also be operated at overtones $n\omega_0$ of the fundamental half-wavelength frequency. At the even overtones $(n = 2, 4, 6, \ldots)$ $\overline{k}d$ is a multiple of 2π, and H_r in (8.235(c)) is zero because of the $(1 - \cos \overline{k}d)$ term in the numerator. There is therefore no conversion at these overtones. At the odd overtones $(n = 3, 5, 7, \ldots)$ $\overline{k}d$ is an odd multiple of π and H_r is unity. Conversion occurs at these frequencies, but with radiation impedance, because

$$\omega_0 \approx n\omega_0$$

in (8.235(a)).

Bandwidth

In (10.238) the overall frequency response of the transducer is governed by both the circuit response function $M_e(\omega)$ and the acoustic resonance response function $M_a(\omega)$. The acoustic bandshape function (8.239(b)), which is just the inverse of H_r in (8.235), has essentially the same frequency response shape as the inverse of the function plotted in Fig. 8.15 and appearing in (8.221). The reason is that the function plotted is just the trigonometric part of H_r. From the derivations in Examples 12 and 13, it is clear that this frequency response arises from interference between the distributed piezoelectric sources and the standing waves of the acoustic resonance.

Figure 8.15 shows that the bandwidth of the response function is increased as the acoustic impedance of the propagation medium Z_{0p} increases relative to the stiffened impedance $\overline{Z}_0$ of the piezoelectric plate. It can be seen, however, from (8.235(b)), that this improvement is paid for by a decrease in the resonant radiation resistance. If this tradeoff is acceptable in the design, the increased load impedance is usually realized by adding one or two quarter-wave acoustic matching transformers (Problems 11 to 15 in Chapter 6), as in Fig. 8.16.

The circuit frequency response is not independent of the acoustic frequency response. As the acoustic bandwidth is increased by raising the acoustic load impedance, the radiation resistance is lowered. This increases the impedance mismatch between $\hat{R}_a$ and R_0, and demands an increased electric impedance transformation ratio to maintain the conversion loss at the same level.

Optimization

Bulk transducer design and optimization is now a highly developed art. The compromises required depend strongly on the operating frequency, the bandwidth required, the impedance of the delay medium, and the operating mode. Optimizing the frequency bandwidth requires balancing the effects of the electric circuit Q and the acoustic radiation Q, which vary inversely with the mechanical load impedance Z_{0p} in Fig. 8.20. For an air-backed transducer with no front-surface $\lambda/4$ transformers, the two

Q's should optimally be equal. Greater bandwidths can be achieved by using wideband matching networks, front-surface $\lambda/4$ transformers, and a lossy backing layer. Details are given in the Ristic book and the Kino book referenced at the beginning of this section.

J.4 Receiving Transducer

Figure 8.20(c) shows the same transducer operating in the receiver mode. In this case modeling requires evaluation of the mechanical input impedance looking into the transducer at terminals 2 in (Fig. 8.17(b)). Calculation of the conversion loss then requires that an incoming acoustic wave in the delay medium be modeled by a mechanical source and source impedance, as in Fig. 8.20(c).

Mechanical Input Impedance

This calculation is completely analogous to the electrical calculation of the previous section. The mechanical input impedance, from Fig. 8.17(b) is defined as

$$Z_{\text{INr}} = \frac{F_2}{v_2} \qquad (8.241(a))$$

and the electrical load impedance looking out from the transducer to the external electrical load is, from Figs. 8.17(b) and 8.20(c),

$$Z_{Lr} = -\frac{V_3}{I_3} = R \qquad (8.241(b))$$

Terminal 1 is, again, unloaded ($F_1 = 0$).

To evaluate (8.241(a))

$$F_1 = 0$$

and V_3 from (8.241(b)) are substituted, with (8.228) , into (8.223). The variables v_1 and I_3 are eliminated from the resulting set of linear equations, and F_2 is found in terms of v_2. This gives the mechanical input impedance

$$Z_{\text{INr}} = \overline{Z}_0 \frac{i \tan \overline{k}d + 2p(1 - \sec \overline{k}d)}{1 + p \tan \overline{k}d} \qquad (8.242(a))$$

with

$$p = \frac{(h_{x5}/\omega)^2}{(R + (i/\omega C_0))\overline{Z}_0} \qquad (8.242(b))$$

the piezoelectric coupling term. In the nonpiezoelectric case ($p = 0$), (8.242(a)) reduces to

$$Z_{\text{INr}} = iZ_0 \tan kd , \qquad (8.243(a))$$

which is just the mechanical impedance looking into the right side of the transmission line in the KLM model (Fig. 8.19), with the left side short circuited ($F_1 = 0$).

At the fundamental resonance ω_0 ($\overline{k}d = \pi$) the mechanical input impedance (8.242(a)) becomes

$$Z_{\text{INr}} = \frac{(h_{x5}/\omega_0)^2}{4R + i/\omega_0 C_0} \qquad (8.243(b))$$

This goes to zero when the electrical terminal is open circuited ($R = \infty$). In other words, disconnecting the electrical load removes the effect of piezoelectric coupling on Z_{INr}. The transducer appears as a purely acoustic load. Since the mechanical impedance (8.242(a)) is defined as A times the acoustic impedance, it can be substituted directly into the reflection coefficient formulas given in Example 3 of Chapter 4 to find the acoustic reflection coefficient of the transducer at terminals 2. Here, the stress reflection coefficient is required. From (4.83) and (4.85) the input reflection coefficient terminals 2 is therefore

$$\Gamma_{\text{INr}} = \frac{Z_{\text{INr}} - Z_{0p}}{Z_{\text{INr}} + Z_{0p}} \qquad (8.244)$$

Just as in the analogous transmitter case, impedance matching (zero reflection) is a desired condition. This is possible only when an electric load is connected, because (8.243(b)) is zero when the electrical terminals are open-circuited and pure imaginary when they are short circuited. In either case the transducer is perfectly reflecting. To acoustically impedance match at the fundamental resonance, C_0 must be tuned out with a series inductor and Z_{INr} must be matched by either adjusting the electrical load resistor R or adding impedance matching transformers to the front face (Fig. 8.16(a)). Note that it is (8.243(b)), not $\overline{Z}_0$, that is being matched to Z_{0p}. As in the case of the transmitting transducer, perfect matching is impossible in practice because of losses in the electric circuit, the transducer plate, and the acoustic transformers.

Conversion Loss

The acoustic-to-electric conversion loss can be calculated from the equivalent circuit (Fig. 8.20(c)), just as in the corresponding transmitting transducer problem. At the acoustic input side, the circuit connection shown in Fig. 8.20(c) can be justified by the input reflection calculation (8.244). By definition, F_{sr} in the figure is the "open-circuit" force—that is, it corresponds to the stress generated by an incoming acoustic wave reflected from an infinite acoustic input impedance. In this situation (8.244) is $+1$. The "open-circuit" force F_{sr} is therefore *twice* the force of the incoming wave. At terminals 2 in the figure, the actual force is the sum of the forces of the incident and reflected waves. That is

$$F_2 = \frac{F_{sr}}{2}(1 + \Gamma_{INr})$$

or

$$F_2 = \frac{F_{sr}Z_{INr}}{Z_{0p} + Z_{INr}} \qquad (8.245(a))$$

The corresponding "current" at terminals 2 is

$$v_{sr} = \frac{F_2}{Z_{INr}} = \frac{F_{sr}}{Z_{0p} + Z_{INr}} \qquad (8.245(b))$$

Comparison with Fig. 8.20(c) shows that (8.245(a)) and (8.245(b)) are exactly the circuit equations for the equivalent acoustic input circuit.

The conversion loss is again calculated as the ratio of the available power from the source into a matched load,

$$P_0 = \frac{F_{sr}^2}{8Z_{0p}}$$

to the power delivered to the electrical load

$$P_L = \frac{1}{2}I_{Lr}^2 R$$

If C_0 is tuned-out with a lossy series inductor, as in Fig. 8.20(b), (8.243(b)) becomes

$$R_{INr} = \frac{(h_{x5}/\omega_0)^2}{R} \qquad (8.246(a))$$

The power absorbed by the transducer at its fundamental resonant frequency is then

$$\begin{aligned} P_{IN} &= \frac{1}{2}v_{sr}^2 R_{INr} \\ &= \frac{1}{2}\frac{F_{sr}^2 R_{INr}}{(Z_{0p} + R_{INr})^2} \end{aligned} \qquad (8.246(b))$$

and this is equal to the power absorbed by R and Re,

$$P_e = \frac{1}{2}I_{Lr}^2(R + R_{se}) \qquad (8.246(c))$$

Using (8.246) and some algebraic manipulation, P_{IN} can be reduced to

$$P_{IN} = \frac{F_{sr}^2}{2}\frac{\hat{R}_e}{(R + R_{se} + \hat{R}_e)^2}\frac{R}{Z_{0p}}$$

with

$$\hat{R}_e = \frac{R_{1Nr}(R + R_{se})}{Z_{0p}} \qquad (8.247)$$

Finally, the conversion loss at ω_0 is

$$L(\omega_0) = \frac{(R + R_{se} + \hat{R}_e)^2}{4R\hat{R}_e} \qquad (8.248)$$

Substituting the explicit expression for k_t^2 into (8.235(b)) shows that

$$\hat{R}_a = \hat{R}_e$$

The transmitting and receiving conversion losses, (8.240) and (8.248), are therefore identical.

Bandwidth and Optimization

Equality of the transmitting and receiving conversion losses has been demonstrated above for operation at ω_0. It is not directly evident from a comparison of (8.231) with (8.242) that the frequency response functions are the same, or that the conversion losses are equal at all frequencies. Fortunately, this can be demonstrated by a much more general, and simpler, argument in the following section. Design rules for optimization of a transmitting transducer therefore also apply to a receiving transducer.

J.5 Reciprocity

In parts (a) and (c) of Fig. 8.20 the box marked $\mathbf{Z}_T$ can be replaced by a T-network, with overall impedance matrix elements

$$\mathcal{Z}_{11}, \mathcal{Z}_{22}, \mathcal{Z}_{21} = \mathcal{Z}_{12}$$

calculated by eliminating v_1 from (8.223), with $F_1 = 0$. The equality of the 12 and 21 matrix elements above follows from the symmetry of $\mathbf{Z}_T$ in (8.228),

$$(Z_T)_{ij} = (Z_T)_{ji} .$$

A straightforward circuit analysis shows that

$$\frac{v_{Lt}}{V_{st}} = \frac{\mathcal{Z}_{21}}{(\mathcal{Z}_{11} + R)(\mathcal{Z}_{22} + Z_{0p}) - \mathcal{Z}_{12}\mathcal{Z}_{21}} \qquad (8.249(a))$$

in part (a) of Fig. 8.20 and

$$\frac{I_{Lr}}{F_{sr}} = \frac{\mathcal{Z}_{12}}{(\mathcal{Z}_{11} + R)(\mathcal{Z}_{22} + Z_{0p}) - \mathcal{Z}_{12}\mathcal{Z}_{21}} \qquad (8.249(b))$$

in part (c) of the figure. The equality of these relations (reciprocity principle) applies at all frequencies, because $\mathcal{Z}_{12}$ and $\mathcal{Z}_{21}$ are always equal. In (a) the conversion loss is

$$L(\omega) = \frac{\frac{1}{8}(V_{st}^2/R)}{\frac{1}{2}V_{Lt}^2 Z_{0p}} = \frac{1}{4RZ_{0p}}\left(\frac{V_{st}}{v_{Lt}}\right)^2$$

and in (b) it is

$$L(\omega) = \frac{\frac{1}{2}(F_{sr}^2/Z_{0p})}{\frac{1}{2}I_{Lr}^2 R} = \frac{1}{4RZ_{0p}}\left(\frac{F_{sr}}{I_{Lr}}\right)^2$$

From (8.249) these two expressions are equal.

The above argument is easily extended to include a series tuning inductor in the electrical circuit. The inductor is simply included in series with $\mathcal{Z}_{11}$. More complicated electrical matching networks and acoustic impedance transform layers may also be added without changing the result. A more general field argument, presented in part 3 of Section 10.L, extends this reciprocity principle to surface wave transducers of arbitrary type.

J.6 Pulsed Operation

Up to this point only the continuous wave (frequency domain) behavior of bulk wave transducers has been considered. But in many applications of acoustics (ranging, medical imaging, nondestructive evaluation of flaws in structures, etc.) the transducers are driven by short pulses, and the time domain response is of more direct interest. This can be calculated from the frequency domain response by using Fourier transform methods. However, more direct physical insight is provided by considering the time domain response directly. Only a brief introduction will be given here. For more detail, the Ristic and Kino books referenced above should be consulted.

Time domain analysis of Fig. 8.17 begins by substituting T_5 from (8.225(b)) into the acoustic field equation

$$\nabla \cdot \mathbf{T} = \rho\frac{\partial \mathbf{v}}{\partial t}$$

and deriving, as in Section 6.A, the wave equation for v_z. The result is

$$\rho\frac{\partial v_z}{\partial t} - c_{44}\frac{\partial^2 v_z}{\partial z^2} = -\frac{\partial}{\partial x}(T_5)_{\text{PE}} \qquad (8.250(a))$$

with

$$(T_5)_{\text{PE}} = -h_{x5}D_x(t) \qquad (8.250(b))$$

From the arguments in part 1 of this section, D_x is spatially constant inside the transducer plate, so that the drive term on the right of (8.250(a)) has spatial δ-functions at $x = 0, d$. That is

$$-\frac{\partial}{\partial x}(T_5)_{\text{PE}} = h_{x5}D_x(t)G(x)$$

$$G(x) = [-\delta(x) + \delta(x - d)]$$

(8.251)

In a physical description, the δ-functions in (8.251) can be said to arise from the unbalanced piezoelectric traction forces acting on elemental layers of thickness dx at the surfaces of the plate (Fig. 8.21(a)). Elemental layers of thickness dx in the interior of the plate have equal and opposite piezoelectric traction forces on its two sides and are therefore not driven piezoelectrically. The transducer plate is, for simplicity, assumed to be loaded on both sides by propagation media having the same mechanical impedance as the plate ($Z_{0p} = Z_0$). There are therefore no acoustic reflections at $x = 0, d$. The transducer is driven by a short voltage impulse at $t = 0$, generating a dielectric displacement

$$D_x(t) = D_x\delta(t)$$

(8.252)

The impulse response of the transducer can be obtained analytically by solving (8.250) to (8.252), but the general features of the solution are easily deduced by a simple physical argument. In (8.250(a)) the source term on the right side, detailed by (8.251) and (8.252), describes the net $+z$-directed forces at the left and right surfaces of the transducer plate. The lefthand force is toward $-z$ and the righthand force is toward $+z$ (Fig. 8.21(a)). In a manner similar to the behavior of an impulse-driven piston (or loudspeaker) in air, each surface layer radiates a positive shear-stress impulse to one side and a negative-shear stress impulse to the other (Fig. 8.21(b)). As shown in the figure, these impulses then propagate away from the surfaces, two impulses to the right and two to the left, without reflection at the interfaces.

The impulse response of a longitudinal transducer is analyzed in detail in the Ristic and Kino books noted above. In these analyses, reflections are included and design criteria for optimum performance are reviewed. Although conditions on the frequency response required for good impulse response can be easily stated (wideband amplitude response and linear phase response), neither these conditions nor the equivalent circuits in Figs. 8.18 and 8.19, give physical insight into the design process for impulse transducers.

J.7 Composite Transducers

Figure 8.15 and part 3 of this section showed that broadband operation of a transducer requires the mechanical impedance of the propagation medium

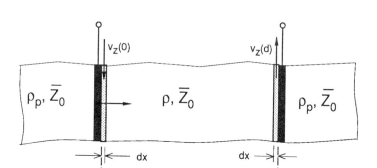

(a) Surface layer generation.

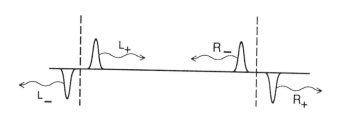

(b) Radiated stress impulses.

FIGURE 8.21. Impulse response of a bulk wave transducer.

to be equal to, or larger than, the mechanical impedance of the transducer plate. Similarly, in part 6, it was seen that good impulse response, without multiple reflections and ringing, requires that the impedances of the propagation medium and the transducer plate be approximately matched. In pulsed ultrasound nondestructive evaluation (NDE), the structural part under test is often immersed in a water bath, which serves as a transmission medium between the transducer and the test piece. Optimization of the transducer performance requires special care because water has a low acoustic impedance ($\rho = 10^3$ kg/m^3, $c_{11} = 0.225 \times 10^{10}$ newton/m^2, $Z_a = 1.5 \times 10^6$ rayl) compared with typical high-coupling piezoelectric materials ($\rho \simeq 6 \times 10^3$, $c_{11} \simeq 15 \times 10^{10}$, $Z_a = 30 \times 10^6$). As has been shown, this large impedance mismatch requires use of front surface matching layers to achieve satisfactory bandwidth; and this increases the conversion loss because of added losses in the transformers. Polyvinyledene fluoride (PVF$_2$), a polymer piezoelectric, has a much closer acoustic impedance match to water ($\rho = 1.79 \times 10^3$, $C_{11} = 0.58 \times 10^{10}$, $Z_a = 3.22 \times 10^6$); but the electromechanical coupling factor is much smaller than the value for piezoceramics such as lead-zirconate-titanate (PZT). Similar problems of mechanical impedance mismatch occur in medical ultrasonics, where the acoustic impedance of tissue is also very low.

The physical constraints imposed by natural piezoelectric materials can be reduced by using composites. These artificial materials are fabricated by inserting into a base (or matrix) material randomly (or periodically) distributed inclusions of one or more different materials. The advantage of using a composite is that the resulting properties can be tailored to give properties superior to those exhibited by the individual material phases of the combined material. Adjustment of the material properties is achieved by suitable choice of the individual phases and by varying their volume fractions in the mixture. Piezoelectric composites are, typically, mixtures of strong coupling piezoceramics with low impedance nonpiezoelectric polymers. For ultrasonic transducer applications, the piezoelectric elements are usually distributed periodically in the nonpiezoelectric matrix. As will be seen, this structuring tends to control spurious acoustic resonances of the transducer.

Optimized Bulk Properties

The main purpose of using composites for piezoelectric transducers is to reduce the mechanical impedance, while maintaining high electromechanical coupling. At the same time, a relatively large dielectric permittivity must be maintained, so that Z_{INt} in (8.231) remains at an appropriate level for electrical matching.

Figure 8.22 shows, in (a), a typical PZT-rod/polymer composite and, in (b), a magnified view of several unit cells in one row of the periodic array. An introduction to the general theory of waves and vibrations in these composites will be given in Section 10.F. Here, attention will be focused

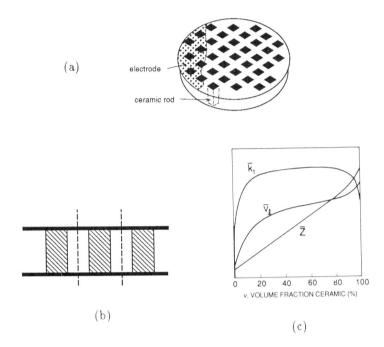

FIGURE 8.22.   Composite transducers.   (a) PZT-rod/polymer disk transducer; (b) Unit cell detail; (c) Transducer parameters versus volume fraction (v) of PZT. (After Supplementary Reference 8.1)

only on longitudinal wave thickness vibrations of a composite longitudinal transducer plate. In this case the piezoelements are all driven in phase by electrodes on the top and bottom of the plate, as in Fig. 8.16, and their spacing is much smaller than an acoustic wavelength at the fundamental transducer resonant frequency ω_0.

The problem is to evaluate the average (or effective medium) mechanical, electrical, and piezoelectric properties of the composite plate by appropriately combining those of the two individual phases. The constitutive relations of the two phases may be written

$$
\begin{aligned}
T_1^p &= (c_{11} + c_{12})S_1^p + c_{12}S_3^p \\
T_3^p &= 2c_{12}S_1^p + c_{11}S_3^p \\
D_z^p &= \epsilon_{xx}E_3^p
\end{aligned}
\qquad (8.253(a))
$$

for the (isotropic) polymer (p), and

$$
\begin{aligned}
T_1^c &= \left(c_{11}^E + c_{12}^E\right) S_1^c + c_{13}^E S_3^c - e_{z1} E_3^c \\
T_3^c &= 2c_{13}^E S_1^c + c_{33}^E S_3^c - e_{z3} E_3^c \\
D_z^c &= 2e_{z1} S_1^c + e_{z3} S_3^c + \epsilon_{zz}^S E_z^c
\end{aligned}
\tag{8.253(b)}
$$

for the ceramic (c). Here, elastic and dielectric constants of the piezoelectric ceramic have the E and S superscripts defined earlier in this chapter.

The two phases are assumed to move together in a uniform thickness vibration (constant strain model), so that

$$
S_3^p(z) = S_3^c(z) = \overline{S}_3(z)
\tag{8.254(a)}
$$

If fringing electric fields are neglected, the electric field is the same in both phases, and

$$
E_z^p(z) = E_z^c(z) = \overline{E}_z(z)
\tag{8.254(b)}
$$

As the piezoceramic rods expand and contract in the thickness direction they also contract and expand in the lateral direction, from the Poisson ratio effect (Problems 3 and 4 of Chapter 6). Each polymer section in part (b) of Fig. 8.22 is therefore subject to equal lateral stresses on either side, and the dashed center lines do not move. Since this constraint applies to each unit cell of the array, the composite as a whole is laterally clamped even though the component phases themselves are not. Therefore

$$
T_1^p(z) = T_1^c(z) = \overline{T}_1(z)
\tag{8.255(a)}
$$

$$
\overline{S}_1(z) = (1 - v)S_1^p(z) + vS_1^c(z) = 0
\tag{8.255(b)}
$$

where v is the volume fraction of ceramic in the composite.

Using (8.255), the lateral strains can be expressed in terms of the thickness strain and electric field as (Supplementary Reference 8.1).

$$
S_1^c = -\frac{c_{13}^E - c_{12}}{c(v)}\overline{S}_3 + \frac{e_{z1}}{c(v)}\overline{E}_z
\tag{8.256(a)}
$$

$$
S_1^p = \frac{\alpha(v)\left(c_{13}^E - c_{12}\right)}{c(v)}\overline{S}_3 - \frac{\alpha(v)e_{z1}}{c(v)}\overline{E}_z
\tag{8.256(b)}
$$

with

$$
c(v) = c_{11}^E + c_{12}^E + \alpha(v)(c_{11} + c_{12})
$$

$$
\alpha(v) = v/(1 - v) \ .
$$

These expressions can then be used to eliminate the lateral strains from (8.253). Finally, the effective thickness stress $\overline{T}_3$ and electric displacement $\overline{D}_3$ are evaluated by averaging over the two phases,

$$\overline{T}_3(z) = vT_3^c(z) + (1-v)T_3^p(z) \qquad (8.257(a))$$

$$\overline{D}_z(z) = vD_z^c(z) + (1-v)D_z^p(z) \qquad (8.257(b))$$

Combining the above results gives the final effective constitutive relations for the composite,

$$\begin{aligned}
\overline{T}_3 &= -\overline{e}_{z3}\overline{E}_z + \overline{c}_{33}^E\overline{S}_3 \\
\overline{D}_z &= \overline{\epsilon}_{zz}^S\overline{E}_z + \overline{e}_{z3}\overline{S}_3
\end{aligned} \qquad (8.258)$$

with

$$\overline{c}_{33}^E = v\left[c_{33}^E - \frac{2\left(c_{13}^E - c_{12}\right)^2}{c(v)}\right] + (1-v)c_{11}$$

$$\overline{e}_{z3} = v\left[e_{z3} - \frac{2e_{z1}\left(c_{13}^E - c_{12}\right)}{c(v)}\right] \qquad (8.259)$$

$$\overline{\epsilon}_{zz}^S = v\left[\epsilon_{zz}^S + \frac{2(e_{z1}^2)}{c(v)}\right] + (1-v)\epsilon_{xx}$$

The average, or effective, composite density is

$$\overline{\rho} = v\rho^c + (1-v)\rho^p \qquad (8.260)$$

To characterize the composite transducer, (8.258) is converted to the form of (8.225(b)). The analysis then proceeds as in parts 1 and 3. In particular, the basic operational parameters of the transducer

$$\overline{k}_t\ ,\ \overline{Z}_\ell\ ,\ \overline{v}_\ell$$

can be directly evaluated. Part (c) of Fig. 8.22 (after Supplementary Reference 8.1) plots these parameters as a function of the volume fraction of ceramic. In this case, the ceramic is PZT-5 (similar to PZT-5H in Appendix A.1) and the polymer is SPURRS epoxy.

The tradeoffs to be made in transducer design are clearly visible from the parameter curves. Over most of the range, $\overline{Z}$ varies linearly with volume fraction v; but it increases suddenly near the 100% point to the laterally clamped impedance of a homogeneous ceramic transducer. The velocity v_ℓ shows corresponding behavior in the same region. For small volume fractions, the variation of v_ℓ is governed by the combined effects of increasing density and stiffness with volume fraction v. The effective

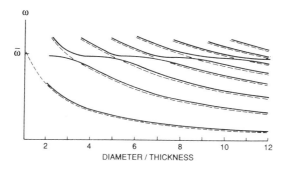

FIGURE 8.23. Typical spurious mode frequency spectrum. (See Supplementary References 8.2 and 8.3.)

electrical coupling $\bar{k}_t$ increases rapidly with v to nearly the value k_{33} for free ceramic rods; then, near 100% composition, it decreases rapidly to the laterally clamped value k_t for a uniform ceramic plate. As the figure shows, a compromise must be made between low impedance and high coupling. For volume fractions in the range of 10%, coupling comparable to k_t can be achieved with significantly reduced $\bar{Z}$.

Spurious Mode Suppression

The infinite plate model (Fig. 8.16(a)) of a thin disk bulk wave transducer ignores the stress-free conditions at the edges of the disk. In that approximate model, the transducer has only one characteristic dimension, its thickness, so that there exists only one family of resonant modes—standing wave resonances of Fabry-Perot-type between the faces of the plate. In fact, the actual finite disk geometry also supports lateral standing waves between the edges of the disk. These do not exist independently of the thickness standing waves. The reason is that the thickness and lateral vibration patterns are coupled at the edges of the disk, through the Poisson ratio effect. As the disk expands and contracts in the thickness direction each volume element contracts and expands to a degree fixed by the Poisson ratio of the material. In the interior of the disk the lateral stresses on each volume element are balanced and no motion ensues. For a volume element at the edge, however, there is a lateral stress on one side only. Consequently, the volume element is driven laterally, exciting a lateral standing wave. This effect is exactly analogous to the surface layer δ-function excitation in Fig. 8.21(a), except that the driving mechanism is the Poisson ratio effect, rather than the piezoelectric stress.

Because of the above Poisson edge coupling mechanism, the resonances of a finite disk longitudinal wave transducer are combinations of thickness and lateral standing waves. This coupling has an extremely strong effect when the thickness and lateral waves are simultaneously resonant.

Frequencies at which this occurs depend on the ratio of the diameter d of the disk to its thickness t. Precise details depend also on the lateral shape of the disk (circular, square, rectangular, etc.). Figure 8.23 shows the form of the resulting mode frequency spectrum, where $\bar{\omega}$ represents the frequency of the fundamental half-wavelength thickness resonance. The dotted hyperbola-like curves represent lateral standing wave resonances. At the crossover points of the thickness curve and the lateral curves, strong interactions occur, and the actual modes become strong admixtures of the two types of standing waves. At these points there is a splitting of the coupled resonances (see Problem 15 in Chapter 12 for coupled waveguide modes). The resulting coupled mode frequency curves are shown as solid lines in the figure. In actuality, the behavior is much more complicated than this, because the lateral waves are plate waveguide modes (Section 10.C), rather than plane waves. The consequences of this fact will be explored more fully in Section 11.B, where spurious mode effects in shear bulk wave transducers are considered.

It can be seen in Fig. 8.23 that the frequency of the desired longitudinal thickness mode is strongly perturbed near the thickness/lateral crossover points, especially for smaller values of d/t. Near these points, the resonant vibrations are strongly mixed thickness and lateral motions, rather than essentially pure thickness vibrations. In designing a finite disk transducer having a small ratio d/t, these crossover points must be avoided, requiring much more complicated design calculations than for the thin plate geometry of part 1 in this section. This design procedure can be simplified for composite transducers. The reason is that the laterally periodic structure of these transducers (parts (a) and (b) of Fig. 8.22) creates stopbands where the lateral waves do not propagate (Section 10.F). In these stopband regions, no lateral standing waves can exist. The problem of avoiding coupling to lateral wave resonances can therefore be eliminated by designing the composite so that the thickness resonant frequency lies inside one of the lateral wave stopbands, where the lateral waves are strongly attenuated.

Figure 8.24 illustrates this stopband phenomenon for a simple transmission line model for the lateral wave resonances (compare Section 11.A). Open circuit boundary conditions are used at the ends, to simulate the stress-free boundaries at the transducer edges. The figure compares the resonant mode spectra of a uniform transmission line (triangles) with those of a line having properties that vary periodically along its length (dots). In either case the resonance condition is

$$\beta(\omega) = n\pi/L$$

where $\beta(\omega)$ for the uniform line is defined by the dashed curve and, for the periodic line, by the solid curve. A derivation of the second dispersion curve will be given in Section 10.F. When the length L is an integral multiple of the period d, resonances of the periodic string occur exactly at the upper

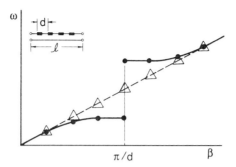

FIGURE 8.24. **Suppression of lateral resonances**

and lower edges of the stopbands. It will be seen in Section 10.F that correctly positioning these band edge resonances is an important step in good composite transducer design.

PROBLEMS

1. The piezoelectric constitutive relations have two dependent variables and two independent variables. In Section B two formulations were given: piezoelectric strain equations with dependent variables $\mathbf{D}$, $\mathbf{S}$ and independent variables $\mathbf{E}$, $\mathbf{T}$; and piezolectric stress equations with dependent variables $\mathbf{D}$, $\mathbf{T}$ and independent variables $\mathbf{E}$, $\mathbf{S}$. Other commonly used formulations are

$$\mathbf{E} = \boldsymbol{\beta}^T \cdot \mathbf{D} - \mathbf{g} : \mathbf{T}$$
$$\mathbf{S} = \underline{\mathbf{g}} \cdot \mathbf{D} + \mathbf{s}^D : \mathbf{T}$$

and

$$\mathbf{E} = \boldsymbol{\beta}^S \cdot \mathbf{D} - \mathbf{h} : \mathbf{S}$$
$$\mathbf{T} = -\underline{\mathbf{h}} \cdot \mathbf{D} + \mathbf{c}^D : \mathbf{S}$$

where $\boldsymbol{\beta}$ is the inverse of the permittivity matrix $\boldsymbol{\epsilon}$. Derive relationships between $\mathbf{g}$, $\mathbf{h}$, $\mathbf{d}$, $\mathbf{e}$, etc.

2. Show that the energy functions appropriate to the constitutive relations in Problem 1 are

$$U - T_I S_I$$

and

$$U$$

respectively. Prove that

$$g_{iJ} = \underline{g}_{Ji} = -\frac{\partial^2}{\partial T_J \partial D_i}(U - T_I S_I)$$
$$h_{iJ} = \underline{h}_{Ji} = -\frac{\partial^2}{\partial S_J \partial D_i}.$$

3. Using the method described in Section E, derive the piezoelectric matrices given in Part B-1 of Appendix 2 for all crystal classes in the hexagonal system.

4. Following the analytical procedure used in Example 4, find the X-polarized X-propagating and X-polarized Y-propagating plane wave solutions for the trigonal crystal classes $3m$ and 32. Find three plane wave solutions for propagation along the Z axis. Compare with Problem 4 in Chapter 3.

5. Find the stiffened Christoffel matrix and the potential equation for wave propagation in the $X = Y, Z$ plane of the cubic crystal classes 23 and $\overline{4}3m$.

6. Derive plane wave dispersion relations for propagation in the $X = Y, Z$ plane of a cubic piezoelectric crystal (Problem 5). Using the material constants given for gallium arsenide in Appendix 2, estimate the maximum error due to the neglect of piezoelectricity in Fig. 3.3 of Part B.2 Appendix 3.

7. Compare the solutions obtained in Problem 6 with (3.4), (3.5), and (3.6) in Part B.2 Appendix 3. Can a single piezoelectric stiffening constant be defined for each wave solution and all values of θ?

8. Suggest some possible methods for measuring the piezoelectric constants of crystals. Compare with the techniques used in practice [see W. P. Mason and H. Jaffe, *Proc. IRE* **42**, 921–930 (1954) and A. W. Warner, M. Onoe, and G. Coquin, *J. Acoust. Soc. Am.* **42**, 1223–1231 (1966)].

9. For fields varying as $e^{i\omega t}$, viscoelastic damping can be accounted for by adding an imaginary part to the elastic stiffness matrix (Problem 11 in Chapter 3). In piezoelectric materials, dielectric and piezoelectric effects may also contribute to the damping (R. Holland and E. P. Eer Nisse, *Design of Resonant Piezoelectric Devices*, pp. 12–16, MIT Press, 1969). These contributions to the damping may be included by allowing the permittivity and piezoelectric matrices to become complex

$$\epsilon_{ij} \rightarrow \epsilon_{ij} + \frac{\sigma_{ij}}{i\omega}$$

$$d_{ij} \rightarrow d'_{iJ} + i\, d''_{iJ}\,.$$

Express the complex Poynting's Theorem given by (8.194) in terms of real and imaginary parts of the constitutive constants, separating the stored energy and power dissipation terms.

10. In (8.208) and (8.209) the body force source terms have the same sign for positive- and negative-traveling waves, while the piezoelectric stress terms have opposite signs. Use the localized body force distribution and the localized piezoelectric stress distribution

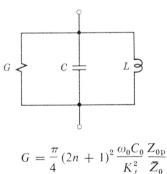

where l is much smaller than the wavelength $2\pi \bar{V}_s/\omega$, to explain this behavior. Describe how a combination of localized sources might be used to radiate acoustic energy in one direction only.

11. Derive impedance equations similar to (8.228) for a Z-cut cadmium sulfide disk, a Z-cut zinc oxide disk, and a Z-cut lithium niobate disk.

12. Derive equations (8.234 and 8.235).

13. Show that the acoustic bandshape function

$$M_a(\omega) = \frac{1}{H_r(\omega)}$$

in (8.239b) is a periodic function of frequency, repeated at frequencies $3\omega_0$, $5\omega_0$, $7\omega_0$, . . . and $M_a(\omega) = \infty$ at frequencies $2\omega_0$, 4ω, 6_0v_0, ...

14. The second (or radiation impedance) term (8.231) can be reduced to the approximate expression

$$Z_a = \frac{1}{\dfrac{\pi}{4}\left(\dfrac{\omega}{\omega_0}\right)\dfrac{\omega C_0}{K_t^2}\dfrac{Z_{0p}}{Z_0} + i\dfrac{\pi}{4}\left(\dfrac{\omega}{\omega_0}\right)\dfrac{\omega C_0}{K_t^2}\dfrac{d}{\bar{V}_s}(\omega - (2n+1)\omega_0)}$$

when $Z_{0p}/Z_0 < 0.1$ and ω is close to $(2n+1)\omega_0$. Prove that this is equivalent to the impedance of the parallel-tuned circuit

$$G = \frac{\pi}{4}(2n+1)^2\frac{\omega_0 C_0}{K_t^2}\frac{Z_{0p}}{Z_0}$$

$$C = \frac{\pi}{4}(2n+1)^2\frac{\omega_0 C_0}{K_t^2}\frac{d}{\bar{V}_s}$$

$$(LC)^{-1/2} = (2n+1)\omega_0,$$

close to its resonant point. Show that the fractional bandwidth between frequencies where

$$|P| = \frac{P_{\max}}{\sqrt{2}}$$

is

$$\frac{\Delta\omega}{(2n+1)\omega_0} = \frac{2}{(2n+1)\pi} \frac{Z_{0\mathrm{p}}}{\bar{Z}_0}.$$

15.   Problem 14 showed that the bandwidth of the radiation impedance term in (8.231) becomes very narrow when $Z_{0\mathrm{p}}/\bar{Z}_0$ is small. This places rather stringent requirements on the thickness of the bond used to attach the

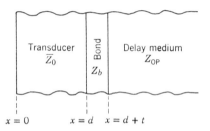

transducer to the delay medium. Use the transformation law for acoustic impedance (Problem 25 in Chapter 7) to show that the impedance at the transducer surface at $x = d$ is

$$Z_{x=d} \simeq Z_b \left[\frac{Z_{0\mathrm{p}} + iZ_b(k_b t)}{Z_b + iZ_{0\mathrm{p}}(k_b t)} \right]$$

where Z_b and k_b are the impedance and wave number in the bonding material and $k_b t \ll \pi/2$. If $Z_b < 0.1\, Z_{0\mathrm{p}}$ (typical for a glued bond), prove that t must be less than $0.1/2\pi$ wavelengths in order to have

$$Z_{0\mathrm{p}} > |Z_{x=d}| > 0.707 Z_{0\mathrm{p}} .$$

16.   The radiation impedance Z_a is usually small compared with $1/\omega C_0$. In this case, show that the transducer input admittance in the vicinity of the fundamental resonance can be approximated by

$$\frac{1}{Z_{\mathrm{IN}}} \simeq i\omega C_0 + (\omega_0 C_0)^2 Z_a$$

and that this can be represented by the equivalent circuit

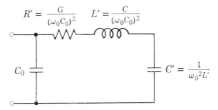

under the conditions applied in Problem 14.

17. Derive equations (8.238 and 8.239).

18. Show that the radiation resistances R_a for the X-cut hexagonal ($6mm$) transducer in Fig. 8.17 and for the transducers considered in Problem 11 are given to a good approximation by

Transducer Type $\qquad (2n + 1)^2 \dfrac{\pi}{4} \dfrac{Z_{0p}}{Z_0} \omega_0 C_0 R_a$

Transducer Type	$(2n + 1)^2 \frac{\pi}{4} \frac{Z_{0p}}{Z_0} \omega_0 C_0 R_a$
X-cut CdS, Shear	$(0.192)^2$
X-cut ZnO, Shear	$(0.268)^2$
Z-cut CdS, Longitudinal	$(0.156)^2$
Z-cut ZnO, Longitudinal	$(0.302)^2$
Z-cut LiNbO$_3$, Longitudinal	$(0.163)^2$

at the resonant frequency $(2n + 1)\omega_0$. The approximation involves use of the coupling constants K given in Table 8.1, rather than the coupling constant K_t that appears in the transducer analysis.

19. Consider a fundamental mode ($n = 0$) Z-cut cadmium sulfide longitudinal wave transducer. Calculate and graph R_a as a function of ω for a Z-cut sapphire delay medium

$$\left(\frac{Z_{0p}}{Z_0} = \frac{44.5}{21.3} \right)$$

and a polyethylene delay medium

$$\left(\frac{Z_{0p}}{Z_0} = \frac{1.75}{21.3} \right).$$

20. In a fundamental mode transducer, the thickness d is related to the resonant frequency by the condition

$$\frac{\omega_{res} d}{\bar{V}_p} = \pi,$$

where $\bar{V}_p$ is the stiffened phase velocity in the transducer medium; and the thin disk analysis applies only if the diameter-to-thickness ratio is large. If

$$\frac{\text{diameter}}{d} = N,$$

show that the geometrical capacitance is

$$C_0 = \frac{\pi^2}{4} \epsilon_{\text{eff}} \frac{N^2 \bar{V}_p}{\omega_{\text{res}}}$$

and the center frequency radiation resistance is

$$R_a = \left(\frac{4}{\pi}\right)^2 \frac{Z_0}{Z_{0p}} \frac{K_t^2}{\pi N^2 \epsilon_{\text{eff}} \bar{V}_p} .$$

Modify these expressions for operation in the $(2n + 1)$th overtone mode, where

$$\frac{\omega_{\text{res}} d}{\bar{V}_p} = (2n + 1)\pi.$$

21. Calculate and tabulate the dimensions and center frequency input impedances

$$Z_{\text{in}} = R_a + \frac{i}{\omega_{\text{res}} C_0}$$

for the zinc oxide transducers of Problem 18. Take the diameter-to-thickness ratio $N = 1000, 100, 10$, and $\omega_{\text{res}} = 1000 \text{ MHz}, 100 \text{ MHz}, 10 \text{ MHz}$. How will the results be modified by operation in the $(2n + 1)$th overtone mode?

22. Derive an expression giving the *acoustic* input impedance at $x = d$ for the transducer show in Fig 8.17. Assume an impedance load Z_L at the electrical terminals.

REFERENCES

Piezoelectric Equations

1. D. A. Berlincourt, D. R. Curran, and H. Jaffe, "Piezoelectric and Piezomagnetic Materials and Their Function in Transducers," pp. 188–189, in *Physical Acoustics*, vol. 1A, W. P. Mason, Ed., Academic Press, New York, 1964.

2. H. Jaffe and D. A. Berlincourt, "Piezoelectric Transducer Materials," *Proc. IEEE* **53**, pp. 1372–1374 (1965).

3. W. P. Mason, *Piezoelectric Crystals and Their Application to Ultrasonics*, Ch. 3, van Nostrand, New York, 1950.

4. W. P. Mason, *Physical Acoustics and the Properties of Solids*, p. 379, van Nostrand, New York, 1958.

5. J. F. Nye, *Physical Properties of Crystals*, pp. 110–116, Oxford, England, 1957.

6. H. F. Tiersten, *Linear Piezoelectric Plate Vibrations*, pp. 34–36, Plenum, New York, 1969.

Thermodynamics of Solids

7. Reference 1, pp. 182–188.

8. Reference 3, pp. 27–40.

9. Reference 4, pp. 373–379.

10. Reference 5, Ch. 10.

Symmetry Relations

11. S. Bhagavantam, *Crystal Symmetry and Physical Properties*, pp. 159–163, Academic Press, New York, 1966.

12. Reference 5, pp. 116–125.

Power Flow and Energy Balance

13. R. Holland, "Representation of Dielectric, Elastic and Piezoelectric Losses by Complex Coefficients," *IEEE Trans.* **SU-14**, pp. 12–16 (1967).

14. R. Holland and Eer Nisse, *Design of Resonant Piezoelectric Devices*, pp. 12–16, MIT Press, 1969.

15. Reference 6, pp. 27–34.

16. J. F. Havlice, W. L. Bond, and L. B. Wigton, "Elastic Poynting Vector in a Piezoelectric Medium," *IEEE Trans* **SU-17**, pp. 246–249 (1970).

Wave Propagation in Piezoelectric Media

17. F. Bardati, G. Barzilai, and G. Gerosa, "Elastic Wave Excitation in Piezoelectric Slabs," *IEEE Trans.* **SU-15**, pp. 193–202 (1968).

18. Reference 1, pp. 193–198.

19. Reference 6, pp. 119–123.

Piezoelectric Stiffening and Electomechanical Coupling Factors

20. Reference 1, pp. 189–193.

21. Reference 14, pp. 16–17.

22. Reference 2, pp. 1374–1375.

Excitation of Plane Waves by Distributed Sources

23. R. F. Mitchell and M. Redwood, "Generation and Detection of Sound by Distributed Piezoelectric Sources", *J. Acous. Soc. Am.* **47**, pp. 701–710 (1969).

Transducers

24. Reference 1, pp. 220–242.

25. Reference 4, pp. 53–74.

26. B. Jaffe, W. R. Cook, Jr., and H. Jaffe, *Piezoelectric Ceramics*, Ch. 12, Academic Press, New York, 1971.

27. L. E. Kinsler and A. R. Frey, *Fundamentals of Acoustics*, Ch. 12, Wiley, New York, 1962.

28. F. A. Fischer, *Fundamentals of Electroacoustics*, Interscience, New York, 1955.

29. O. E. Mattiat, *Ultrasonic Transducer Materials*, Plenum, New York, 1971.

30. T. M. Reeder and D. K. Winslow, "Characteristics of Microwave Acoustic Transducers for Volume Wave Excitation," *IEEE Trans.* **MTT-17,** pp. 927–941 (1969).

31. R. Krimholtz, D. A. Leedom, and G. L. Matthaei, "New Equivalent Circuits for Elementary Piezoelectric Transducers," *Electron. Lett.* **6,** pp. 398-399 (1970).

Appendix 1

CYLINDRICAL AND SPHERICAL COORDINATES

A. BASIC DERIVATIVE OPERATORS

Cylindrical and spherical coordinate representations of the basic *vector* derivative operators are derived in a number of elementary texts, and only the final results will be given here.

A.1 Cylindrical Coordinates

$$\nabla \Phi = \hat{\mathbf{r}} \frac{\partial \Phi}{\partial r} + \hat{\boldsymbol{\phi}} \frac{1}{r} \frac{\partial \Phi}{\partial \phi} + \hat{\mathbf{z}} \frac{\partial \Phi}{\partial z}$$

$$\nabla \cdot \mathbf{D} = \frac{1}{r} \frac{\partial}{\partial r} (r D_r) + \frac{1}{r} \frac{\partial D_\phi}{\partial \phi} + \frac{\partial D_z}{\partial z}$$

$$\nabla \times \mathbf{A} = \hat{\mathbf{r}} \left[\frac{1}{r} \frac{\partial A_z}{\partial \phi} - \frac{\partial A_\phi}{\partial z} \right] + \hat{\boldsymbol{\phi}} \left[\frac{\partial A_r}{\partial z} - \frac{\partial A_z}{\partial r} \right] + \hat{\mathbf{z}} \left[\frac{1}{r} \frac{\partial (r A_\phi)}{\partial r} - \frac{1}{r} \frac{\partial A_r}{\partial \phi} \right] \quad (1.1)$$

$$\nabla^2 \Phi = \frac{1}{r} \frac{\partial}{\partial r} \left(r \frac{\partial \Phi}{\partial r} \right) + \frac{1}{r^2} \frac{\partial^2 \Phi}{\partial \phi^2} + \frac{\partial^2 \Phi}{\partial z^2}$$

357

A.2   Spherical Coordinates

$$\nabla\Phi = \hat{\mathbf{r}}\frac{\partial\Phi}{\partial r} + \hat{\boldsymbol{\theta}}\frac{1}{r}\frac{\partial\Phi}{\partial\theta} + \frac{\hat{\boldsymbol{\phi}}}{r\sin\theta}\frac{\partial\Phi}{\partial\theta}$$

$$\nabla\cdot\mathbf{D} = \frac{1}{r^2}\frac{\partial}{\partial r}(r^2 D_r) + \frac{1}{r\sin\theta}\frac{\partial}{\partial\theta}(\sin\theta\, D_\theta) + \frac{1}{r\sin\theta}\frac{\partial D_\phi}{\partial\phi}$$

$$\nabla\times\mathbf{A} = \frac{\hat{\mathbf{r}}}{r\sin\theta}\left[\frac{\partial}{\partial\theta}(A_\phi\sin\theta) - \frac{\partial A_\theta}{\partial\phi}\right]$$

$$+ \frac{\hat{\boldsymbol{\theta}}}{r}\left[\frac{1}{\sin\theta}\frac{\partial A_r}{\partial\phi} - \frac{\partial}{\partial r}(rA_\phi)\right] + \frac{\hat{\boldsymbol{\phi}}}{r}\left[\frac{\partial}{\partial r}(rA_\theta) - \frac{\partial A_r}{\partial\theta}\right] \tag{1.2}$$

$$\nabla^2\Phi = \frac{1}{r^2}\frac{\partial}{\partial r}\left(r^2\frac{\partial\Phi}{\partial r}\right) + \frac{1}{r^2\sin\theta}\frac{\partial}{\partial\theta}\left(\sin\theta\frac{\partial\Phi}{\partial\theta}\right) + \frac{1}{r^2\sin^2\theta}\frac{\partial^2\Phi}{\partial\phi^2}$$

B.   GRADIENT AND SYMMETRIC GRADIENT OF A VECTOR

In Chapter 1 the gradient of the particle displacement field was defined by the relation

$$d\mathbf{u} = \nabla\mathbf{u}\cdot d\mathbf{L} = \frac{\partial\mathbf{u}}{\partial L_1}dL_1 + \frac{\partial\mathbf{u}}{\partial L_2}dL_2 + \frac{\partial\mathbf{u}}{\partial L_3}dL_3, \tag{1.3}$$

where dL_1, dL_2, dL_3 are components of $d\mathbf{L}$ in some orthogonal coordinate system. For a rectangular Cartesian coordinate system

$$\mathbf{u} = \hat{\mathbf{x}}u_x + \hat{\mathbf{y}}u_y + \hat{\mathbf{z}}u_z,$$

where the unit vectors $\hat{\mathbf{x}}$, $\hat{\mathbf{y}}$, $\hat{\mathbf{z}}$ are *constant*, and performance of the differential operations in (1.3) does not affect the unit vectors. Cylindrical and spherical coordinate systems, on the other hand, have unit vector directions that vary with coordinate position, and this must be taken into account in evaluating (1.3).

B.1   Cylindrical Coordinates

In this case, the components of $d\mathbf{L}$ are $dL_1 = dr$, $dL_2 = rd\phi$, $dL_3 = dz$, and (1.3) takes the form

$$d\mathbf{u} = \frac{\partial\mathbf{u}}{\partial r}(dr) + \frac{1}{r}\frac{\partial\mathbf{u}}{\partial\phi}(rd\phi) + \frac{\partial\mathbf{u}}{\partial z}dz. \tag{1.4}$$

By considering small increments in r, ϕ, and z one can show that derivatives of the unit vectors are

$$\frac{\partial}{\partial r}\hat{\mathbf{r}} = 0, \qquad \frac{\partial}{\partial r}\hat{\boldsymbol{\phi}} = 0, \qquad \frac{\partial}{\partial r}\hat{\mathbf{z}} = 0$$

$$\frac{\partial}{\partial \phi}\hat{\mathbf{r}} = \hat{\boldsymbol{\phi}}, \qquad \frac{\partial}{\partial \phi}\hat{\boldsymbol{\phi}} = -\hat{\mathbf{r}}, \qquad \frac{\partial}{\partial \phi}\hat{\mathbf{z}} = 0 \qquad (1.5)$$

$$\frac{\partial}{\partial z}\hat{\mathbf{r}} = 0, \qquad \frac{\partial}{\partial z}\hat{\boldsymbol{\phi}} = 0, \qquad \frac{\partial}{\partial z}\hat{\mathbf{z}} = 0.$$

Substitution of

$$\mathbf{u} = \hat{\mathbf{r}}u_r + \hat{\boldsymbol{\phi}}u_\phi + \hat{\mathbf{z}}u_z$$

into (1.4) then gives

$$d\mathbf{u} = \left(\frac{\partial u_r}{\partial r}\hat{\mathbf{r}} + \frac{\partial u_\phi}{\partial r}\hat{\boldsymbol{\phi}} + \frac{\partial u_z}{\partial r}\hat{\mathbf{z}}\right) dr$$

$$+ \frac{1}{r}\left(\left(\frac{\partial u_r}{\partial \phi} - u_\phi\right)\hat{\mathbf{r}} + \left(\frac{\partial u_\phi}{\partial \phi} + u_r\right)\hat{\boldsymbol{\phi}} + \frac{\partial u_z}{\partial \phi}\hat{\mathbf{z}}\right) r d\phi$$

$$+ \left(\frac{\partial u_r}{\partial z}\hat{\mathbf{r}} + \frac{\partial u_\phi}{\partial z}\hat{\boldsymbol{\phi}} + \frac{\partial u_z}{\partial z}\hat{\mathbf{z}}\right) dz,$$

and the gradient therefore has the matrix representation

$$\nabla \mathbf{u} \rightarrow \begin{bmatrix} \dfrac{\partial u_r}{\partial r} & \dfrac{1}{r}\dfrac{\partial u_r}{\partial \phi} - \dfrac{u_\phi}{r} & \dfrac{\partial u_r}{\partial z} \\[2ex] \dfrac{\partial u_\phi}{\partial r} & \dfrac{1}{r}\dfrac{\partial u_\phi}{\partial \phi} + \dfrac{u_r}{r} & \dfrac{\partial u_\phi}{\partial z} \\[2ex] \dfrac{\partial u_z}{\partial r} & \dfrac{1}{r}\dfrac{\partial u_z}{\partial \phi} & \dfrac{\partial u_z}{\partial z} \end{bmatrix} \qquad (1.6)$$

in cylindrical coordinates. The strain matrix, which is defined by the relation

$$\mathbf{S} = \tfrac{1}{2}(\nabla \mathbf{u} + \widetilde{\nabla \mathbf{u}}) = \nabla_s \mathbf{u},$$

is then obtained by adding the matrix in (1.6) to its transpose and dividing by two. In abbreviated subscript notation the strain is represented by a six-element column vector with elements

$$\begin{aligned} S_1 &= S_{rr} & S_4 &= 2S_{\phi z} \\ S_2 &= S_{\phi\phi} & S_5 &= 2S_{rz} & \qquad (1.7) \\ S_3 &= S_{zz} & S_6 &= 2S_{r\phi}\,, \end{aligned}$$

and the symmetric gradient operation has the matrix representation

$$
\nabla_s \rightarrow
\begin{bmatrix}
\dfrac{\partial}{\partial r} & 0 & 0 \\[2mm]
\dfrac{1}{r} & \dfrac{1}{r}\dfrac{\partial}{\partial \phi} & 0 \\[2mm]
0 & 0 & \dfrac{\partial}{\partial z} \\[2mm]
0 & \dfrac{\partial}{\partial z} & \dfrac{1}{r}\dfrac{\partial}{\partial \phi} \\[2mm]
\dfrac{\partial}{\partial z} & 0 & \dfrac{\partial}{\partial r} \\[2mm]
\dfrac{1}{r}\dfrac{\partial}{\partial \phi} & \dfrac{\partial}{\partial r}-\dfrac{1}{r} & 0
\end{bmatrix}
\tag{1.8}
$$

B.2   Spherical Coordinates

In spherical coordinates $dL_1 = dr$, $dL_\theta = rd\theta$, $dL_3 = r\sin\theta\,d\phi$, and (1.3) takes the form

$$
d\mathbf{u} = \frac{\partial \mathbf{u}}{\partial r}(dr) + \frac{1}{r}\frac{\partial \mathbf{u}}{\partial \theta}(rd\theta) + \frac{1}{r\sin\theta}\frac{\partial \mathbf{u}}{\partial \phi}(r\sin\theta\,d\phi).
\tag{1.9}
$$

Derivatives of the unit vectors are now

$$
\frac{\partial \hat{\mathbf{r}}}{\partial r} = 0, \qquad \frac{\partial \hat{\boldsymbol{\theta}}}{\partial r} = 0, \qquad \frac{\partial \hat{\boldsymbol{\phi}}}{\partial r} = 0
$$

$$
\frac{\partial \hat{\mathbf{r}}}{\partial \theta} = \hat{\boldsymbol{\theta}}, \qquad \frac{\partial \hat{\boldsymbol{\theta}}}{\partial \theta} = -\hat{\mathbf{r}}, \qquad \frac{\partial \hat{\boldsymbol{\phi}}}{\partial \theta} = 0
\tag{1.10}
$$

$$
\frac{\partial \hat{\mathbf{r}}}{\partial \phi} = \hat{\boldsymbol{\phi}}\sin\theta, \qquad \frac{\partial \hat{\boldsymbol{\theta}}}{\partial \phi} = \hat{\boldsymbol{\phi}}\cos\theta, \qquad \frac{\partial \hat{\boldsymbol{\phi}}}{\partial \phi} = -\hat{\mathbf{r}}\sin\theta - \hat{\boldsymbol{\theta}}\cos\theta.
$$

The derivation proceeds as above, with

$$
\mathbf{u} = \hat{\mathbf{r}}u_r + \hat{\boldsymbol{\theta}}u_\theta + \hat{\boldsymbol{\phi}}u_\phi
$$

substituted into (1.9). In abbreviated subscript notation the strain components are

$$
\begin{aligned}
&S_1 = S_{rr} &\quad &S_4 = 2S_{\theta\phi} \\
&S_2 = S_{\theta\theta} &\quad &S_5 = 2S_{r\phi} \\
&S_3 = S_{\phi\phi} &\quad &S_6 = 2S_{r\theta}
\end{aligned}
\tag{1.11}
$$

and the symmetric gradient operation has the matrix representation

$$\nabla_s \rightarrow \begin{bmatrix} \dfrac{\partial}{\partial r} & 0 & 0 \\[2mm] \dfrac{1}{r} & \dfrac{1}{r}\dfrac{\partial}{\partial \theta} & 0 \\[2mm] \dfrac{1}{r} & \dfrac{\cot\theta}{r} & \dfrac{1}{r\sin\theta}\dfrac{\partial}{\partial\phi} \\[2mm] 0 & \dfrac{1}{r\sin\theta}\dfrac{\partial}{\partial\phi} & \dfrac{1}{r}\dfrac{\partial}{\partial\theta} - \dfrac{\cot\theta}{r} \\[2mm] \dfrac{1}{r\sin\theta}\dfrac{\partial}{\partial\phi} & 0 & \dfrac{\partial}{\partial r} - \dfrac{1}{r} \\[2mm] \dfrac{1}{r}\dfrac{\partial}{\partial\theta} & \dfrac{\partial}{\partial r} - \dfrac{1}{r} & 0 \end{bmatrix} \tag{1.12}$$

C. DIVERGENCE OF STRESS

In Chapter 2 divergence of the stress field was defined by the relation

$$\nabla \cdot \mathbf{T} = \lim_{\delta V \to 0} \frac{\displaystyle\int_{\delta S} \mathbf{T} \cdot \hat{\mathbf{n}} \, dS}{\delta V} \tag{1.13}$$

where δV is an elemental volume in some orthogonal coordinate system and δS is the surface enclosing this volume. To evaluate $\mathbf{T} \cdot \hat{\mathbf{n}}$ at points on δS one must take derivatives of the traction forces, which contain the unit coordinate vectors.† In the rectangular Cartesian coordinate system treated in Chapter 2 complications do not arise, because the unit vectors $\hat{\mathbf{x}}$, $\hat{\mathbf{y}}$, $\hat{\mathbf{z}}$ are *constant*. For cylindrical and spherical coordinates, the unit vector directions vary with coordinate position and this must be taken into account in taking derivatives of the traction forces.

C.1 Cylindrical Coordinates

In this case the elemental volume is as shown in Fig. 1.1. If the stress components at the center of this element are T_{rr}^0, $T_{\phi\phi}^0$, etc., then the traction forces

† See, for example, (2.19) in Section C of Chapter 2.

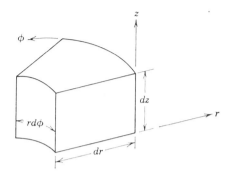

FIGURE 1.1.   **Elemental volume in cylindrical coordinates.**

acting on the *outside* of the $+r$, $+\phi$, $+z$ faces are

$$\mathbf{T}_r^0 + \delta\mathbf{T}_r = \mathbf{T}_r^0 + \frac{\partial}{\partial r}(\hat{\mathbf{r}}T_{rr} + \hat{\boldsymbol{\phi}}T_{\phi r} + \hat{\mathbf{z}}T_{zr})\frac{\delta r}{2}$$

$$\mathbf{T}_\phi^0 + \delta\mathbf{T}_\phi = \mathbf{T}_\phi^0 + \frac{\partial}{\partial\phi}(\hat{\mathbf{r}}T_{r\phi} + \hat{\boldsymbol{\phi}}T_{\phi\phi} + \hat{\mathbf{z}}T_{z\phi})\frac{\delta\phi}{2} \qquad (1.14)$$

$$\mathbf{T}^0 + \delta\mathbf{T}_z = \mathbf{T}_z^0 + \frac{\partial}{\partial z}(\hat{\mathbf{r}}T_{rz} + \hat{\boldsymbol{\phi}}T_{\phi z} + \hat{\mathbf{z}}T_{zz})\frac{\delta z}{2}.$$

The derivation proceeds in exactly the same way as in Section 2.C, but the derivatives of the unit vectors (given by (1.5)) must now be taken into account. In abbreviated subscript notation

$$\begin{aligned}
T_1 &= T_{rr} & T_4 &= T_{\phi z} \\
T_2 &= T_{\phi\phi} & T_5 &= T_{rz} \\
T_3 &= T_{zz} & T_6 &= T_{r\phi}.
\end{aligned} \qquad (1.15)$$

and the divergence-of-stress operator is

$$\nabla\cdot \rightarrow
\begin{bmatrix}
\dfrac{\partial}{\partial r} + \dfrac{1}{r} & -\dfrac{1}{r} & 0 & 0 & \dfrac{\partial}{\partial z} & \dfrac{1}{r}\dfrac{\partial}{\partial\phi} \\[2ex]
0 & \dfrac{1}{r}\dfrac{\partial}{\partial\phi} & 0 & \dfrac{\partial}{\partial z} & 0 & \dfrac{\partial}{\partial r} + \dfrac{2}{r} \\[2ex]
0 & 0 & \dfrac{\partial}{\partial z} & \dfrac{1}{r}\dfrac{\partial}{\partial\phi} & \dfrac{\partial}{\partial r} + \dfrac{1}{r} & 0
\end{bmatrix} \qquad (1.16)$$

C.2 Spherical Coordinates

The derivation follows the same steps as in the cylindrical case. In abbreviated subscript notation

$$
\begin{aligned}
T_1 &= T_{rr} & T_4 &= T_{\theta\phi} \\
T_2 &= T_{\theta\theta} & T_5 &= T_{r\phi} \\
T_3 &= T_{\phi\phi} & T_6 &= T_{r\theta}
\end{aligned}
\tag{1.17}
$$

and the divergence-of-stress operator is

$$
\begin{bmatrix}
\dfrac{2}{r}+\dfrac{\partial}{\partial r} & -\dfrac{1}{r} & -\dfrac{1}{r} & 0 & \dfrac{1}{r\sin\theta}\dfrac{\partial}{\partial\phi} & \dfrac{\cot\theta}{r}+\dfrac{1}{r}\dfrac{\partial}{\partial\theta} \\[2ex]
0 & \dfrac{\cot\theta}{r}+\dfrac{1}{r}\dfrac{\partial}{\partial\theta} & -\dfrac{\cot\theta}{r} & \dfrac{1}{r\sin\theta}\dfrac{\partial}{\partial\phi} & 0 & \dfrac{3}{r}+\dfrac{\partial}{\partial r} \\[2ex]
0 & 0 & \dfrac{1}{r\sin\theta}\dfrac{\partial}{\partial\phi} & \dfrac{2\cot\theta}{r}+\dfrac{1}{r}\dfrac{\partial}{\partial\theta} & \dfrac{3}{r}+\dfrac{\partial}{\partial r} & 0
\end{bmatrix}
\tag{1.18}
$$

D. CONSTITUTIVE MATRICES

Coordinate transformation laws for compliance and stiffness (Sections B and D of Chapter 3), permittivity (Example 2 of Chapter 4), and the piezoelectric constants (Section B of Chapter 8) can be applied to cylindrical and spherical coordinates by using the appropriate transformation matrices [a]. These are

$$
[a] = \begin{bmatrix}
\cos\phi & \sin\phi & 0 \\
-\sin\phi & \cos\phi & 0 \\
0 & 0 & 1
\end{bmatrix}
\tag{1.19}
$$

for transforming from rectangular to cylindrical coordinates, and

$$
[a] = \begin{bmatrix}
\sin\theta\cos\phi & \sin\theta\sin\phi & \cos\theta \\
\cos\theta\cos\phi & \cos\theta\sin\phi & -\sin\theta \\
-\sin\phi & \cos\phi & 0
\end{bmatrix}
\tag{1.20}
$$

for transforming from rectangular to spherical coordinates. The transformation matrices (1.19) and (1.20) can also be used to convert strain and stress field components from rectangular coordinates to cylindrical and spherical coordinates. (See P. M. Morse and H. Feshbach, *Methods of Theoretical*

Physics, pp. 29–31, 54–55, 66–71, McGraw-Hill, New York, 1953, and note that the unit vectors in (1.5) and (1.10) have *unit length*.)

Since the transformation matrix (1.19) is the same as for a rotation of rectangular coordinates about the z axis, it is clear that the isotropic stiffness matrix (defined in rectangular coordinates by (6.12) and (6.13)) is unchanged by transforming to cylindrical coordinates. A similar argument can be used to show that the isotropic stiffness matrix in spherical coordinates is also given by (6.12) and (6.13).

PROPERTIES OF MATERIALS†

† Additional information on the properties of materials may be found in the following references:

1. K. S. Aleksandrov and T. V. Ryzhova, "The Elastic Properties of Crystals," *Sov. Phys.-Crystall.* **6**, 228–252 (1961).
2. H. B. Huntington, "The Elastic Constants of Crystals," *Solid State Physics*, F. Seitz and D. Turnbull, eds., Vol. 7, 213–351, Academic Press, New York, 1958.
3. Landolt-Börnstein, *Numerical Data and Functional Relationships in Science and Technology*, K.-H. Hellwege and A. M. Hellvege, eds., Group III—Crystal and Solid State Physics, Vol. 1, Springer, Berlin, 1966.
4. G. Simmons, "Single Crystal Constants and Crystal Aggregate Properties," *J. Grad. Res. Center* **34**, pp. 1–152, Department of Geol. and Geophys., Southern Methodist University, Dallas (1965).
5. *American Institute of Physics Handbook*, 3rd Ed., D. E. Gray, ed., Sections 2b, 2e, 3f, 3g, 9f, McGraw-Hill, New York, 1972.
6. The Geological Society of America, *Memoir 97, Handbook of Physical Constants*, S. P. Clark, ed., Sections 7–9, New York, 1966.
7. References 1, 2, 26, and 29 in Chapter 8.

MECHANICAL PROPERTIES   A.1   Crystal Symmetry Class and Mass Density

Material	Chemical Formula	Symmetry Class	Density (kg/m^3)§
INSULATORS			
Ammonium Dihydrogen Phosphate (ADP)†	$NH_4H_2PO_4$	Tetr. $\bar{4}2m$	1803
Barium Fluoride	BaF_2	Cubic $m3m$	4886
Barium Sodium Niobate†	$Ba_2NaNb_5O_{15}$	Orth. $2mm$	5300
Barium Titanate†	$BaTiO_3$	Tetr. $4mm$	6020
Barium Titanate†	$BaTiO_3$	Uniaxial ‖	5700
Bismuth Germanate†	$Bi_4Ge_3O_{12}$	Cubic $\bar{4}3m$	7095
Bismuth Germanium Oxide†	$Bi_{12}GeO_{20}$	Cubic 23	9200
Calcium Molybdate	$CaMoO_4$	Tetr. $4/m$	4255
Diamond	C	Cubic $m3m$	3512
Europium Iron Garnet‡	$Eu_3Fe_5O_{12}$	Cubic $m3m$	6280
Fused Silica	SiO_2	Isotropic	2200
Glass—Heavy Silicate Flint	—	Isotropic	3879
Glass—Light Borate Crown	—	Isotropic	2243
Glass—Pyrex	—	Isotropic	2320
Glass—T-40	—	Isotropic	3390
Iodic Acid†	HIO_3	Orth. 222	4629
Lead Molybdate	$PbMoO_4$	Tetr. $4/m$	6950
Lead Titanate-Zirconate (PZT-2)†	—	Uniaxial ‖	7600
Lead Titanate-Zirconate (PZT-5H)†	—	Uniaxial ‖	7500
Lithium Fluoride	LiF	Cubic $m3m$	2601
Lithium Gallate†	$LiGaO_2$	Orth. $2mm$	4187
Lithium Niobate†	$LiNbO_3$	Trig. $3m$	4700
Lithium Tantalate†	$LiTaO_3$	Trig. $3m$	7450
Lucite	—	Isotropic	1182
Magnesium Oxide	MgO	Cubic $m3m$	3650
Polyethylene	—	Isotropic	900
Polystyrene	—	Isotropic	1056
Potassium Dihydrogen Phosphate (KDP)†	KH_2PO_4	Tetr. $\bar{4}2m$	2338
Quartz†	SiO_2	Trig. 32	2651

Material	Chemical Formula	Symmetry Class	Density (kg/m^3)§
Rochelle Salt†	$NaKC_4H_4O_6 \cdot 4H_2O$	Orth. *222*	1767
Rutile	TiO_2	Tetr. *4/mmm*	4260
Sapphire	Al_2O_3	Trig. *$\bar{3}m$*	3986
Sodium Fluoride	NaF	Cubic *m3m*	2790
Strontium Titanate	$SrTiO_3$	Cubic *m3m*	5110
Yttrium Aluminum Garnet (YAG)	$Y_3Al_5O_{12}$	Cubic *m3m*	4550
Yttrium Gallium Garnet (YGaG)	$Y_3Ga_5O_{12}$	Cubic *m3m*	5790
Yttrium Iron Garnet (YIG)‡	$Y_3Fe_5O_{12}$	Cubic *m3m*	5170

SEMICONDUCTORS

Material	Chemical Formula	Symmetry Class	Density (kg/m^3)§
Beryllium Oxide†	BeO	Hex. *6mm*	3009
Cadmium Selenide†	CdSe	Hex. *6mm*	5684
Cadmium Sulfide†	CdS	Hex. *6mm*	4820
Gallium Arsenide†	GaAs	Cubic *$\bar{4}3m$*	5307
Gallium Phosphide†	GaP	Cubic *$\bar{4}3m$*	4130
Germanium	Ge	Cubic *m3m*	5322
Indium Antimonide†	InSb	Cubic *$\bar{4}3m$*	5770
Indium Arsenide†	InAs	Cubic *$\bar{4}3m$*	5672
Indium Phosphide†	InP	Cubic *$\bar{4}3m$*	4787
Silicon	Si	Cubic *m3m*	2332
Tellurium Dioxide†	TeO_2	Tetr. *422*	5990
Zinc Oxide†	ZnO	Hex. *6mm*	5680
Zinc Sulfide†	ZnS	Hex. *6mm*	3980

METALS

Material	Chemical Formula	Symmetry Class	Density (kg/m^3)§
Aluminum	Al	Cubic *m3m*	2695
Gold	Au	Cubic *m3m*	19,300
Indium	In	Tetr. *4/mmm*	7300
Iron‡	Fe	Cubic *m3m*	7870
Nickel‡	Ni	Cubic *m3m*	8905
Silver	Ag	Cubic *m3m*	10,490
Tellurium	Te	Trig. *32*	6260
Titanium	Ti	Hex. *6/mmm*	4500
Tungsten	W	Cubic *m3m*	19,200

† Piezoelectric.
‡ Ferromagnets and ferrimagnets.
§ Units and conversion ratios on back endpaper.
‖ Poled ceramic.

MECHANICAL PROPERTIES   A.2   Symmetry Characteristics of Compliance and Stiffness. The matrix elements are referred to coordinate axes x, y, z that coincide with the crystal axes X, Y, Z in Table 7.1. Except where noted, the stiffness matrices have the same form as the compliance matrices.

Triclinic System

21 constants

$$
\begin{bmatrix}
s_{11} & s_{12} & s_{13} & s_{14} & s_{15} & s_{16} \\
s_{12} & s_{22} & s_{23} & s_{24} & s_{25} & s_{26} \\
s_{13} & s_{23} & s_{33} & s_{34} & s_{35} & s_{36} \\
s_{14} & s_{24} & s_{34} & s_{44} & s_{45} & s_{46} \\
s_{15} & s_{25} & s_{35} & s_{45} & s_{55} & s_{56} \\
s_{16} & s_{26} & s_{36} & s_{46} & s_{56} & s_{66}
\end{bmatrix}
$$

Monoclinic System

13 constants

$$
\begin{bmatrix}
s_{11} & s_{12} & s_{13} & 0 & s_{15} & 0 \\
s_{12} & s_{22} & s_{23} & 0 & s_{25} & 0 \\
s_{13} & s_{23} & s_{33} & 0 & s_{35} & 0 \\
0 & 0 & 0 & s_{44} & 0 & s_{46} \\
s_{15} & s_{25} & s_{35} & 0 & s_{55} & 0 \\
0 & 0 & 0 & s_{46} & 0 & s_{66}
\end{bmatrix}
$$

Orthorhombic System

9 constants

$$
\begin{bmatrix}
s_{11} & s_{12} & s_{13} & 0 & 0 & 0 \\
s_{12} & s_{22} & s_{23} & 0 & 0 & 0 \\
s_{13} & s_{23} & s_{33} & 0 & 0 & 0 \\
0 & 0 & 0 & s_{44} & 0 & 0 \\
0 & 0 & 0 & 0 & s_{55} & 0 \\
0 & 0 & 0 & 0 & 0 & s_{66}
\end{bmatrix}
$$

Tetragonal System

Classes 4, $\bar{4}$, $4/m$

7 constants

$$\begin{bmatrix} s_{11} & s_{12} & s_{13} & 0 & 0 & s_{16} \\ s_{12} & s_{11} & s_{13} & 0 & 0 & -s_{16} \\ s_{13} & s_{13} & s_{33} & 0 & 0 & 0 \\ 0 & 0 & 0 & s_{44} & 0 & 0 \\ 0 & 0 & 0 & 0 & s_{44} & 0 \\ s_{16} & -s_{16} & 0 & 0 & 0 & s_{66} \end{bmatrix}$$

Tetragonal System

Classes $4mm$, 422, $\bar{4}2m$, $4/mmm$

6 constants

$$\begin{bmatrix} s_{11} & s_{12} & s_{13} & 0 & 0 & 0 \\ s_{12} & s_{11} & s_{13} & 0 & 0 & 0 \\ s_{13} & s_{13} & s_{33} & 0 & 0 & 0 \\ 0 & 0 & 0 & s_{44} & 0 & 0 \\ 0 & 0 & 0 & 0 & s_{44} & 0 \\ 0 & 0 & 0 & 0 & 0 & s_{66} \end{bmatrix}$$

Trigonal System

Classes 3, $\bar{3}$

7 constants

$$\begin{bmatrix} s_{11} & s_{12} & s_{13} & s_{14} & -s_{25} & 0 \\ s_{12} & s_{11} & s_{13} & -s_{14} & s_{25} & 0 \\ s_{13} & s_{13} & s_{33} & 0 & 0 & 0 \\ s_{14} & -s_{14} & 0 & s_{44} & 0 & 2s_{25} \\ -s_{25} & s_{25} & 0 & 0 & s_{44} & 2s_{14} \\ 0 & 0 & 0 & 2s_{25} & 2s_{14} & 2(s_{11} - s_{12}) \end{bmatrix}$$

$c_{46} = c_{25}$
$c_{56} = c_{14}$
$c_{66} = \frac{1}{2}$
$(c_{11} - c_{12})$

369

Trigonal System

32, 3m, $\bar{3}m$

6 constants

$$\begin{bmatrix} s_{11} & s_{12} & s_{13} & s_{14} & 0 & 0 \\ s_{12} & s_{11} & s_{13} & -s_{14} & 0 & 0 \\ s_{13} & s_{13} & s_{33} & 0 & 0 & 0 \\ s_{14} & -s_{14} & 0 & s_{44} & 0 & 0 \\ 0 & 0 & 0 & 0 & s_{44} & 2s_{14} \\ 0 & 0 & 0 & 0 & 2s_{14} & 2(s_{11} - s_{12}) \end{bmatrix}$$

$c_{56} = c_{14}$
$c_{66} = \frac{1}{2}(c_{11} - c_{12})$

Hexagonal System

5 constants

$$\begin{bmatrix} s_{11} & s_{12} & s_{13} & 0 & 0 & 0 \\ s_{12} & s_{11} & s_{13} & 0 & 0 & 0 \\ s_{13} & s_{13} & s_{33} & 0 & 0 & 0 \\ 0 & 0 & 0 & s_{44} & 0 & 0 \\ 0 & 0 & 0 & 0 & s_{44} & 0 \\ 0 & 0 & 0 & 0 & 0 & 2(s_{11} - s_{12}) \end{bmatrix}$$

$c_{66} = \frac{1}{2}(c_{11} - c_{12})$

Cubic System

3 constants

$$\begin{bmatrix} s_{11} & s_{12} & s_{12} & 0 & 0 & 0 \\ s_{12} & s_{11} & s_{12} & 0 & 0 & 0 \\ s_{12} & s_{12} & s_{11} & 0 & 0 & 0 \\ 0 & 0 & 0 & s_{44} & 0 & 0 \\ 0 & 0 & 0 & 0 & s_{44} & 0 \\ 0 & 0 & 0 & 0 & 0 & s_{44} \end{bmatrix}$$

Isotropic

$$\begin{bmatrix} s_{11} & s_{12} & s_{12} & 0 & 0 & 0 \\ s_{12} & s_{11} & s_{12} & 0 & 0 & 0 \\ s_{12} & s_{12} & s_{11} & 0 & 0 & 0 \\ 0 & 0 & 0 & s_{44} & 0 & 0 \\ 0 & 0 & 0 & 0 & s_{44} & 0 \\ 0 & 0 & 0 & 0 & 0 & s_{44} \end{bmatrix}$$

2 constants

$$s_{12} = s_{11} - \tfrac{1}{2}s_{44}$$
$$c_{12} = c_{11} - 2c_{44}$$

MECHANICAL PROPERTIES A.3 Relations between Stiffness and Compliance Constants. For the more symmetric crystal classes there are simple relations between the c_{IJ}'s and the s_{IJ}'s. The equations given below are for converting from $[s]$ to $[c]$. Equations for converting from $[c]$ to $[s]$ are obtained by interchanging all c's and s's. That is,

$$s_{11} + s_{12} = \frac{c_{33}}{c}, \text{ etc.}$$

with

$$c = c_{33}(c_{11} + c_{12}) - 2c_{13}^2,$$

and so on.

TETRAGONAL SYSTEM (Classes *4mm, 422, $\bar{4}2m$, 4/mmm* only)

$$c_{11} + c_{12} = \frac{s_{33}}{s}, \qquad c_{11} - c_{12} = \frac{1}{s_{11} - s_{12}}, \qquad c_{13} = -\frac{s_{13}}{s}$$

$$c_{33} = \frac{s_{11} + s_{12}}{s}, \qquad c_{44} = \frac{1}{s_{44}}, \qquad c_{66} = \frac{1}{s_{66}}$$

with

$$s = s_{33}(s_{11} + s_{12}) - 2s_{13}^2$$

TRIGONAL SYSTEM (Classes *3m, 32, $\bar{3}m$* only)

$$c_{11} + c_{12} = \frac{s_{33}}{s}, \qquad c_{11} - c_{12} = \frac{s_{44}}{s'}, \qquad c_{13} = -\frac{s_{13}}{s}$$

$$c_{14} = -\frac{s_{14}}{s'}, \qquad c_{33} = \frac{s_{11} + s_{12}}{s}, \qquad c_{44} = \frac{s_{11} - s_{12}}{s'}$$

with

$$s = s_{33}(s_{11} + s_{12}) - 2s_{13}^2$$
$$s' = s_{44}(s_{11} - s_{12}) - 2s_{14}^2$$

HEXAGONAL SYSTEM

$$c_{11} + c_{12} = \frac{s_{33}}{s}, \qquad c_{11} - c_{12} = \frac{1}{s_{11} - s_{12}}, \qquad c_{13} = -\frac{s_{13}}{s}$$

$$c_{33} = \frac{s_{11} + s_{12}}{s}, \qquad c_{44} = \frac{1}{s_{44}}$$

with

$$s = s_{33}(s_{11} + s_{12}) - 2s_{13}^2$$

CUBIC SYSTEM

$$c_{11} = \frac{s_{11} + s_{12}}{(s_{11} - s_{12})(s_{11} + 2s_{12})}$$

$$c_{12} = \frac{-s_{12}}{(s_{11} - s_{12})(s_{11} + 2s_{12})}$$

$$c_{44} = \frac{1}{s_{44}}$$

ISOTROPIC

$$c_{11} = \frac{s_{11} + s_{12}}{(s_{11} - s_{12})(s_{11} + 2s_{12})}$$

$$c_{44} = \frac{1}{s_{44}}$$

MECHANICAL PROPERTIES A.4 Compliance Constants s_{IJ}

Material						Compliance (10^{-12} m²/newton)§						
	s_{11}	s_{22}	s_{33}	s_{44}	s_{55}	s_{66}	s_{12}	s_{13}	s_{14}	s_{16}	s_{23}	s_{25}
INSULATORS												
Ammonium Dihydrogen Phosphate (ADP)†	18.1		43.5	116		166	1.9	−11.8				
Barium Fluoride	—			—			—					
Barium Sodium Niobate†	5.3	5.14	8.33	15.4	15.2	13.2	−1.98	−1.20			−1.25	
Barium Titanate†	8.05		15.7	18.4		8.84	−2.35	−5.24				
Barium Titanate†‖	9.1		9.5	22.8			−2.7	−2.9				
Bismuth Germanate†	9.4			23			1.8					
Bismuth Germanium Oxide†				—								
Calcium Molybdate	—		—	—			—	—			—	
Diamond	1.12		—	2.07			−0.22					
Europium Iron Garnet‡												
Fused Silica												
Glass—Heavy Silicate, Flint												
Glass—Light Borate, Crown												
Glass—Pyrex												
Glass—T-40												
Iodic Acid†		—	—	—			—	—				
Lead Molybdate		—	—	—			—	—		—	—	
Lead Titanate-Zirconate (PZT-2)†‖	11.6		14.8	45.0			−3.33	−4.97				
Lead Titanate-Zirconate (PZT-5H)†‖	16.5		20.7	43.5			−4.78	−8.45				
Lithium Fluoride	11.35			15.9			−3.1					

TABLE A.4 *(Cont.)*

Material					Compliance (10^{-12} m²/newton)§							
	s_{11}	s_{22}	s_{33}	s_{44}	s_{55}	s_{66}	s_{12}	s_{13}	s_{14}	s_{16}	s_{23}	s_{25}
INSULATORS *(Cont.)*												
Lithium Gallate†	7.3	9.2	—	—	—	—	—	—			—	
Lithium Niobate†	5.78		5.02	17.0			−1.01	−1.47	−1.02			
Lithium Tantalate†	4.87		4.36	10.8			−0.58	−1.25	0.64			
Lucite	—			—								
Magnesium Oxide	4.08			6.76			−0.95					
Polyethylene	—			—								
Polystryrene	—			—								
Potassium Dihydrogen Phosphate (KDP)†	17.5		20	77.7		161	−4	−7.5				
Quartz†	12.77		9.60	20.04			−1.79	−1.22	−4.50			
Rochelle Salt†	52.0	36.8	35.9	150.2	350.3	104.2	−16.3	−11.6			−12.2	
Rutile	6.788		2.592	8.072		5.302	−4.017	−0.799				
Sapphire	2.38		2.20	7.05			−0.70	−0.38	0.49			
Sodium Fluoride	11.5			35.6			−2.3					
Strontium Titanate	3.73			8.09			−0.91					
Yttrium Aluminum Garnet (YAG)	—			—			—					
Yttrium Gallium Garnet	—			—			—					
Yttrium Iron Garnet (YIG)‡	—			—			—					

SEMICONDUCTORS

Beryllium Oxide†	—	—	—	—	—	—
Cadmium Selenide†	23.38	17.35	75.95	−11.22		−5.72
Cadmium Sulfide†	20.69	16.97	66.49	−9.99		−5.81
Gallium Arsenide†	12.64	—	18.6	−4.234		—
Gallium Phosphide†	—	—	—	—		—
Germanium	9.78	—	14.90	−2.66		—
Indium Antimonide†	—	—	—	—		—
Indium Arsenide†	—	—	—	—		—
Indium Phosphide†	—	—	—	—		—
Silicon	7.68	—	12.56	−2.14		—
Tellurium Dioxide†	115.0	10.3	37.7	−104.8	15.2	−2.11
Zinc Oxide†	7.858	6.940	23.57	−3.432		−2.206
Zinc Sulfide†	11.12	8.47	34.4	−4.56		−1.4
METALS						
Aluminum	15.9	—	35.2	−5.8		—
Gold	23.3	—	23.8	−10.7		—
Indium	—	—	—	—		—
Iron‡	7.57	—	8.62	−2.82		—
Nickel‡	8.0	—	8.44	−3.12		—
Silver	23.2	—	22.9	−9.93		—
Tellurium	48.7	23.4	58.1	−6.9		−13.8
Titanium	9.59	6.99	21.4	−4.62		−1.90
Tungsten	2.4	—	6.2	−0.6		—

† Piezoelectric. Values given are s^E.
‡ Ferromagnets and ferrimagnets.
§ Only the independent constants are given. Units conversion ratios on back endpaper.
‖ Poled ceramic. The compliance matrix has the same form as for the hexagonal crystal system, with Z along the poling direction.

MECHANICAL PROPERTIES A.5   Stiffness Constants c_{IJ}

Material	Stiffness (10^{10} newton/m²)§											
	c_{11}	c_{22}	c_{33}	c_{44}	c_{55}	c_{66}	c_{12}	c_{13}	c_{14}	c_{16}	c_{23}	c_{25}
INSULATORS												
Ammonium Dihydrogen Phosphate (ADP)†	6.76		3.38	0.867		0.608	0.59	2.0				
Barium Fluoride	9.20			2.57			4.18					
Barium Sodium Niobate†	23.9	24.7	13.5	6.5	6.6	7.6	10.4	5.0			5.2	
Barium Titanate†	27.5		16.5	5.43		11.3	17.9	15.1				
Barium Titanate†‖	15.0		14.6	4.4			6.6	6.6				
Bismuth Germanate†	11.58			4.36			2.70					
Bismuth Germanium Oxide†	12.80			2.55			3.05					
Calcium Molybdate	14.47		12.65	3.69		4.51	6.64	4.46		1.34		
Diamond	102			49.2			25					
Europium Iron Garnet‡	25.1			7.62			10.70					
Fused Silica	7.85			3.12								
Glass—Heavy Silicate, Flint	6.13			2.18								
Glass—Light Borate, Crown	5.82			1.81								
Glass—Pyrex	7.3			2.5								
Glass—T-40	6.30			2.26								
Iodic Acid†	3.03	5.45	4.36	1.84	2.19	1.74	1.19	1.17			0.55	
Lead Molybdate	10.92		9.17	2.67		3.37	6.83	5.28		1.36		

Material							
Lead Titanate-Zirconate (PZT-2)†‖	13.5	11.3	2.22		6.79	6.81	
Lead Titanate-Zirconate (PZT-5H)†‖	12.6 / 11.12	11.7	2.30 / 6.28		7.95 / 4.20	8.41	
Lithium Fluoride	—			4.45	—	—	—
Lithium Gallate†	20.3	14.7	5.42				
Lithium Niobate†	23.3	24.5	6.0		5.3	7.5	0.9
Lithium Tantalate†		27.5	9.4		4.7	8.0	−1.1
Lucite	0.848		0.143				
Magnesium Oxide	28.6		14.8		8.7		
Polyethylene	0.340		0.026				
Polystyrene	0.58		0.12				
Potassium Dihydrogen Phosphate (KDP)†	7.85	7.63	1.23		3.2	3.87	
Quartz†	8.674 / 4.14	10.72	5.794	0.61	0.699	1.191	+1.791
Rochelle Salt†	2.8	3.94	0.666	0.285 / 0.96	1.74	1.50	
Rutile	26.60	46.99	12.39	18.86	17.33	13.62	1.97
Sapphire	49.4	49.6	14.5		15.8	11.4	−2.3
Sodium Fluoride	9.7		2.81		2.44		
Strontium Titanate	31.8		12.4		10.2		
Yttrium Aluminum Garnet (YAG)	33.32		11.50		11.07		
Yttrium Gallium Garnet	29.03		9.55		11.73		
Yttrium Iron Garnet (YIG)‡	26.8		7.66		11.06		

TABLE A.5 (*Cont.*)

Material	c_{11}	c_{22}	c_{33}	c_{44}	c_{55}	c_{66}	c_{12}	c_{13}	c_{14}	c_{16}	c_{23}	c_{25}
					Stiffness (10^{10} newton/m^2)§							

SEMICONDUCTORS

Material	c_{11}	c_{22}	c_{33}	c_{44}	c_{55}	c_{66}	c_{12}	c_{13}	c_{14}	c_{16}	c_{23}	c_{25}
Beryllium Oxide†	46.06		49.16	14.77			12.65	8.848				
Cadmium Selenide†	7.41		8.36	1.317			4.52	3.93				
Cadmium Sulfide†	9.07		9.38	1.504			5.81	5.10				
Gallium Arsenide†	11.88			5.94			5.38					
Gallium Phosphide†	14.12			7.047			6.253					
Germanium	12.89			6.71			4.83					
Indium Antimonide†	6.72			3.02			3.67					
Indium Arsenide†	8.329			3.959			4.526					
Indium Phosphide†	10.22			4.60			5.76					
Silicon	16.57			7.956			6.39					
Tellurium Dioxide	5.57		10.58	2.65		6.59	5.12	2.18				
Zinc Oxide†	20.97		21.09	4.247			12.11	10.51				
Zinc Sulfide†	12.04		12.76	2.28			6.92	6.20				

METALS

Aluminium, crystal	10.80		2.85		6.13	
Aluminum, polycrystal	11.1		2.5			
Gold, crystal	18.6		4.20		15.7	
Gold, polycrystal	20.7		2.85			
Indium	4.535	4.515	0.646	1.207	4.006	4.151
Iron‡, crystal	23.7		11.6		14.1	
Iron‡, polycrystal	27.7		8.2			
Nickel‡, crystal	25.0		11.85		16.0	
Nickel‡, polycrystal	32.4		8.0			
Silver, crystal	11.9		4.37		8.94	
Silver, polycrystal	13.95		2.7			
Tellurium	—	7.00	—		—	2.31
Titanium, crystal	16.2	18.1	4.67		9.2	6.9
Titanium, polycrystal	16.59		4.4			
Tungsten, crystal	50.2		15.2		19.9	
Tungsten, polycrystal	58.1		13.4			

† Piezoelectric. Values given are c^E.
‡ Ferromagnets and ferrimagnets.
§ Only the independent constants are given. Units conversion ratios on back end paper.
‖ Poled ceramic. The stiffness matrix has the same form as for the hexagonal crystal system, with Z along the poling axis.

PIEZOELECTRIC PROPERTIES **B.1** **Symmetry Characteristics of $[d_{iJ}]$ and $[e_{iJ}]$.** The matrix elements are referred to coordinate axes x, y, z that coincide with the crystal axes X, Y, Z in Table 7.1. Except where noted, the piezoelectric stress matrices $[e_{iJ}]$ have the same form as the piezoelectric strain matrices $[d_{iJ}]$.

Triclinic *1*

$$\begin{bmatrix} d_{x1} & d_{x2} & d_{x3} & d_{x4} & d_{x5} & d_{x6} \\ d_{y1} & d_{y2} & d_{y3} & d_{y4} & d_{y5} & d_{y6} \\ d_{z1} & d_{z2} & d_{z3} & d_{z4} & d_{z5} & d_{z6} \end{bmatrix}$$

Monoclinic *2*

$$\begin{bmatrix} 0 & 0 & 0 & d_{x4} & 0 & d_{x6} \\ d_{y1} & d_{y2} & d_{y3} & 0 & d_{y5} & 0 \\ 0 & 0 & 0 & d_{z4} & 0 & d_{z6} \end{bmatrix}$$

Monoclinic *m*

$$\begin{bmatrix} d_{x1} & d_{x2} & d_{x3} & 0 & d_{x5} & 0 \\ 0 & 0 & 0 & d_{y4} & 0 & d_{y6} \\ d_{z1} & d_{z2} & d_{z3} & 0 & d_{z5} & 0 \end{bmatrix}$$

Orthorhombic *222*

$$\begin{bmatrix} 0 & 0 & 0 & d_{x4} & 0 & 0 \\ 0 & 0 & 0 & 0 & d_{y5} & 0 \\ 0 & 0 & 0 & 0 & 0 & d_{z6} \end{bmatrix}$$

Orthorhombic *2mm*

$$\begin{bmatrix} 0 & 0 & 0 & 0 & d_{x5} & 0 \\ 0 & 0 & 0 & d_{y4} & 0 & 0 \\ d_{z1} & d_{z2} & d_{z3} & 0 & 0 & 0 \end{bmatrix}$$

Tetragonal $\bar{4}$

$$\begin{bmatrix} 0 & 0 & 0 & d_{x4} & d_{x5} & 0 \\ 0 & 0 & 0 & -d_{x5} & d_{x4} & 0 \\ d_{z1} & -d_{z1} & 0 & 0 & 0 & d_{z6} \end{bmatrix}$$

Tetragonal *4*

$$\begin{bmatrix} 0 & 0 & 0 & d_{x4} & d_{x5} & 0 \\ 0 & 0 & 0 & d_{x5} & -d_{x4} & 0 \\ d_{z1} & d_{z1} & d_{z3} & 0 & 0 & 0 \end{bmatrix}$$

Tetragonal $\bar{4}2m$

$$\begin{bmatrix} 0 & 0 & 0 & d_{x4} & 0 & 0 \\ 0 & 0 & 0 & 0 & d_{x4} & 0 \\ 0 & 0 & 0 & 0 & 0 & d_{z6} \end{bmatrix}$$

Tetragonal 422

$$\begin{bmatrix} 0 & 0 & 0 & d_{x4} & 0 & 0 \\ 0 & 0 & 0 & 0 & -d_{x4} & 0 \\ 0 & 0 & 0 & 0 & 0 & 0 \end{bmatrix}$$

Tetragonal $4mm$

$$\begin{bmatrix} 0 & 0 & 0 & 0 & d_{x5} & 0 \\ 0 & 0 & 0 & d_{x5} & 0 & 0 \\ d_{z1} & d_{z1} & d_{z3} & 0 & 0 & 0 \end{bmatrix}$$

Trigonal 3

$$\begin{bmatrix} d_{x1} & -d_{x1} & 0 & d_{x4} & d_{x5} & -2d_{y2} \\ -d_{y2} & d_{y2} & 0 & d_{x5} & -d_{x4} & -2d_{x1} \\ d_{z1} & d_{z1} & d_{z3} & 0 & 0 & 0 \end{bmatrix}$$

$e_{x6} = -e_{y2}$
$e_{y6} = -e_{x1}$

Trigonal 32

$$\begin{bmatrix} d_{x1} & -d_{x1} & 0 & d_{x4} & 0 & 0 \\ 0 & 0 & 0 & 0 & -d_{x4} & -2d_{x1} \\ 0 & 0 & 0 & 0 & 0 & 0 \end{bmatrix}$$

$e_{y6} = -e_{x1}$

Trigonal $3m$

$$\begin{bmatrix} 0 & 0 & 0 & 0 & d_{x5} & -2d_{y2} \\ -d_{y2} & d_{y2} & 0 & d_{x5} & 0 & 0 \\ d_{z1} & d_{z1} & d_{z3} & 0 & 0 & 0 \end{bmatrix}$$

$e_{x6} = -e_{y2}$

Hexagonal 6

$$\begin{bmatrix} 0 & 0 & 0 & d_{x4} & d_{x5} & 0 \\ 0 & 0 & 0 & d_{x5} & -d_{x4} & 0 \\ d_{z1} & d_{z1} & d_{z3} & 0 & 0 & 0 \end{bmatrix}$$

Hexagonal 622

$$\begin{bmatrix} 0 & 0 & 0 & d_{x4} & 0 & 0 \\ 0 & 0 & 0 & 0 & -d_{x4} & 0 \\ 0 & 0 & 0 & 0 & 0 & 0 \end{bmatrix}$$

Hexagonal $6mm$

$$\begin{bmatrix} 0 & 0 & 0 & 0 & d_{x5} & 0 \\ 0 & 0 & 0 & d_{x5} & 0 & 0 \\ d_{z1} & d_{z1} & d_{z3} & 0 & 0 & 0 \end{bmatrix}$$

381

Hexagonal $\bar{6}$
$$\begin{bmatrix} d_{x1} & -d_{x1} & 0 & 0 & 0 & -2d_{y2} \\ -d_{y2} & d_{y2} & 0 & 0 & 0 & -2d_{x1} \\ 0 & 0 & 0 & 0 & 0 & 0 \end{bmatrix}$$
$$e_{x6} = -e_{y2}$$
$$e_{y6} = -e_{x1}$$

Hexagonal $\bar{6}m2$
$$\begin{bmatrix} d_{x1} & -d_{x1} & 0 & 0 & 0 & 0 \\ 0 & 0 & 0 & 0 & 0 & -2d_{x1} \\ 0 & 0 & 0 & 0 & 0 & 0 \end{bmatrix}$$
$$e_{y6} = -e_{x1}$$

Cubic 23 and $\bar{4}3m$
$$\begin{bmatrix} 0 & 0 & 0 & d_{x4} & 0 & 0 \\ 0 & 0 & 0 & 0 & d_{x4} & 0 \\ 0 & 0 & 0 & 0 & 0 & d_{x4} \end{bmatrix}$$

PIEZOELECTRIC PROPERTIES   B.2   Piezoelectric Strain Constants d_{iJ}

Material	Symmetry	Piezoelectric Strain Constants, $d_{iJ} = d_{Ji}$ (10^{-12} coulomb/newton)†									
		d_{x1}	d_{x4}	d_{x5}	d_{y2}	d_{y4}	d_{y5}	d_{z1}	d_{z2}	d_{z3}	d_{z6}
INSULATORS											
Ammonium Dihydrogen Phosphate (ADP)	Tetr. $\bar{4}2m$		11.7								51.7
Barium Sodium Niobate	Orth. 2mm			42		52		−7	−6	37	
Barium Titanate	Tetr. 4mm			392				−34.5		85.6	
Barium Titanate	Uniaxial‡			260				−78		190	
Bismuth Germanate	Cub. $\bar{4}3m$		0.87								
Bismuth Germanium Oxide	Cub. 23		—								
Lead Titanate-Zirconate (PZT-2)	Uniaxial‡			440				−60		152	
Lead Titanate-Zirconate (PZT-5H)	Uniaxial‡			741				−274		593	
Lithium Gallate	Orth. 2mm			−6.4	21	−4.7		−2.8	−4.0	7.7	
Lithium Niobate	Trig. 3m			68	21			−1		6	
Lithium Tantalate	Trig. 3m			26	7			−2		8	
Potassium Dihydrogen Phosphate (KDP)	Tetr. $\bar{4}2m$		1.3								21
Quartz	Trig. 32	−2.3	−0.67								
Rochelle Salt	Orth. 222		345				54				12

383

TABLE B.2 (Cont.)

Material	Symmetry	Piezoelectric Strain Constants, $d_{iJ} = d_{Ji}$ (10^{-12} coulomb/newton)†									
		d_{x1}	d_{x4}	d_{x5}	d_{y2}	d_{y4}	d_{y5}	d_{z1}	d_{z2}	d_{z3}	d_{z6}
SEMICONDUCTORS											
Beryllium Oxide	Hex. 6mm			—				−0.12		0.24	
Cadmium Sulfide	Hex. 6mm			−14				−5		10.3	
Gallium Arsenide	Cub. $\overline{4}3m$		2.6								
Gallium Phosphide	Cub. $\overline{4}3m$		—								
Indium Antimonide	Cub. $\overline{4}3m$		—								
Indium Phosphide	Cub. $\overline{4}3m$		—								
Tellurium Dioxide	Tetr. 422		8.13								
Zinc Oxide	Hex. 6mm			−11.34				−5.43		11.67	
Zinc Sulfide	Hex. 6mm			−2.8				−1.13		3.23	

† Only the independent constants are given. Units conversion ratios on back endpaper.
‡ Piezoelectric ceramics are poled during preparation and are isotropic in the plane transverse to the poling (or Z) axis. The piezoelectric strain matrix is of the form

$$\begin{bmatrix} 0 & 0 & 0 & 0 & d_{x5} & 0 \\ 0 & 0 & 0 & d_{x5} & 0 & 0 \\ d_{z1} & d_{z1} & d_{z3} & 0 & 0 & 0 \end{bmatrix}$$

PIEZOELECTRIC PROPERTIES B.3 Piezoelectric Stress Constants e_{iJ}

Material	Symmetry	Piezoelectric Stress Constants, $e_{iJ} = e_{Ji}$ (coulomb/m²)†									
		e_{x1}	e_{x4}	e_{x5}	e_{y2}	e_{y4}	e_{y5}	e_{z1}	e_{z2}	e_{z3}	e_{z6}
INSULATORS											
Ammonium Dihydrogen Phosphate (ADP)	Tetr. $\bar{4}2m$		0.101								0.134
Barium Sodium Niobate	Orth. $2mm$			2.8		3.4		−0.4	−0.3	4.3	
Barium Titanate	Tetr. $4mm$			21.3				−2.74		3.70	
Barium Titanate	Uniaxial‡			11.4				−4.35		17.5	
Bismuth Germanate	Cub. $\bar{4}3m$		0.0376								
Bismuth Germanium Oxide	Cub. 23		0.99								
Lead Titanate-Zirconate (PZT-2)	Uniaxial‡			9.8				−1.9		9.0	
Lead Titanate-Zirconate (PZT-5H)	Uniaxial‡			17.0				−6.5		23.3	
Lithium Gallate	Orth. $2mm$			−0.28		0.25		—	—	0.88	
Lithium Niobate	Trig. $3m$			3.7	2.5			0.2		1.3	
Lithium Tantalate	Trig. $3m$			2.6	1.6			0.0		1.9	
Potassium Dihydrogen Phosphate (KDP)	Tetr. $\bar{4}2m$		0.0172								
Quartz	Trig. 32	0.171	−0.0436								0.14
Rochelle Salt	Orth. 222		2.23				0.154				0.115

TABLE B.2 (*Cont.*)

Material	Symmetry	Piezoelectric Stress Constants, $e_{i,J} = e_{Ji}$ (coulomb/m²)†									
		e_{x1}	e_{x4}	e_{x5}	e_{y2}	e_{y4}	e_{y5}	e_{z1}	e_{z2}	e_{z3}	e_{z6}
SEMICONDUCTORS											
Beryllium Oxide	Hex. $6mm$			—				−0.051		0.092	
Cadmium Sulfide	Hex. $6mm$			−0.21				−0.24		0.44	
Gallium Arsenide	Cub. $\bar{4}3m$		0.154								
Gallium Phosphide	Cub. $\bar{4}3m$		−0.10								
Indium Antimonide	Cub. $\bar{4}3m$		0.71								
Indium Phosphide	Cub. $\bar{4}3m$		—								
Tellurium Dioxide	Tetr. 422		0.216								
Zinc Oxide	Hex. $6mm$			−0.48				−0.573		1.32	
Zinc Sulfide	Hex. $6mm$			−0.0638				−0.0140		0.272	

† Only the independent constants are given. Units conversion ratios on back endpaper.
‡ Piezoelectric ceramics are poled during preparation and are isotropic in the plane transverse to the poling (or Z) axis. The piezoelectric stress matrix is of the form

$$
\begin{bmatrix}
0 & 0 & 0 & 0 & e_{z5} & 0 \\
0 & 0 & 0 & e_{z5} & 0 & 0 \\
e_{z1} & e_{z1} & e_{z3} & 0 & 0 & 0
\end{bmatrix}
$$

ELECTRICAL PROPERTIES C.1 Symmetry Characteristics of $[\epsilon^S_{iJ}]$.

The matrix elements are referred to coordinate axes x, y, z that coincide with the crystal axes X, Y, Z in Table 7.1 The matrix forms given also apply to $[\epsilon^T_{iJ}]$ and to $[\epsilon_{ij}]$ for nonpiezoelectric materials.

TRICLINIC

$$\begin{bmatrix} \epsilon^S_{xx} & \epsilon^S_{xy} & \epsilon^S_{xz} \\ \epsilon^S_{xy} & \epsilon^S_{yy} & \epsilon^S_{yz} \\ \epsilon^S_{xz} & \epsilon^S_{yz} & \epsilon^S_{zz} \end{bmatrix}$$

MONOCLINIC

$$\begin{bmatrix} \epsilon^S_{xx} & 0 & \epsilon^S_{xz} \\ 0 & \epsilon^S_{yy} & 0 \\ \epsilon^S_{xz} & 0 & \epsilon^S_{zz} \end{bmatrix}$$

ORTHORHOMBIC

$$\begin{bmatrix} \epsilon^S_{xx} & 0 & 0 \\ 0 & \epsilon^S_{yy} & 0 \\ 0 & 0 & \epsilon^S_{zz} \end{bmatrix}$$

HEXAGONAL, TRIGONAL, TETRAGONAL

$$\begin{bmatrix} \epsilon^S_{xx} & 0 & 0 \\ 0 & \epsilon^S_{xx} & 0 \\ 0 & 0 & \epsilon^S_{zz} \end{bmatrix}$$

CUBIC, ISOTROPIC

$$\begin{bmatrix} \epsilon^S_{xx} & 0 & 0 \\ 0 & \epsilon^S_{xx} & 0 \\ 0 & 0 & \epsilon^S_{xx} \end{bmatrix}$$

ELECTRICAL PROPERTIES C.2 Relative Permittivity Constants for Piezoelectric Materials

Material	Symmetry	Constant Strain†			Constant Stress†		
		$\epsilon^S_{xx}/\epsilon_0$	$\epsilon^S_{yy}/\epsilon_0$	$\epsilon^S_{zz}/\epsilon_0$	$\epsilon^T_{xx}/\epsilon_0$	$\epsilon^T_{yy}/\epsilon_0$	$\epsilon^T_{zz}/\epsilon_0$
INSULATORS							
Ammonium Dihydrogen Phosphate (ADP)	Tetr. $\bar{4}2m$	—		—	56		15.4
Barium Sodium Niobate	Orth. 2mm	222	227	32	235	247	51
Barium Titanate	Tetr. 4mm	—		—	2920		168
Barium Titanate	Uniaxial‡	1115		1260	1450		1700
Bismuth Germanate	Cub. $\bar{4}3m$	16			16		
Bismuth Germanium Oxide	Cub. 23	38			—		
Lead Titanate-Zirconate (PZT-2)	Uniaxial‡	504		260	990		450
Lead Titanate-Zirconate (PZT-5H)	Uniaxial‡	1700		1470	3130		3400
Lithium Gallate	Orth. 2mm	7.0	—	—	7.18	6.18	8.8
Lithium Niobate	Trig. 3m	44		29	84		30
Lithium Tantalate	Trig. 3m	41		43	51		45

Potassium Dihydrogen Phosphate (KDP)	Tetr. $\bar{4}2m$	—	42	—	21
Quartz	Trig. 32	4.5	4.52	4.6	4.68
Rochelle Salt	Orth. 222	—	205	9.6	9.5

SEMICONDUCTORS

Beryllium Oxide	Hex. 6mm	—	—	—	7.66
Cadmium Sulfide	Hex. 6mm	9.02	9.35	9.53	10.3
Gallium Arsenide	Cub. $\bar{4}3m$	—	12.5		
Gallium Phosphide	Cub. $\bar{4}3m$	—	11.1		
Indium Antimonide	Cub. $\bar{4}3m$	—	17.7		
Indium Phosphide	Cub. $\bar{4}3m$	—	12.35		
Tellurium Dioxide	Tetr. 422	—	22.9	—	24.7
Zinc Oxide	Hex. 6mm	8.55	9.16	10.2	12.64
Zinc Sulfide	Hex. 6mm	—	8.7	—	8.7

† $\epsilon_0 = 8.854 \times 10^{-12}$ farads/m. Units conversion ratios on back endpaper. Only the independent constants are shown.

‡ Poled ceramic material. The permittivity matrix has the same form as for the hexagonal crystal system, with Z along the poling axis.

C.3 Relative Permittivity Constants for
Nonpiezoelectric Materials

Material	Symmetry	$\dfrac{\epsilon_{xx}}{\epsilon_0}$	$\dfrac{\epsilon_{yy}}{\epsilon_0}$	$\dfrac{\epsilon_{zz}}{\epsilon_0}$†
INSULATORS				
Calcium Molybdate	Tetr. $4/m$	24.0		20.0
Diamond	Cubic $m3m$	5.67		
Europium Iron Garnet‡	Cubic $m3m$	—		
Fused Silica	Isotropic	3.78		
Glass—Heavy Silicate Flint	Isotropic	—		
Glass—Light Borate Crown	Isotropic	—		
Glass—Pyrex	Isotropic	4.5		
Glass—T40	Isotropic	—		
Lead Molybdate	Tetr. $4/m$	34.0		40.6
Lithium Fluoride	Cubic $m3m$	9.0		
Lucite	Isotropic	—		
Magnesium Oxide	Cubic $m3m$	9.6		
Polyethylene	Isotropic	2.25		
Polystyrene	Isotropic	—		
Rutile	Tetr. $4/mmm$	89		173
Sapphire	Trig. $\bar{3}m$	9.34		11.54
Sodium Fluoride	Cubic $m3m$	6.0		
Strontium Titanate	Cubic $m3m$	302		
Yttrium Aluminum Garnet (YAG)	Cubic $m3m$	—		
Yttrium Gallium Garnet (YGaG)	Cubic $m3m$	—		
Yttrium Iron Garnet (YIG)	Cubic $m3m$	—		
SEMICONDUCTORS				
Germanium	Cubic $m3m$	15.8		
Silicon	Cubic $m3m$	11.7		

† $\epsilon_0 = 8.854 \times 10^{-12}$ farads/m. Units conversion ratios on back endpaper. Only the independent constants are shown.
‡ Ferrimagnetic materials.

Appendix 3

ACOUSTIC PLANE WAVE PROPERTIES

A. CHRISTOFFEL EQUATIONS FOR ISOTROPIC AND ANISOTROPIC SOLIDS†

ISOTROPIC AND CUBIC

$$k^2 \begin{bmatrix} c_{11}l_x^2 + c_{44}(1 - l_x^2) & (c_{12} + c_{44})l_x l_y & (c_{12} + c_{44})l_x l_z \\ (c_{12} + c_{44})l_x l_y & c_{11}l_y^2 + c_{44}(1 - l_y^2) & (c_{12} + c_{44})l_y l_z \\ (c_{12} + c_{44})l_x l_z & (c_{12} + c_{44})l_y l_z & c_{11}l_z^2 + c_{44}(1 - l_z^2) \end{bmatrix} \begin{bmatrix} v_x \\ v_y \\ v_z \end{bmatrix} = \rho\omega^2 \begin{bmatrix} v_x \\ v_y \\ v_z \end{bmatrix}$$

$c_{11} = c_{12} + 2c_{44}$ for the isotropic case.

† The wave vector *direction* is defined by the unit vector

$$\hat{\mathbf{l}} = \hat{\mathbf{x}}l_x + \hat{\mathbf{y}}l_y + \hat{\mathbf{z}}l_z$$

where the coordinate axes x, y, z coincide with the crystal axes X, Y, Z in Table 7.1.

HEXAGONAL

$$k^2 \begin{bmatrix} c_{11}l_x^2 + c_{66}l_y^2 + c_{44}l_z^2 & (c_{12} + c_{66})l_x l_y & (c_{13} + c_{44})l_x l_z \\ (c_{12} + c_{66})l_x l_y & c_{66}l_x^2 + c_{11}l_y^2 + c_{44}l_z^2 & (c_{13} + c_{44})l_y l_z \\ (c_{13} + c_{44})l_x l_z & (c_{13} + c_{44})l_y l_z & c_{44}l_x^2 + c_{44}l_y^2 + c_{33}l_z^2 \end{bmatrix} \begin{bmatrix} v_x \\ v_y \\ v_z \end{bmatrix} = \rho\omega^2 \begin{bmatrix} v_x \\ v_y \\ v_z \end{bmatrix}$$

$$c_{66} = \tfrac{1}{2}(c_{11} - c_{12}).$$

TRIGONAL

Classes 32, 3m, $\bar{3}m$

$$k^2 \begin{bmatrix} c_{11}l_x^2 + c_{66}l_y^2 + c_{44}l_z^2 + 2c_{14}l_y l_z & (c_{12} + c_{66})l_x l_y + 2c_{14}l_x l_z & (c_{13} + c_{44})l_x l_z + 2c_{14}l_x l_y \\ (c_{12} + c_{66})l_x l_y + 2c_{14}l_x l_z & c_{66}l_x^2 + c_{11}l_y^2 + c_{44}l_z^2 - 2c_{14}l_y l_z & (c_{13} + c_{44})l_y l_z + c_{14}(l_x^2 - l_y^2) \\ (c_{13} + c_{44})l_x l_z + 2c_{14}l_x l_y & (c_{13} + c_{44})l_y l_z + c_{14}(l_x^2 - l_y^2) & c_{44}(l_x^2 + l_y^2) + c_{33}l_z^2 \end{bmatrix}$$

$$\times \begin{bmatrix} v_x \\ v_y \\ v_z \end{bmatrix} = \rho\omega^2 \begin{bmatrix} v_x \\ v_y \\ v_z \end{bmatrix}$$

$$c_{66} = \tfrac{1}{2}(c_{11} - c_{12}).$$

Classes 3, $\bar{3}$.   In 7.11 the elements of the Christoffel matrix are:

$$\alpha = c_{11}l_x^2 + c_{66}l_y^2 + c_{44}l_z^2 + 2c_{14}l_y l_z - 2c_{25}l_z l_x$$
$$\beta = c_{66}l_x^2 + c_{11}l_y^2 + c_{44}l_z^2 - 2c_{14}l_y l_z + 2c_{25}l_z l_x$$
$$\gamma = c_{44}(l_x^2 + l_y^2) + c_{33}l_z^2$$
$$\delta = (c_{12} + c_{66})l_x l_y + 2c_{14}l_z l_x + 2c_{25}l_y l_z$$
$$\epsilon = -c_{25}(l_x^2 - l_y^2) + (c_{13} + c_{44})l_z l_x + 2c_{14}l_x l_y$$
$$\zeta = c_{14}(l_x^2 - l_y^2) + (c_{13} + c_{44})l_y l_z + 2c_{25}l_x l_z.$$

TETRAGONAL

Classes _4mm, 422, $\bar{4}2m$, 4/mmm._   Same as hexagonal, but with arbitrary c_{66}.

Classes 4, $\bar{4}$, 4/m

$$k^2 \begin{bmatrix} c_{11}l_x^2 + c_{66}l_y^2 + c_{44}l_z^2 + 2c_{16}l_x l_y & c_{16}(l_x^2 - l_y^2) + (c_{12} + c_{66})l_x l_y & (c_{13} + c_{44})l_z l_x \\ c_{16}(l_x^2 - l_y^2) + (c_{12} + c_{66})l_x l_y & c_{66}l_x^2 + c_{11}l_y^2 + c_{44}l_z^2 - 2c_{16}l_x l_y & (c_{13} + c_{44})l_y l_z \\ (c_{13} + c_{44})l_z l_x & (c_{13} + c_{44})l_y l_z & c_{44}(l_x^2 + l_y^2) + c_{33}l_z^2 \end{bmatrix}$$

$$\times \begin{bmatrix} v_x \\ v_y \\ v_z \end{bmatrix} = \rho\omega^2 \begin{bmatrix} v_x \\ v_y \\ v_z \end{bmatrix}.$$

ORTHORHOMBIC

$$k^2 \begin{bmatrix} c_{11}l_x^2 + c_{66}l_y^2 + c_{55}l_z^2 & (c_{12} + c_{66})l_x l_y & (c_{13} + c_{55})l_z l_x \\ (c_{12} + c_{66})l_x l_y & c_{66}l_x^2 + c_{22}l_y^2 + c_{44}l_z^2 & (c_{23} + c_{44})l_y l_z \\ (c_{13} + c_{55})l_z l_x & (c_{23} + c_{44})l_y l_z & c_{55}l_x^2 + c_{44}l_y^2 + c_{33}l_z^2 \end{bmatrix} \begin{bmatrix} v_x \\ v_y \\ v_z \end{bmatrix} = \rho \omega^2 \begin{bmatrix} v_x \\ v_y \\ v_z \end{bmatrix}$$

B. SLOWNESS SURFACES FOR ISOTROPIC AND ANISOTROPIC SOLIDS

This section gives slowness curves for the principal planes of all crystal classes except triclinic and monoclinic. Representative dimensions on these curves are stated in terms of the material parameters, neglecting the piezo-electric effect. Using the material constants given in Appendix 2, these dimensions can be easily evaluated for any material and then used to sketch the general shape of the curves.

B.1 Isotropic

In Section C of Chapter 6 it was seen that the phase velocity $V_p = \omega/k$ is independent of direction and has two values,

$$V_l = \sqrt{\frac{c_{11}}{\rho}} = \sqrt{\frac{\lambda + 2\mu}{\rho}}$$

for the longitudinal wave, and

$$V_s = \sqrt{\frac{c_{44}}{\rho}} = \sqrt{\frac{\mu}{\rho}}$$

for the two shear waves. The slowness surface consists of two concentric spheres (Fig. 3.1). Since the Lamé constants λ and μ are both positive, the outer sphere is for shear waves and the inner one for longitudinal waves.

B.2 Cubic

The case of a cubic material was treated in Examples 4 and 5 of Chapter 7. For propagation in a cube face the characteristic equation factors into a linear term and a quadratic term. There is a pure shear wave polarized normal to the cube face,

$$\left(\frac{k}{\omega}\right)_1 = \left(\frac{\rho}{c_{44}}\right)^{1/2}, \tag{3.1}$$

a quasishear wave,

$$\left(\frac{k}{\omega}\right)_2 = (2\rho)^{1/2} \left| c_{11} + c_{44} - \sqrt{(c_{11} - c_{44})^2 \cos^2 2\phi + (c_{12} + c_{44})^2 \sin^2 2\phi} \right|^{-1/2}$$

$$\tag{3.2}$$

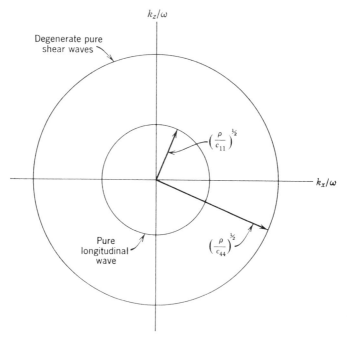

FIGURE 3.1.   **Isotropic. Propagation in an arbitrary plane.**

and a quasilongitudinal wave,

$$\left(\frac{k}{\omega}\right)_3 = (2\rho)^{1/2}\left\{c_{11} + c_{44} + \sqrt{(c_{11} - c_{44})^2 \cos^2 2\phi + (c_{12} + c_{44})^2 \sin^2 2\phi}\right\}^{-1/2},$$

$$(3.3)$$

where $\cos \phi = l_x$. Inverse velocity curves are shown in Fig. 3.2 for gallium arsenide. Representative dimensions are given in terms of material parameters. From these, the general shape of the curves can be deduced for any cubic material.

For propagation in a plane passing through a cube face diagonal the characteristic equation again factors into linear and quadratic terms. There is a pure shear mode polarized normal to the plane of propagation,

$$\left(\frac{k}{\omega}\right)_2 = \left(\frac{\rho}{\dfrac{c_{11} - c_{12}}{2} \cos^2 \theta + c_{44} \sin^2 \theta}\right)^{1/2},$$

$$(3.4)$$

a quasishear wave,

$$\left(\frac{k}{\omega}\right)_1 = \left(\frac{2\rho}{B - \sqrt{B^2 - C}}\right)^{1/2},$$

$$(3.5)$$

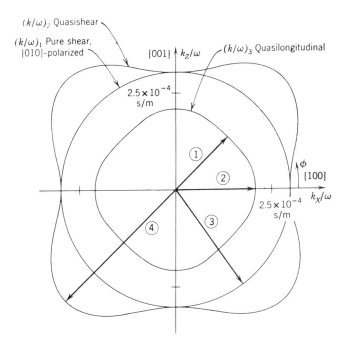

① $\left(\dfrac{\rho}{c_{11} + c_{44}(1 - 1/A)}\right)^{1/2}$

③ $(\rho/c_{44})^{1/2}$

② $(\rho/c_{11})^{1/2}$

④ $(\rho A/c_{44})^{1/2}$

Anisotropy factor $A = \dfrac{2c_{44}}{c_{11} - c_{12}}$

FIGURE 3.2.   Cubic crystal classes. Propagation in a cube face. Curves shown are for GaAs, with the piezoelectric effect ignored. Sketches may be made for other materials, using the key dimensions shown.

and a quasilongitudinal wave,

$$\left(\frac{k}{\omega}\right)_3 = \left(\frac{2\rho}{B + \sqrt{B^2 - C}}\right)^{1/2} \tag{3.6}$$

Here

$$B = (c_{11} + c_{12} + 4c_{44})\frac{\cos^2 \theta}{2} + (c_{11} + c_{44}) \sin^2 \theta$$

and

$$C = (c_{11}c_{11}' - c_{12}^2 - 2c_{12}c_{44}) \sin^2 2\theta + 4c_{44}(c_{11}' \cos^4 \theta + c_{11} \sin^4 \theta)$$

with

$$c'_{11} = \frac{c_{11} + c_{12} + 2c_{44}}{2}.$$

The direction angle θ is defined in Fig. 3.3, which gives curves for gallium arsenide and also some representative dimensions.

B.3   Hexagonal

For hexagonal materials the characteristic equation factors when propagation is in the XY plane, that is, normal to the Z-axis. The dispersion relation is

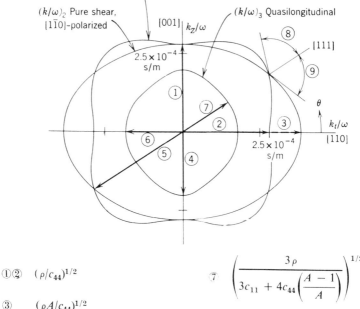

①② $(\rho/c_{44})^{1/2}$

③ $(\rho A/c_{44})^{1/2}$

④ $(\rho/c_{11})^{1/2}$

⑤ $\left(\dfrac{3\rho A}{(2+A)c_{44}}\right)^{1/2}$

⑥ $\left(\dfrac{\rho}{c_{11} + c_{44}\left(\dfrac{A-1}{A}\right)}\right)^{1/2}$

⑦ $\left(\dfrac{3\rho}{3c_{11} + 4c_{44}\left(\dfrac{A-1}{A}\right)}\right)^{1/2}$

⑧⑨ $\cot^{-1} \sqrt{2}\left(\dfrac{A-1}{A+2}\right)$

Anisotropy factor

$$A = \frac{2c_{44}}{c_{11} - c_{12}}$$

FIGURE 3.3.   **Cubic crystal classes. Propagation in a cube diagonal plane. Curves shown are for GaAs, with the piezoelectric effect ignored.**

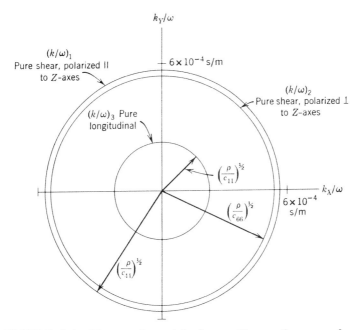

FIGURE 3.4. **Hexagonal crystal classes. Propagation normal to the Z-axis. Curves are for CdS, with the piezoelectric effect ignored.**

independent of the propagation direction in this plane. There is one pure shear mode

$$\left(\frac{k}{\omega}\right)_2 = \left(\frac{\rho}{c_{66}}\right)^{1/2}, \tag{3.7}$$

polarized normal to the Z axis, another pure shear mode

$$\left(\frac{k}{\omega}\right)_1 = \left(\frac{\rho}{c_{44}}\right)^{1/2} \tag{3.8}$$

polarized parallel to the Z axis, and a pure longitudinal mode

$$\left(\frac{k}{\omega}\right)_3 = \left(\frac{\rho}{c_{11}}\right)^{1/2}. \tag{3.9}$$

Figure 3.4 shows curves for cadmium sulfide, neglecting the piezoelectric effect.

The characteristic equation also factors for propagation in the XZ plane. Since the Christoffel equation can be shown to be symmetric with respect to an arbitrary rotation about the Z-axis the same dispersion relation will apply for any meridian plane. That is, the wave slowness surface is always rotationally symmetric. There is one pure shear mode

$$\left(\frac{k}{\omega}\right)_2 = \left(\frac{\rho}{c_{66} \sin^2 \theta + c_{44} \cos^2 \theta}\right)^{1/2}, \tag{3.10}$$

polarized normal to the meridian plane containing **k**. The other solutions are a quasishear wave

$$\left(\frac{k}{\omega}\right)_1 = (2\rho)^{1/2} \Big\{ c_{11} \sin^2 \theta + c_{33} \cos^2 \theta + c_{44}$$

$$- \sqrt{[(c_{11} - c_{44}) \sin^2 \theta + (c_{44} - c_{33}) \cos^2 \theta]^2 + (c_{13} + c_{44})^2 \sin^2 2\theta} \Big\}^{-1/2}. \tag{3.11}$$

and a quasilongitudinal wave

$$\left(\frac{k}{\omega}\right)_3 = (2\rho)^{1/2} \Big\{ c_{11} \sin^2 \theta + c_{33} \cos^2 \theta + c_{44}$$

$$+ \sqrt{[(c_{11} - c_{44}) \sin^2 \theta + (c_{44} - c_{33}) \cos^2 \theta]^2 + (c_{13} + c_{44})^2 \sin^2 2\theta} \Big\}^{-1/2}. \tag{3.12}$$

The direction angle θ is measured from the Z axis (Fig. 3.5).

B.4   Trigonal

B.4a   Classes *32, 3m, $\bar{3}m$.*   In this case, the characteristic equation factors only for propagation in the YZ plane or for propagation along the X, Y, and Z axes.

For X propagation there are two pure shear waves

$$\left(\frac{k}{\omega}\right)_1 = (2\rho)^{1/2} \Big\{ c_{44} + c_{66} + \sqrt{(c_{66} - c_{44})^2 + 4c_{14}^2} \Big\}^{-1/2} \tag{3.13}$$

$$\left(\frac{k}{\omega}\right)_2 = (2\rho)^{1/2} \Big\{ c_{44} + c_{66} - \sqrt{(c_{66} - c_{44})^2 + 4c_{14}^2} \Big\}^{-1/2} \tag{3.14}$$

and a pure longitudinal wave

$$\left(\frac{k}{\omega}\right)_3 = \left(\frac{\rho}{c_{11}}\right)^{1/2}. \tag{3.15}$$

Along Y there is one pure shear wave

$$\left(\frac{k}{\omega}\right)_2 = \left(\frac{\rho}{c_{66}}\right)^{1/2} \tag{3.16}$$

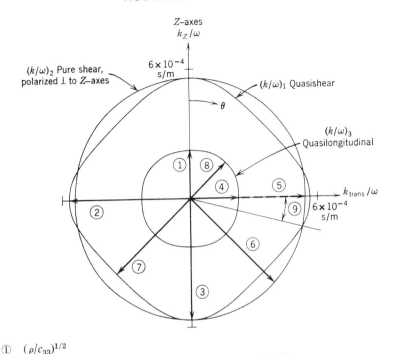

① $(\rho/c_{33})^{1/2}$

② $(\rho/c_{44})^{1/2}$

③ $(\rho/c_{44})^{1/2}$

④ $(\rho/c_{11})^{1/2}$

⑤ $(\rho/c_{66})^{1/2}$

⑥ $\left(\dfrac{2\rho}{c_{66} + c_{44}}\right)^{1/2}$

⑦ $\dfrac{(4\rho)^{1/2}}{\{c_{11} + c_{33} + 2c_{44} - \sqrt{(c_{11} - c_{33})^2 + 4(c_{13} + c_{44})^2}\}^{1/2}}$

⑧ $\dfrac{(4\rho)^{1/2}}{\{c_{11} + c_{33} + 2c_{44} + \sqrt{(c_{11} - c_{33})^2 + 4(c_{13} + c_{44})^2}\}^{1/2}}$

⑨ $\cot^{-1}\sqrt{\dfrac{(c_{13} + c_{44})^2 - (c_{11} - c_{66})(c_{33} - c_{44})}{(c_{44} - c_{66})(c_{11} - c_{66})}}$

FIGURE 3.5. **Hexagonal crystal classes. Propagation in a meridian plane. Curves shown are for CdS, with the piezoelectric effect ignored.**

polarized along X. The quasishear wave is

$$\left(\frac{k}{\omega}\right)_1 = (2\rho)^{1/2}\left\{(c_{44} + c_{11}) - \sqrt{(c_{11} - c_{44})^2 + 4c_{14}^2}\right\}^{-1/2} \tag{3.17}$$

and the quasilongitudinal wave is

$$\left(\frac{k}{\omega}\right)_3 = (2\rho)^{1/2}\left\{(c_{44} + c_{11}) + \sqrt{(c_{11} - c_{44})^2 + 4c_{14}^2}\right\}^{-1/2} \tag{3.18}$$

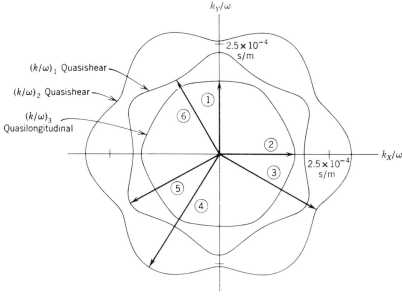

① $\left(\dfrac{2\rho}{c_{11} + c_{44} + c'}\right)^{1/2}$

② $(\rho/c_{11})^{1/2}$

③ $(\rho/c_{66})^{1/2}$

④ $\left(\dfrac{2\rho}{c_{44} + c_{66} - c''}\right)^{1/2}$

⑤ $\left(\dfrac{2\rho}{c_{11} + c_{44} - c'}\right)^{1/2}$

⑥ $\left(\dfrac{2\rho}{c_{44} + c_{66} + c''}\right)^{1/2}$

$c' = \{(c_{11} - c_{44})^2 + 4c_{14}^2\}^{1/2}$

$c'' = \{(c_{66} - c_{44})^2 + 4c_{14}^2\}^{1/2}$

FIGURE 3.6.   **Trigonal crystal classes** $32, 3m, \bar{3}m$. **Propagation normal to the Z-axis. Curves shown are for quartz, with the piezoelectric effect ignored.**

Because of the threefold symmetry about the Z-axis and the inherent inversion symmetry of the elastic stiffness matrix the wave vector curves have a sixfold pattern in the XY plane (Fig. 3.6). Piezoelectric stiffening (Part 4 of Section 8.F) is usually only a small correction to the dispersion curves and has been neglected in calculating the curves shown for quartz.

In the YZ plane a pure shear wave, with polarization along the X axis, is governed by the relation

$$\left(\frac{k}{\omega}\right)_2 = \rho^{1/2}\{c_{66} \sin^2 \theta + c_{44} \cos^2 \theta + c_{14} \sin 2\theta\}^{1/2}. \tag{3.19}$$

The quasishear wave is described by

$$\left(\frac{k}{\omega}\right)_1 = (2\rho)^{1/2}\left\{A - \sqrt{B^2 + C}\right\}^{-1/2} \tag{3.20}$$

and the quasilongitudinal wave by

$$\left(\frac{k}{\omega}\right)_3 = (2\rho)^{1/2}\left\{A + \sqrt{B^2 + C}\right\}^{-1/2}, \tag{3.21}$$

where

$$A = c_{44} + c_{11} \sin^2 \theta + c_{33} \cos^2 \theta - c_{14} \sin 2\theta$$
$$B = (c_{44} - c_{11}) \sin^2 \theta + (c_{33} - c_{44}) \cos^2 \theta + c_{14} \sin 2\theta$$
$$C = ((c_{13} + c_{44}) \sin 2\theta - 2c_{14} \sin^2 \theta)^2.$$

Figure 3.7 shows curves for quartz. The characteristic equation does not factor for propagation in the XZ plane and little information can be obtained without numerical computation. For propagation along Z the pure shear waves are degenerate,

$$\left(\frac{k}{\omega}\right)_1 = \left(\frac{k}{\omega}\right)_2 = \left(\frac{\rho}{c_{44}}\right)^{1/2}, \tag{3.22}$$

and the pure longitudinal wave is governed by

$$\left(\frac{k}{\omega}\right)_3 = \left(\frac{\rho}{c_{33}}\right)^{1/2}. \tag{3.23}$$

Along X there are pure shear waves (3.13) and (3.14), and a pure longitudinal wave (3.15). Curves for quartz are given in Fig. 3.8.

B.4b   Classes $3, \bar{3}$.   In this case it is possible to deduce the general shape of the slowness curves for these crystal classes by noting that the stiffness matrix (given in Part A.5 of Appendix 2) assumes exactly the same form as for classes 32, $3m$, $\bar{3}m$ when it is referred to a set of coordinate axes that are rotated clockwise through an angle ξ, with

$$\tan 3\xi = \frac{c_{25}}{c_{14}}, \tag{3.24}$$

about the Z axis. The general shape of the slowness curves for classes 3 and $\bar{3}$, referred to the crystal axis directions, is thus obtained by rotating the curves in Fig. 3.6 to 3.8 *counterclockwise* through the angle ξ about the Z axis.[†] Because there are no measured stiffness constants for materials of this kind,

---

† It should be noted, however, that the dimensions given in these figures cannot be calculated *directly* from the stiffness constants referred to the crystal axes. The *transformed* stiffness constants must be evaluated first.

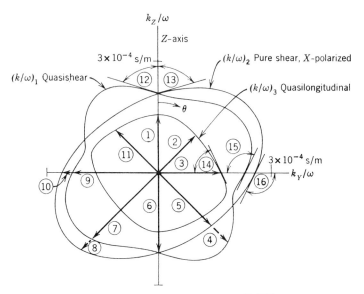

① $(\rho/c_{33})^{1/2}$

② $\left(\dfrac{4\rho}{c_{11} + c_{33} + 2(c_{44} - c_{14}) + c'''}\right)^{1/2}$

③ $\dfrac{(2\rho)^{1/2}}{(c_{11} + c_{44} + c')^{1/2}}$

④ $\left(\dfrac{4\rho}{c_{11} + c_{33} + 2(c_{44} + c_{14}) - c''''}\right)^{1/2}$

⑤ $\dfrac{(2\rho)^{1/2}}{(c_{44} + c_{66} - 2c_{14})^{1/2}}$

⑥ $(\rho/c_{44})^{1/2}$

⑦ $\left(\dfrac{4\rho}{c_{11} + c_{33} + 2(c_{44} - c_{14}) - c'''}\right)^{1/2}$

⑧ $\dfrac{(2\rho)^{1/2}}{(c_{44} + c_{66} + 2c_{14})^{1/2}}$

⑨ $\dfrac{(2\rho)^{1/2}}{(c_{11} + c_{44} - c')^{1/2}}$

⑩ $(\rho/c_{66})^{1/2}$

⑪ $\left(\dfrac{4\rho}{c_{11} + c_{33} + 2(c_{44} + c_{14}) + c'''}\right)^{1/2}$

⑫ ⑬ $\cot^{-1} - c_{14}/c_{44}$

⑭ $\cot^{-1} \dfrac{-c_{14}(c_{11} + c_{44} + 2c_{13} + c')}{c'(c_{11} + c_{44} + c')}$

⑮ $\cot^{-1} \dfrac{c_{14}(c_{11} + c_{44} + 2c_{13} - c')}{c'(c_{11} + c_{44} - c')}$

⑯ $\cot^{-1} c_{14}/c_{66}$

$c' = ((c_{11} - c_{44})^2 + 4c_{14}^2)^{1/2}$

c'''
$c'''' = \{(c_{33} - c_{11} \pm 2c_{14})^2 + 4(c_{44} + c_{13} \mp c_{14})^2\}^{1/2}$

FIGURE 3.7. **Trigonal crystal classes** *32*, *3m*, $\bar{3}m$. **Propagation in the** *YZ* **plane. Curves shown are for quartz, with the piezoelectric effect ignored.**

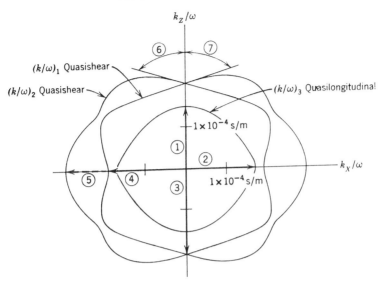

$$① \ (\rho/c_{33})^{1/2}$$

$$② \ (\rho/c_{11})^{1/2}$$

$$③ \ (\rho/c_{44})^{1/2}$$

$$④ \ \left(\frac{2\rho}{c_{44} + c_{66} + c''}\right)^{1/2}$$

$$⑤ \ \left(\frac{2\rho}{c_{44} + c_{66} - c''}\right)^{1/2}$$

$$⑥⑦ \ \cot^{-1} - c_{14}/c_{44}$$

$$c'' = \{(c_{66} - c_{44})^2 + 4c_{14}^2\}^{1/2}$$

FIGURE 3.8. **Trigonal crystal classes** *32, 3m, 3̄m.* **Propagation in the**
XZ **plane. Curves shown are for quartz, with the piezoelectric effect**
ignored.

sample curves cannot be given. However, the same phenomenon occurs in
the tetragonal classes and will be illustrated below.

B.5   Tetragonal

B.5a   Classes *4mm, 422, 4̄2m, 4/mmm.*   The characteristic equation factors
for propagation in the *XY* and *XZ* (or *YZ*) planes.

In the *XY* plane

$$\left(\frac{k}{\omega}\right)_1 = \left(\frac{\rho}{c_{44}}\right)^{1/2} \tag{3.25}$$

is a pure shear wave polarized along *Z*. The quasishear wave is

$$\left(\frac{k}{\omega}\right)_2 = (2\rho)^{1/2}\left|c_{11} + c_{66} - \sqrt{(c_{11} - c_{66})^2 \cos^2 2\phi + (c_{12} + c_{66})^2 \sin^2 2\phi}\right|^{-1/2} \tag{3.26}$$

and the quasilongitudinal wave is

$$\left(\frac{k}{\omega}\right)_3 = (2\rho)^{1/2}\left|c_{11} + c_{66} + \sqrt{(c_{11} - c_{66})^2 \cos^2 2\phi + (c_{12} + c_{66})^2 \sin^2 2\phi}\right|^{-1/2},$$

(3.27)

where the angle ϕ is defined in Fig. 3.9. These expressions are the same as for hexagonal crystals, except that there is now no restriction on c_{66}. The consequences of this are (1) the slowness surfaces are no longer rotationally symmetric and (2) there is only one pure mode propagating in the XY plane.

In the XZ plane there is a pure shear wave polarized along Y,

$$\left(\frac{k}{\omega}\right)_2 = \rho^{1/2}\{c_{66} \sin^2 \theta + c_{44} \cos^2 \theta\}^{-1/2}.$$

(3.28)

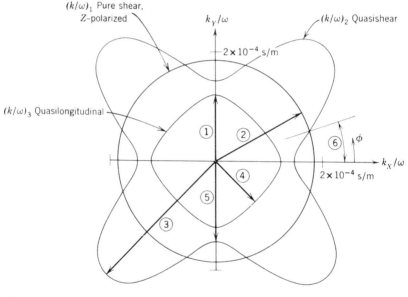

① $(\rho/c_{11})^{1/2}$

② $(\rho/c_{44})^{1/2}$

③ $\left(\dfrac{2\rho}{c_{11} - c_{12}}\right)^{1/2}$

④ $\left(\dfrac{2\rho}{c_{11} + 2c_{66} + c_{12}}\right)^{1/2}$

⑤ $(\rho/c_{66})^{1/2}$

⑥ $\sin^2 \phi = \dfrac{1}{2} \pm \sqrt{\dfrac{1}{4} + F}$

$$F = \frac{(c_{11} - c_{44})(c_{66} - c_{44})}{(c_{11} - c_{66})^2 - (c_{12} + c_{66})^2}$$

FIGURE 3.9. **Tetragonal crystal classes** *4mm, 422, 42m, 4/mmm.* **Propagation in the** XY **plane. Curves shown are for rutile.**

The quasishear wave is described by

$$\left(\frac{k}{\omega}\right)_1 = (2\rho)^{1/2}\left|A - \sqrt{B^2 + C}\right|^{-1/2} \tag{3.29}$$

and the quasilongitudinal wave by

$$\left(\frac{k}{\omega}\right)_3 = (2\rho)^{1/2}\left|A + \sqrt{B^2 + C}\right|^{-1/2}, \tag{3.30}$$

where

$$A = c_{11}\sin^2\theta + c_{33}\cos^2\theta + c_{44}$$
$$B = (c_{11} - c_{44})\sin^2\theta + (c_{44} - c_{33})\cos^2\theta$$
$$C = (c_{13} + c_{44})^2\sin^2 2\theta$$

Curves for rutile are shown in Fig. 3.10.

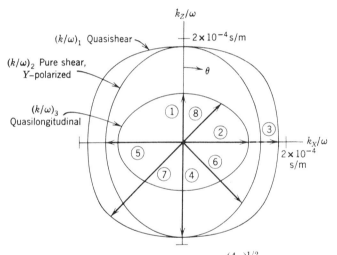

① $(\rho/c_{33})^{1/2}$

② $(\rho/c_{11})^{1/2}$

③④ $(\rho/c_{44})^{1/2}$

⑤ $(\rho/c_{66})^{1/2}$

⑥ $\left(\dfrac{2\rho}{c_{66} + c_{44}}\right)^{1/2}$

⑦ $\left[\dfrac{(4\rho)^{1/2}}{c_{11} + c_{33} + 2c_{44} - \sqrt{(c_{11} - c_{33})^2 + 4(c_{13} + c_{44})^2}}\right]^{1/2}$

⑧ $\left[\dfrac{(4\rho)^{1/2}}{c_{11} + c_{33} + 2c_{44} + \sqrt{(c_{11} - c_{33})^2 + 4(c_{13} + c_{44})^2}}\right]^{1/2}$

FIGURE 3.10. Tetragonal crystal classes $4mm$, 422, $\bar{4}2m$, $4/mmm$. **Propagation in the XZ plane. Curves shown are for rutile.**

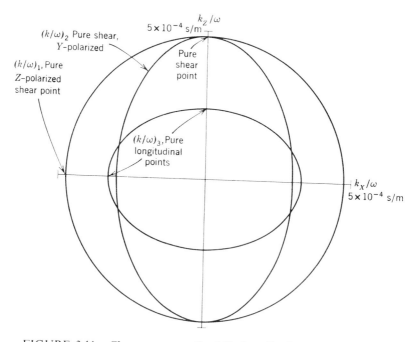

FIGURE 3.11.   **Slowness curves for tellurium dioxide, corresponding to Fig. 3.10.**

One material of this crystal class (tellurium dioxide) has the interesting property that $c_{66} > c_{11}$, while c_{44} satisfies the usual condition $c_{44} < c_{11}$. This means that the X-intercept ⑤ in Fig. 3.10 falls *inside* the X-intercept ②. From this it appears that the $(k/\omega)_2$ and $(k/\omega)_3$ curves should cross each other. Figure 3.11 shows that this does occur. Along the X axis, the longitudinal wave has the unusual property of being slower than one of the shear waves; but normal conditions are restored after the curves have crossed. Figure 3.12 shows slowness curves in the XY plane for the same material. In this case, the solution $(k/\omega)_3$ changes from pure longitudinal to pure shear as ϕ increases from 0° to 45°, and the solution $(k/\omega)_2$ changes from pure shear to pure longitudinal.

B.5b   Classes *4, $\bar{4}$, 4/m.*   The characteristic equation factors only for propagation in the XY plane and along the Z axis.

In the XY plane there is a pure shear wave polarized along Z,

$$\left(\frac{k}{\omega}\right)_1 = \left(\frac{\rho}{c_{44}}\right)^{1/2}. \tag{3.31}$$

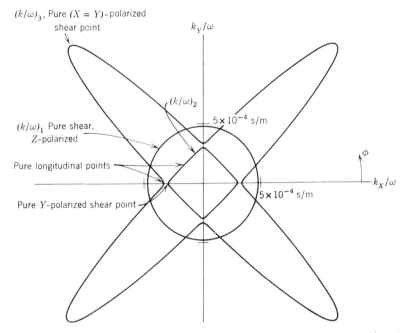

FIGURE 3.12. **Slowness curves for tellurium dioxide, corresponding to Fig. 3.9.**

For the quasishear wave

$$\left(\frac{k}{\omega}\right)_2 = (2\rho)^{1/2}\left\{c_{11} + c_{66} - \sqrt{(c_{11} + c_{66})^2 - 4C}\right\}^{-1/2} \qquad (3.32)$$

and for the quasilongitudinal wave

$$\left(\frac{k}{\omega}\right)_3 = (2\rho)^{1/2}\left\{c_{11} + c_{66} + \sqrt{(c_{11} + c_{66})^2 - 4C}\right\}^{-1/2}. \qquad (3.33)$$

where

$$C = (c_{11}\cos^2\phi + c_{66}\sin^2\phi + c_{16}\sin 2\phi)(c_{11}\sin^2\phi + c_{66}\cos^2\phi$$
$$- c_{16}\sin 2\phi) - (c_{16}\cos 2\phi + (c_{12} + c_{66})\sin\phi\cos\phi)^2.$$

Along the Z axis there are two degenerate shear waves

$$\left(\frac{k}{\omega}\right)_1 = \left(\frac{k}{\omega}\right)_2 = \left(\frac{\rho}{c_{44}}\right)^{1/2}$$

and a pure longitudinal wave.

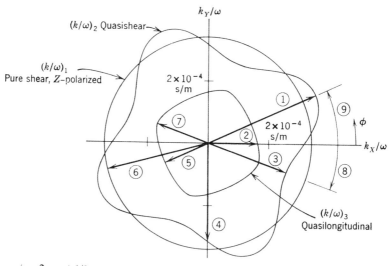

$$\textcircled{1} \quad \left(\frac{2\rho}{c'_{11} - c'_{12}}\right)^{1/2}$$

$$\textcircled{2} \quad (2\rho)^{1/2}\left\{c_{11} + c_{66} + \sqrt{(c_{11} - c_{66})^2 + 4c_{16}^2}\right\}^{-1/2}$$

$$\textcircled{3} \quad \left(\frac{\rho}{c'_{66}}\right)^{1/2}$$

$$\textcircled{4} \quad (2\rho)^{1/2}\left\{c_{11} + c_{66} - \sqrt{(c_{11} - c_{66})^2 + 4c_{16}^2}\right\}^{-1/2}$$

$$\textcircled{5} \quad \left(\frac{2\rho}{c'_{11} + 2c'_{66} + c'_{12}}\right)^{1/2}$$

$$\textcircled{6} \quad \left(\frac{\rho}{c_{44}}\right)^{1/2}$$

$$\textcircled{7} \quad \left(\frac{\rho}{c'_{11}}\right)^{1/2}$$

$$\textcircled{8} \quad \tan 4\phi_{ma} = \frac{-4c_{16}}{c_{11} - c_{12} - 2c_{16}}$$

$$\textcircled{9} \quad \phi_{mi} = \frac{\pi}{4} - \phi_{ma}$$

$$c'_{11} = c_{11}(\cos^4 \phi_{ma} + \sin^4 \phi_{ma}) + \tfrac{1}{2}c_{12} \sin^2 2\phi_{ma} - c_{16} \sin 4\phi_{ma} + c_{66} \sin^2 2\phi_{ma}$$

$$c'_{12} = \tfrac{1}{2}c_{11} \sin^2 2\phi_{ma} + c_{12}(\cos^4 \phi_{ma} + \sin^4 \phi_{ma}) + c_{16} \sin 4\phi_{ma}$$
$$- c_{66} \sin^2 2\phi_{ma}$$

$$c'_{66} = (c_{11} - c_{12}) \sin^2 2\phi_{ma} + c_{16} \sin 4\phi_{ma} + c_{66} \cos^2 2\phi_{ma}$$

FIGURE 3.13. **Tetragonal classes** *4*, *4̄*, *4/m*. **Propagation in the** *XY* **plane. Curves are for calcium molybdate.**

408

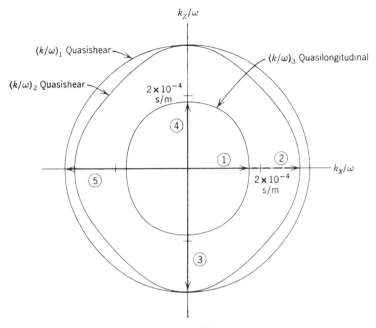

$$① \quad (2\rho)^{1/2}\left\{c_{11} + c_{66} + \sqrt{(c_{11} - c_{66})^2 + 4c_{16}^2}\right\}^{-1/2}$$

$$④ \quad (\rho/c_{33})^{1/2}$$

$$② \quad (2\rho)^{1/2}\left\{c_{11} + c_{66} - \sqrt{(c_{11} - c_{66})^2 + 4c_{16}^2}\right\}^{-1/2}$$

$$⑤ \quad (\rho/c_{44})^{1/2}$$

$$③ \quad (\rho/c_{44})^{1/2}$$

FIGURE 3.14.   **Tetragonal classes** 4, $\bar{4}$, $4/m$. **Propagation in the** XZ **plane. Curves are for calcium molybdate.**

The stiffness matrix for materials of this kind can be converted to the same form as for classes $4mm$, 422, $\bar{4}2m$, $4/mmm$ by performing an appropriate rotation of coordinates about the Z axis. Slowness curves therefore have the same general appearance as Fig. 3.9, but are rotated about the Z axis. Curves for calcium molybdate are shown in Figs. 3.13 to 3.15.

B.6   Orthorhombic

The characteristic equation is of the same form as for the hexagonal classes, but with more general stiffness coefficients. It factors for propagation in the XY, XZ, and YZ planes.

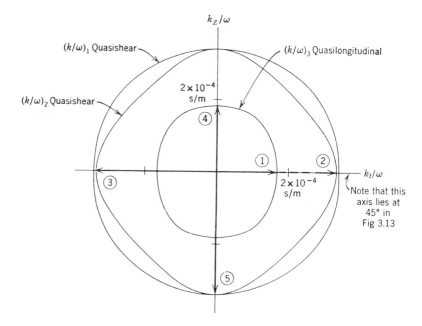

FIGURE 3.15. **Tetragonal classes** *4*, *$\bar{4}$*, *4/m*. **Propagation in the plane**
$X = Y, Z$. **Curves are for calcium molybdate.**

In the XY plane there is a pure shear wave polarized along the Z axis,

$$\left(\frac{k}{\omega}\right)_1 = \rho^{1/2}\{c_{44} \cos^2 \phi + c_{55} \sin^2 \phi\}^{-1/2}. \tag{3.34}$$

The quasishear wave is

$$\left(\frac{k}{\omega}\right)_2 = (2\rho)^{1/2}\Big\{c_{66} + c_{11} \cos^2 \phi + c_{22} \sin^2 \phi$$
$$- \sqrt{(c_{66} + c_{11} \cos^2 \phi + c_{22} \sin^2 \phi)^2 - 4C}\Big\}^{-1/2} \tag{3.35}$$

and the quasilongitudinal wave is

$$\left(\frac{k}{\omega}\right)_3 = (2\rho)^{1/2}\bigg|c_{66} + c_{11}\cos^2\phi + c_{22}\sin^2\phi$$
$$+ \sqrt{(c_{66} + c_{11}\cos^2\phi + c_{22}\sin^2\phi)^2 - 4C}\,\bigg|^{-1/2} \quad (3.36)$$

where

$$C = (c_{11}\cos^2\phi + c_{66}\sin^2\phi)(c_{66}\cos^2\phi + c_{22}\sin^2\phi)$$
$$- (c_{12} + c_{66})^2\cos^2\phi\sin^2\phi.$$

The direction angle ϕ is defined in Fig. 3.16, which shows curves for barium sodium niobate, neglecting the piezoelectric effect.

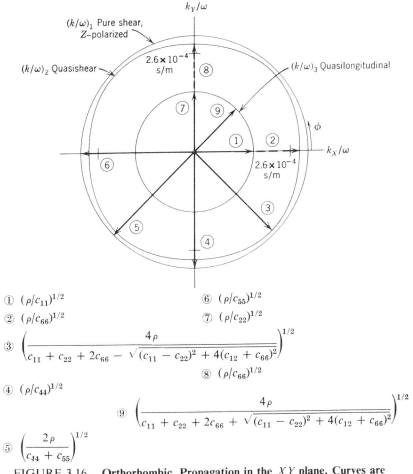

$$① \quad (\rho/c_{11})^{1/2}$$

$$② \quad (\rho/c_{66})^{1/2}$$

$$③ \quad \left(\frac{4\rho}{c_{11} + c_{22} + 2c_{66} - \sqrt{(c_{11} - c_{22})^2 + 4(c_{12} + c_{66})^2}}\right)^{1/2}$$

$$④ \quad (\rho/c_{44})^{1/2}$$

$$⑤ \quad \left(\frac{2\rho}{c_{44} + c_{55}}\right)^{1/2}$$

$$⑥ \quad (\rho/c_{55})^{1/2}$$

$$⑦ \quad (\rho/c_{22})^{1/2}$$

$$⑧ \quad (\rho/c_{66})^{1/2}$$

$$⑨ \quad \left(\frac{4\rho}{c_{11} + c_{22} + 2c_{66} + \sqrt{(c_{11} - c_{22})^2 + 4(c_{12} + c_{66})^2}}\right)^{1/2}$$

FIGURE 3.16.   Orthorhombic. Propagation in the XY plane. Curves are for barium sodium niobate, with the piezoelectric effect neglected.

For propagation in the XZ plane the pure shear wave is polarized along the Y axis,

$$\left(\frac{k}{\omega}\right)_2 = \rho^{1/2}\{c_{66}\sin^2\theta + c_{44}\cos^2\theta\}^{-1/2}. \tag{3.37}$$

The quasishear wave is

$$\left(\frac{k}{\omega}\right)_1 = (2\rho)^{1/2}\Big\{c_{55} + c_{11}\sin^2\theta + c_{33}\cos^2\theta$$

$$- \sqrt{(c_{55} + c_{11}\sin^2\theta + c_{33}\cos^2\theta)^2 - 4\bar{C}}\Big\}^{-1/2} \tag{3.38}$$

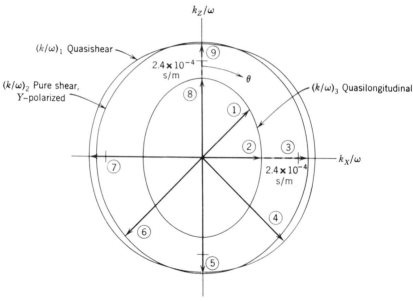

$(k/\omega)_1$ Quasishear

$(k/\omega)_2$ Pure shear, Y–polarized

$(k/\omega)_3$ Quasilongitudinal

k_Z/ω

2.4×10^{-4} s/m

θ

k_X/ω

2.4×10^{-4} s/m

① $\left(\dfrac{4\rho}{c_{11} + c_{33} + 2c_{55} + \sqrt{(c_{11} - c_{33})^2 + 4(c_{13} + c_{55})^2}}\right)^{1/2}$

② $(\rho/c_{11})^{1/2}$ ③ $(\rho/c_{66})^{1/2}$

④ $\left(\dfrac{4\rho}{c_{11} + c_{33} + 2c_{55} - \sqrt{(c_{11} - c_{33})^2 + 4(c_{13} + c_{55})^2}}\right)^{1/2}$

⑤ $(\rho/c_{44})^{1/2}$ ⑥ $\left(\dfrac{2\rho}{c_{44} + c_{66}}\right)^{1/2}$

⑦⑨ $(\rho/c_{55})^{1/2}$ ⑧ $(\rho/c_{33})^{1/2}$

FIGURE 3.17. **Orthorhombic. Propagation in the** XZ **plane. Curves are for barium sodium niobate, with the piezoelectric effect neglected.**

and the quasilongitudinal wave is

$$\left(\frac{k}{\omega}\right)_3 = (2\rho)^{1/2}\Big\{c_{55} + c_{11}\sin^2\theta + c_{33}\cos^2\theta$$

$$+ \sqrt{(c_{55} + c_{11}\sin^2\theta + c_{33}\cos^2\theta)^2 - 4\bar{C}}\Big\}^{-1/2} \quad (3.39)$$

where

$$\bar{C} = (c_{11}\sin^2\theta + c_{55}\cos^2\theta)(c_{55}\sin^2\theta + c_{33}\cos^2\theta)$$
$$- (c_{13} + c_{55})^2\sin^2\theta\cos^2\theta.$$

Curves are shown in Fig 3.17.

In the YZ plane the pure shear wave is polarized along the X axis

$$\left(\frac{k}{\omega}\right)_1 = \rho^{1/2}\{c_{66}\sin^2\theta + c_{55}\cos^2\theta\}^{-1/2}. \quad (3.40)$$

The quasishear wave is

$$\left(\frac{k}{\omega}\right)_2 = (2\rho)^{1/2}\Big\{c_{44} + c_{22}\sin^2\theta + c_{33}\cos^2\theta$$

$$- \sqrt{(c_{44} + c_{22}\sin^2\theta + c_{33}\cos^2\theta)^2 - 4\bar{\bar{C}}}\Big\}^{-1/2} \quad (3.41)$$

and the quasilongitudinal wave is

$$\left(\frac{k}{\omega}\right)_3 = (2\rho)^{1/2}\Big\{c_{44} + c_{22}\sin^2\theta + c_{33}\cos^2\theta$$

$$+ \sqrt{(c_{44} + c_{22}\sin^2\theta + c_{33}\cos^2\theta)^2 - 4\bar{\bar{C}}}\Big\}^{-1/2}, \quad (3.42)$$

where

$$\bar{\bar{C}} = (c_{22}\sin^2\theta + c_{44}\cos^2\theta)(c_{44}\sin^2\theta + c_{33}\cos^2\theta)$$
$$- (c_{23} + c_{44})^2\sin^2\theta\cos^2\theta.$$

Curves are shown in Fig. 3.18.

C. PURE MODE DIRECTIONS

In anisotropic media acoustic plane waves do not usually have particle motion polarized either parallel or normal to the wave vector. Pure longitudinal and pure transverse polarizations occur only for certain wave vector directions, called *pure mode directions*. These may occur in both symmetry and nonsymmetry directions.

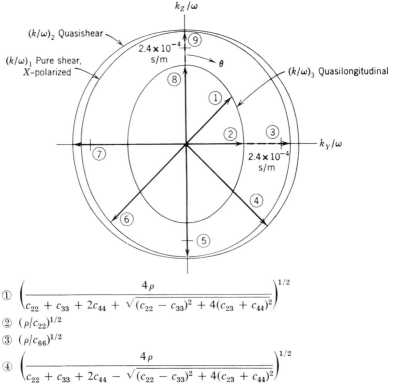

$$\text{①} \left(\frac{4\rho}{c_{22} + c_{33} + 2c_{44} + \sqrt{(c_{22} - c_{33})^2 + 4(c_{23} + c_{44})^2}}\right)^{1/2}$$

② $(\rho/c_{22})^{1/2}$

③ $(\rho/c_{66})^{1/2}$

$$\text{④} \left(\frac{4\rho}{c_{22} + c_{33} + 2c_{44} - \sqrt{(c_{22} - c_{33})^2 + 4(c_{23} + c_{44})^2}}\right)^{1/2}$$

⑤ $(\rho/c_{44})^{1/2}$ ⑦ $(\rho/c_{44})^{1/2}$

$$\text{⑥} \left(\frac{2\rho}{c_{55} + c_{66}}\right)^{1/2}$$ ⑧ $(\rho/c_{33})^{1/2}$

⑨ $(\rho/c_{55})^{1/2}$

FIGURE 3.18.   Orthorhombic. Propagation in the YZ plane. Curves are for barium sodium niobate, with the piezoelectric effect neglected.

C.1   Symmetry Directions

Propagation in a Symmetry Plane
One pure shear mode, polarized normal to plane.

Propagation Normal to a 2-fold, 4-fold or 6-fold Axis
One pure shear mode, polarized parallel to axis.

Propagation Along A 2-fold Axis
All modes are pure.

Propagation Along a 3-fold, 4-fold or 6-fold Axis
All modes are pure.

Shear modes are degenerate.

C.2   Nonsymmetry Directions†

C.2a   Hexagonal

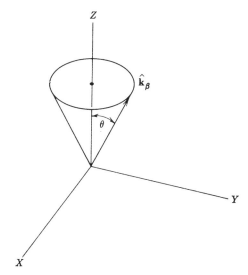

$$\cot^2 \theta = \frac{c_{11} - 2c_{44} - c_{13}}{c_{33} - 2c_{44} - c_{13}}.$$

C.2b   Trigonal (32, $3m$, $\bar{3}m$)

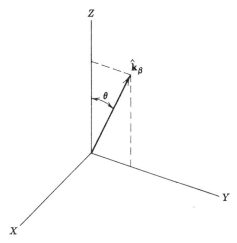

$$(c_{33} - 2c_{44} - c_{13}) \cot^3 \theta + 3c_{14} \cot^2 \theta - (c_{11} - 2c_{44} - c_{13}) \cot \theta - c_{14} = 0.$$

† After K. Brugger, "Pure Modes for Elastic Waves in Crystals," *J. Appl. Phys.* **36**, 759–768 (1965).

C.2c   Trigonal $(3, \bar{3})$

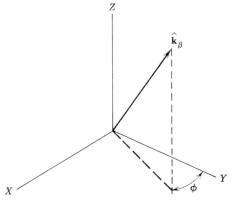

$$\tan 3\phi = -\frac{c_{25}}{c_{14}}$$

$$q = \frac{(\hat{k}_\beta)_Z}{(\hat{k}_\beta)_Y}$$

$(c_{33} - 2c_{44} - c_{13})q^3$

$$-\frac{3(\tan^2 \phi + 1)^2}{(3\tan^2 \phi - 1)} c_{14}q^2 - (\tan^2 \phi + 1)(c_{11} - 2c_{44} - c_{13})q$$

$$+\frac{(\tan^2 \phi + 1)^3}{3\tan^2 \phi - 1}c_{14} = 0.$$

C.2d   Tetragonal $(4mm, 422, \bar{4}2m, 4/mmm)$

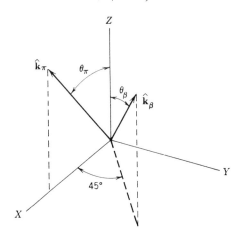

$$\cot \theta_\beta = \left\{ \frac{c_{11} - 4c_{44} - 2(c_{13} - c_{66}) - c_{12}}{2(c_{33} - 2c_{44} - c_{13})} \right\}^{1/2}$$

$$\cot \theta_\pi = \left(\frac{c_{11} - 2c_{44} - c_{13}}{c_{33} - 2c_{44} - c_{13}} \right)^{1/2}.$$

C.2e   Tetragonal ($4, \bar{4}, 4/m$)

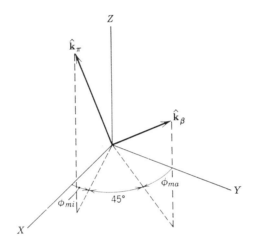

$$\tan 4\phi_{ma} = \frac{-4c_{16}}{c_{11} - c_{12} - 2c_{66}}$$

$$\phi_{mi} = \frac{\pi}{4} - \phi_{ma}.$$

$$\frac{(\hat{k}_\pi)_Z}{(\hat{k}_\pi)_X}$$

$$= \left[\frac{\tan^2 \phi_{mi} + 1}{c_{33} - 2c_{44} - c_{13}} \left\{ c_{11} - 2c_{44} - c_{13} + \frac{(c_{11} - 2c_{66} - c_{12})2 \tan^2 \phi_{mi}}{\tan^4 \phi_{mi} - 6 \tan^2 \phi_{mi} + 1} \right\} \right]^{1/2}.$$

$$\frac{(\hat{k}_\beta)_Z}{(\hat{k}_\beta)_X}$$

$$= \left[\frac{\tan^2 \phi_{ma} + 1}{c_{33} - 2c_{44} - c_{13}} \left\{ c_{11} - 2c_{44} - c_{13} + \frac{(c_{11} - 2c_{66} - c_{12})2 \tan^2 \phi_{ma}}{\tan^4 \phi_{ma} - 6 \tan^2 \phi_{ma} + 1} \right\} \right]^{1/2}.$$

Note that ϕ_{ma} and ϕ_{mi} are the same as ⑧ and ⑨ in Fig. 3.13.

C.2f   Orthorhombic

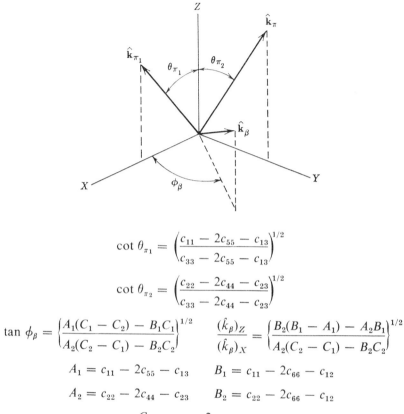

$$\cot \theta_{\pi_1} = \left(\frac{c_{11} - 2c_{55} - c_{13}}{c_{33} - 2c_{55} - c_{13}}\right)^{1/2}$$

$$\cot \theta_{\pi_2} = \left(\frac{c_{22} - 2c_{44} - c_{23}}{c_{33} - 2c_{44} - c_{23}}\right)^{1/2}$$

$$\tan \phi_\beta = \left\{\frac{A_1(C_1 - C_2) - B_1 C_1}{A_2(C_2 - C_1) - B_2 C_2}\right\}^{1/2} \qquad \frac{(\hat{k}_\beta)_Z}{(\hat{k}_\beta)_X} = \left\{\frac{B_2(B_1 - A_1) - A_2 B_1}{A_2(C_2 - C_1) - B_2 C_2}\right\}^{1/2}$$

$$A_1 = c_{11} - 2c_{55} - c_{13} \qquad B_1 = c_{11} - 2c_{66} - c_{12}$$

$$A_2 = c_{22} - 2c_{44} - c_{23} \qquad B_2 = c_{22} - 2c_{66} - c_{12}$$

$$C_1 = c_{33} - 2c_{55} - c_{13}$$

$$C_2 = c_{33} - 2c_{44} - c_{23}.$$

SUPPLEMENTARY REFERENCES

General

G.1. V. M. Ristic, *Principles of Acoustic Devices*, Wiley-Interscience, New York, 1983.

G.2. E. Dieulesaint and D. Royer, *Elastic Waves in Solids*, Wiley, New York, 1980.

G.3. G. S. Kino, *Acoustic Waves—Devices, Imaging and Analogue Signal Processing*, Prentice-Hall, Englewood Cliffs, N.J., 1987.

G.4. J. F. Rosenbaum, *Bulk Acoustic Wave Theory and Devices*, Artech, Norwood, MA., 1988.

Chapter 3

3.1. *IEEE Standard on Piezoelectricity*, The Institute of Electrical and Electronics Engineers, Inc., ANSI/IEEE Std., 176–1978 (1978).

Chapter 7

7.1. A. G. Every, "General Closed Firm Expressions for Acoustic Waves in Elastically Anisotropic Solids," *Phys. Rev. B* **22**, 1746 (1980).

7.2. A. G. Every, "Ballistic Phonons and the Shape of the Ray Surface in Cubic Crystals," *Phys. Rev. B.* **24**, 3456 (1981).

7.3. M. J. P. Musgrave, "Elastodynamic Classification of Orthorhombic Media," *Proc. R. Soc. Lond.* **A374**, 401 (1981).

7.4. H. M. Ledbetter and R. Kriz, "Elastic Wave Surfaces in Solids," *Physica Status Solidi* **114**, 475 (1982).

7.5. C. M. Sayers, "Ultrasonic Velocities in Anisotropic Polycrystalline Aggregates," *J.Phys. D: Appl. Phys.* **15**, 2157 (1982).

7.6. A. G. Every, G. L. Koos, and J. P. Wolfe, "Ballistic Phonon Imaging in Sapphire: Bulk Focusing and Critical Cone Channeling Effects," *Phys. Rev. B* **29**, 2190 (1984).

7.7. F. E. Stanke and G. S. Kino, "A Unified Theory for Elastic Wave Propagation in Polycrystalline Materials," *J. Acous. Soc. Am.* **75**, 665 (1984).

7.8. C-K. Jen, G. W. Farnell, E. L. Adler, and J. E. B. Oliveira, "Interactive Computer-Aided Analysis for Bulk Acoustic Waves in Materials of Arbitrary Anisotropy and Piezoelectricity," *IEEE Trans.* SU–32, 56 (1985).

7.9. A. G. Every, "Formation of Phonon-Focusing Caustics in Crystals," *Phys. Rev. B* **34**, 2852 (1986).

7.10. K. Helbig and M. Schoenberg, "Anomalous Propagation of Elastic Waves in Transversely Isotropic Media," *J. Acous. Soc. Am.* **81**, 1235 (1987).

Chapter 8

8.1. W. A. Smith, A. Shaulov, and B. A. Auld, "Tailoring the Properties of Composite Piezoelectric Materials for Medical Ultrasonic Transducers," 1985 IEEE Ultrasonics Symposium Proceedings, 642 (1985).

8.2. S. Ikegami, I. Ueda, and S. Kobayashi, "Frequency Spectra of Resonant Vibration in Disk Plates of PbTiO$_3$ Piezoelectric Ceramics," *J. Acous. Soc. Am.* **55**, 339 (1974).

8.3. S. Ueha, S. Sakuma, and E. Mori, " Measurement of Vibration Velocity Distributions and Mode Analysis in Thick Disks of Pb(Zr.Ti)O$_3$," *J. Acous. Soc. Am.* **73**, 1842 (1983).

8.4. H. Takeuchi, S. Jyomura, Y. Ishikawa, and E. Yamamoto, "A 7.5 MHz Linear Array Ultrasonic Probe Using Modified PbTiO$_3$ Ceramics," 1982 IEEE Ultrasonics Symposium Proceedings, 849 (1982).

8.5. T. R. Gururaja, W. A. Schulze, L. E. Cross, B. A. Auld, Y. A. Shui, and Y. Wang, "Resonant Modes of Vibration in Piezoelectric PZT-Polymer Composites with Two Dimensional Periodicity," *Ferroelectrics* **54**, 183 (1983).

8.6. W. A. Smith, A. A. Shaulov, and B. M. Singer, "Properties of Composite Piezoelectric Materials for Ultrasonic Transducers," 1984 IEEE Ultrasonics Symposium Proceedings, 539 (1984).

8.7. A. A. Shaulov, W. A. Smith, and B. M. Singer, "Performance of Ultrasonic Transducers Made from Composite Piezoelectric Materials," 1984 IEEE Ultrasonics Symposium Proceedings, 545 (1984).

8.8 T. R. Gururaja, W. A. Schulze, L. E. Cross, R. E. Newnham, B. A. Auld, and Y. Wang, " Piezoelectric Composite Materials for Ultrasonic Transducer Applications, Part I: Resonant Modes of Vibration of PZT Rod-Polymer Composites," *IEEE Trans. Son. and Ultrason.* **SU–32**, 481 (1985).

8.9. A. Shaulov and W. A. Smith, "Ultrasonic Transducer Arrays Made from Composite Piezoelectric Materials," 1985 IEEE Ultrasonics Symposium Proceedings, 648 (1985).

8.10. C. Nakaya, H. Takeuchi, K. Katakura, and A. Sakamoto, "Ultrasonic Probe Using Composite Piezoelectric Materials," 1985 IEEE Ultrasonics Symposium Proceedings, 634 (1985).

8.11. B. A. Auld, "High Frequency Piezoelectric Resonators," Proc. 6th IEEE Int. Symp. Appl. Ferroelectrics, IEEE (CH2358-0/86-0076), 288 (1986).

8.12. G. E. Martin, "Dielectric, Elastic, and Piezoelectric Losses in Piezoelectric Materials," 1974 IEEE Ultrasonics Symposium Proceedings, 613 (1975).

8.13. G. S. Kino and C. S. DeSilets, "Design of Slotted Transducer Arrays with Matched Backings," *Ultrasonic Imaging* **1**, 189 (1979).

8.14. H. C. Tuan, A. R. Selfridge, J. E. Bowers, B. T. Khuri-Yakub, and G. S. Kino, "An Edge-Bonded Surface Acoustic Wave Transducer Array," 1979 IEEE Ultrasonics Symposium Proceedings, 221 (1980).

8.15. B. A. Auld, "Wave Propagation and Resonance in Piezoelectric Materials," *J. Acoust. Soc. Am.* **70**, 1577 (December 1981).

8.16. H. Honda, Y. Yamashita, and K. Uchida, "Array Transducer Using New Modified $PbTiO_3$ Ceramics," 1982 IEEE Ultrasonics Symposium Proceedings, 845 (1982).

8.17. IEEE Standard on Piezoelectricity, *IEEE Trans.* **S–31**, part 2, March 1984.

8.18. C-K. Jen, G. W. Farnell, E. L. Adler, and J. E. B. Oliveira, "Interactive Computer-Aided Analysis for Bulk Acoustic Waves in Materials of Arbitrary Anisotropy and Piezoelectricity," *IEEE Trans.* **SU–32**, 56 (1985).

8.19. A. G. Every, "Electroacoustic Waves in Piezoelectric Crystals: Certain Limiting Cases," *Wave Motion* **9**, 493 (1987).

8.20. T. Inoue, M. Ohta, and S. Takahashi, "Design of Ultrasonic Transducers with Multiple Acoustic Matching Layers for Medical Appliction," *IEEE Trans.* **UFFC–34**, 8 (1987).

INDEX

4. VECTOR AND TENSOR IDENTITIES

$$\nabla \cdot (\mathbf{E} \times \mathbf{H}) = \mathbf{H} \cdot \nabla \times \mathbf{E} - \mathbf{E} \cdot \nabla \times \mathbf{H}$$

$$\nabla \cdot (\mathbf{v} \cdot \mathbf{T}) = \mathbf{v} \cdot (\nabla \cdot \mathbf{T}) + \mathbf{T} : \nabla_s \mathbf{v}$$

$$\nabla \cdot (\Phi \mathbf{D}) = \mathbf{D} \cdot \nabla \Phi + \Phi \nabla \cdot \mathbf{D}$$

$$\nabla \times \nabla \times \mathbf{A} = \nabla(\nabla \cdot \mathbf{A}) - \nabla^2 \mathbf{A}$$

$$\nabla \times \nabla \Phi = 0$$

$$\nabla \cdot \nabla \times \mathbf{A} = 0$$

$$\int_V \nabla \cdot \mathbf{A} \, dV = \oint_S \mathbf{A} \cdot \hat{\mathbf{n}} \, dS \qquad \text{(Divergence Theorem)}$$

$$\mathbf{A} \cdot (\mathbf{B} \times \mathbf{C}) = \mathbf{B} \cdot (\mathbf{C} \times \mathbf{A}) = \mathbf{C} \cdot (\mathbf{A} \times \mathbf{B})$$